DIELECTRIC MATERIAL

Edited by **Marius Alexandru Silaghi**

Dielectric Material
http://dx.doi.org/10.5772/2781
Edited by Marius Alexandru Silaghi

Contributors

Zulkifli Ahmad, L. Huitema, T. Monediere, He Seung Lee, Albert. S. Lee, Kyung-Youl Baek, Seung Sang Hwang, D.A. Hoble, M.A. Silaghi, Wee Fwen Hoon, Soh Ping Jack, Mohd Fareq Abd Malek, Nornikman Hasssan, A.G. Belous, Daniel Vázquez-Moliní, Antonio Álvarez Fernández-Balbuena, Berta García-Fernández, N.D. Tran , N. Harada, T. Sasaki, T. Kikuchi, Sonia M. Holik, Salvador Dueñas, Helena Castán, Héctor García, Luis Bailón, I. Mladenovic, Ch. Weindl, Yuriy Prokopenko, Yuriy Poplavko, Victor Kazmirenko, Irina Golubeva

Published by InTech
Janeza Trdine 9, 51000 Rijeka, Croatia

Publishing Process Manager Marija Radja
Technical Editor InTech Prepress, Novi Sad
Cover InTech Design team

Additional hard copies can be obtained from orders@intechopen.com

Dielectric Material, Edited by Marius Alexandru Silaghi
p. cm.
ISBN 978-953-51-0764-4

For EU product safety concerns:
IN TECH d.o.o., Prolaz Marije Krucifikse Kozulić 3, 51000 Rijeka, Croatia,
info@intechopen.com or visit our website at intechopen.com.

Meet the editor

Marius Alexandru SILAGHI received the M.Sc. in 1990, in electrical engineering at the POLITEHNICA University of Timisoara and the Ph.D. degree in electrical engineering, in 1999 at the University of Oradea, Romania. He is a professor at Electrical Engineering Department of Electrical Engineering and Information Technology Faculty, University of Oradea, responsible for teaching electromagnetic field theory, industrial electrotehnics , microwaves technique, microwave technology, electromagnetic compatibility and electrical energy quality. He is a member of AMPERE(1997), IEEE MTT Society(1998), ASLA(2002), AGIR(2002), ACER(2009), IEEE EMC Society (2009). His research interests include electrical engineering and microwave technology, energy quality in electrical systems, electromagnetic field simulations and analysis, biomedical engineering. He has published more than 100 papers in Journals and Conference Proceedings, authored 22 books, text books, and application guides. He conducted more than 20 research and development projects, grants and contracts in the field of electrical engineering. Prof. Silaghi has been involved in setting up national and international conferences as a reviewer and/or member of organizing committee or board of sections. He was a visiting professor at TU Kosice, Slovakia and U Modena, Italy.
In the last 18 years was appointed as Head of the Electrical Engineering Department, Vice Dean and Dean of Electrical Engineering and Information Technology Faculty , University of Oradea.

Contents

Preface

This book attempts to bring together the theory and practice of dielectric materials for different kind of industrial applications.

Fragmented information on dielectric theory and properties of materials, design of equipment and state of the art in applications relevant to the manufacturing industry should be collated and updated and presented as a single reference volume.

In this book relevant and useful information is presented in the quoted literature and covered by our key patent applications.

As an Editor and also Author in this field, I am honored to edit this book written by a selected group of researchers.

Marius Alexandru Silaghi
University of Oradea, Oradea,
Romania

Section 1

Polymeric Dielectric Materials

Chapter 1

Polymeric Dielectric Materials

Zulkifli Ahmad

Additional information is available at the end of the chapter

http://dx.doi.org/10.5772/50638

1. Introduction

1.1. Background and brief history

The definition for dielectric constant relates to the permittivity of the material (symbol use here ε). The permittivity expresses the ability of a material to polarise in response to an applied field. It is the ratio of the permittivity of the dielectric to the permittivity of a vacuum. Physically it means the greater the polarisation developed by a material in an applied field of given strength, the greater the dielectric constant will be. Traditionally dielectric materials are made from inorganic substances eg. mica and silicon dioxide. However polymers are gaining wider use as dielectric materials. This is due to the easier processing, flexibility, able to tailor made for specific uses and better resistance to chemical attack. As early as mid-60's polymers eg polyvinyl fluoride [1] and aromatic-containing polymers [2] are used as dielectric materials in capacitors. Further improvement in organic film fabrication was established as revealed in US Patent 4153925. Polymers can be fabricated fairly easily into thin film by solution casting and spin coating, immersion in organic substrate, electron or UV radiation and glow discharge methods. This is mainly due to lower thermal properties such as glass transition and melting temperature which contribute to a lower temperature processing windows. Their solubility is controllable without offsetting their intrinsic properties. In the case of inorganic material and ceramic, they have much higher thermal properties hence temperature requirement leads to an extreme end of processing temperatures. On the other hand polymers cannot stand too high a temperature. Their coefficient of thermal expansion is relatively larger than ceramic materials and susceptible to atmospheric and hydrolytic degradation. Table 1 shows the values of dielectric properties of several polymers with comparisons with several inorganic materials.

Inorganic/ceramics materials have higher dielectric constant than polymers. Water has a relatively high dielectric constant. This is quite cumbersome as any traces of moisture trapped or absorb will dramatically alter the desired dielectric properties. Inorganic materials generally have higher dielectric constant compared to polymeric materials. Intrinsically they contains ions and polar groups which contribute to their high dielectric constant. Air having a dielectric constant of 1.02 is taken as reference dielectric.

Materials	Dielectric constant, ε	Materials	Dielectric constant, ε
TiO_2	100	Fluorinated polyimide	2.5 – 2.9
H_2O	78	Methylsilsesquioxane	2.6 – 2.8
neoprene	9.8	Polyarelene ether	2.8 – 2.9
PVDF	6.0	Polyethylene	2.3 – 2.7
SiO2	3.9 – 4.5	Polystyrene	2.5 – 2.9
Fluorosilicate glass	3.2 – 4.0	Teflon AF	2.1
Polyimide	2.8 – 3.2	Air	1.02

Table 1. Dielectric constant of several polymers and inorganic materials. (Adapted from Ref 3)

1.2. Application of polymeric dielectric materials.

Both dielectrics with low and high dielectric constant are essential in electronic industries. Low dielectric constant is required basically as insulators. They are known as passivation materials. Their applications ranged in isolating signal-carrying conductors from each other, fast signal propagation, interlayer dielectric to reduce the resistance-capacitance (RC) time delays, crosstalk and power dissipation in the high density and high speed integration [4]. They are of necessity in very dense multi-layered IC's, wherein coupling between very close metal lines need to be suppressed to prevent degradation in device performance. This role involve packaging and encapsulation. In electronic packaging, they separate interlayers and provide isolated pathways for electronic devices connection in multilayer printed circuit boards. As the trends are towards miniaturization in microprocessor fabrication, any decrease in relative permittivity will reduces the deleterious effect of stray and coupling capacitances. Dielectric naterials are also employed to encapsulate the balls which bridged the die and substrate. This encapsulation is specifically called underfill which helps to protect any circuitary failures as well as reducing thermal mismatch between the bridging layers.(Figure 1) In LED encapsulation low dielectric materials is used for insulation at the lead frame housing.

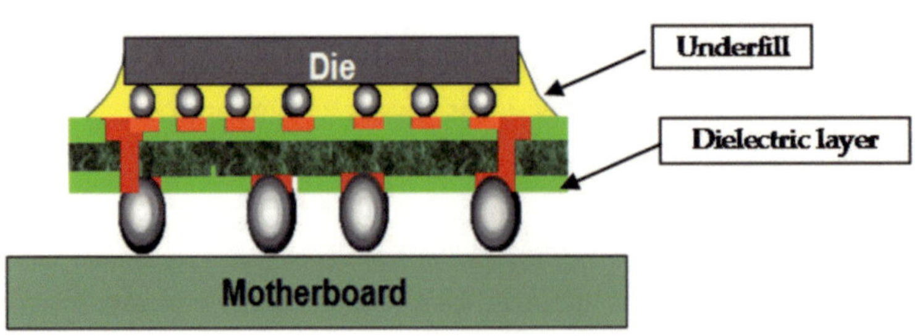

Figure 1. Application of dielectric polymers in IC packaging

As an active components, designing is geared towards high ε value and are used as polarizable media for capacitors, in apparatus used for the propagation or reflection of electromagnetic waves, and for a variety of artifacts, such as rectifiers and semiconductor devices, piezoelectric transducers, dielectric amplifiers, and memory elements. Despite being insulators, hence non-polar, these materials can be made polar by introducing small amount of impurities. In this state, the material is able to store large amount of charges at small applied electrical field. This is the case with polyvinylidene fluoride when introduced with impurities chlorotrifluoroethylene.[5] Indeed several works have been performed on polymers like polyimide with added Al_2O_3, $BaTiO_3$ and ZrO_3 'impurities' [6,7,8] which showed an improved dielectric constant. Once there is large charge storage, it can be readily released on demand. In a rectifier, a capacitor is used to smooth off the pulsating direct current.

2. Theory of dielectric properties in polymer

2.1. Mechanism of interaction with electric field

Quantitative treatment of a dielectric in an electric field can be summarized using Clausius–Mossotti equation (1).

$$\mathrm{P} = \frac{\varepsilon_r - 1}{\varepsilon_r + 2} \cdot \frac{M}{\rho} = \frac{N_A \alpha}{3\varepsilon_0} \tag{1}$$

P is the molar polarisability, ε_r is the relative permittivity, ε_o is the permittivity in vacuum, M is molecular weight of a repeat unit, ϱ is density, α is polarisability, N_a is the Avogadro constant. This equation shows that dielectric constant is dependent on polarisability and free volume of the constituents element present in the materials. Polarisability refer to the proportionality constant for the formation of dipole under the influence of electric field. Thus its value is typical for each different type of atom or molecule.[9] The relation between polarizability with the permittivity of the dielectric material can be shown as in Equation (2):

$$\varepsilon_r = 1 + \frac{N\alpha}{\varepsilon_0} \tag{2}$$

It shows that relative permittivity ε_r is the ratio of total permittivity of one mole of material with that in vacuum. The dependency of free volume of relative permittivity thus originate from the volume involved in one mole of the material. Again the molar volume is characteristic of each different type of atom or molecule. Molar polarization therefore is obtained if the molar volume is introduced into these derivations leading to Clausius–Mossotti equation.

Physically, polarisability is induced when there is electric field applied onto the materials. In the absence of electric field, the electrons are distributed evenly around the nuclei.

When the electric field is applied the electron cloud is displaced from the nuclei in the direction opposite to the applied field. This result in separation of positive and negative charges and the molecules behave like an electric dipole. There are three mode of polarizations [10]:

i. Electronic polarization – slight displacement of electrons with respect to the nucleus.
ii. Atomic polarization – distortion of atomic position in a molecule or lattice
iii. Orientational polarization – For polar molecules, there is a tendency for permanent dipole to align by the electric field to give a net polarization in that direction

When a static electric field is applied on to these materials the dipoles become permanently polarized giving a dielectric constant as ε_{static}. However if the field changes as when alternating electric current is applied, polarization will also oscillate with the changing electric field. All three modes of polarization contributing to the overall dielectric constant will be dependent on the frequency of the oscillating electric field. Obviously the electronic polarization is instantaneous as it is able to follow in phase with the changing electric field compared to atomic polarization which in turn better able to follow the oscillating electric field compared to the orientational polarization. Certain structures and elements display a higher polarisibility than the others. Aromatic rings, sulphur, iodine and bromine are considered as highly polarisable. The present of these groups induced an increase in dielectric constant. The π bond in the aromatic rings is loosely attached compared to the sigma bond. Therefore it is easily polarized. For large size atoms like bromine and iodine, the electron cloud is so large and further apart from the influence of electrostatic attraction of the positive nucleus. It is expected to display a high polarisibility. This is as oppose to fluorine which has small atomic radius and concentrated negative charge. It is able to hold the electron cloud much tightly resulted in a low polarisability. This will induce a lower dielectric constant.

Free volume is also an important factor in determining the dielectric constant. Free volume is defined as the volume which is not occupied by the polymeric material. The free volume associated with one mole of repeat units of the polymer may be estimated by subtracting the occupied molar volume of a repeat unit, V_0, from the total molar volume, M/ρ, where M is the molar mass of the repeat unit. [10] The fractional free volume V_n is given by Equation (3):

$$V_n = \frac{M/\rho - V_0}{M/\rho} \tag{3}$$

The addition of pendant groups, flexible bridging units, and bulky groups which limit chain packing density have been utilised to enhance free volume. [21] The presence of free volume in the form of pores will similarly result in a decrease in dielectric constant as it being occupied by air whose relative permittivity is about one. A higher fractional free volume means that the density of the material will be lower resulting in a lower polarisible group per unit volume. Replacement of hydrogen with fluorine result in lowering of dielectric constant since fluorine occupies higher volume. Thus beside being low polarisability, introduction of fluorine induce a significant decrease of dielectric constant through an increase in free volume.

2.2. Effect of chain polarity

Polymers can be polar or non-polar. This feature affect significantly the dielectric properties. Examples of polar polymers include PMMA, PVC, PA (Nylon), PC while non-polar polymers include PTFE (and many other fluoropolymers), PE, PP and PS. Under alternating electric field, polar polymers require sometime to aligned the dipoles. At very low frequencies the dipoles have sufficient time to align with the field before it changes direction. At very high frequencies the dipoles do not have time to align before the field changes direction. At intermediate frequencies the dipoles move but have not completed their movement before the field changes direction and they must realign with the changed field. The electronic polarization and to some extent atomic polarization, is instantaneous weather at high or low frequency for both polar and non polar polymers. Therefore, polar polymers at low frequencies (eg 60 Hz) generally have dielectric constants of between 3 and 9 and at high frequencies (eg 100 Hz) generally have dielectric constants of between 3 and 5. For non-polar polymer the dielectric constant is independent of the alternating current frequency because the electron polarization is effectively instantaneous hence they always have dielectric constants of less than 3. The chain geometry determines whether a polymer is polar or non-polar. If the polymer is held in a fix confirmation, the resulting dipole will depend whether their dipole moments reinforce or cancell each other. In the case of extended configuration of PTFE, the high dipole moment of $-CF_2-$ units at each alternating carbon backbone cancelled each other since their vector are in opposite directions. Its dielectric constant therefore is low (2.1). On the other hand, PVC has its dipole moment directing parallel to each other resulting in reinforcement of dipole. Its dielectric constant is 4.5. This is illustrated as in Figure 2.

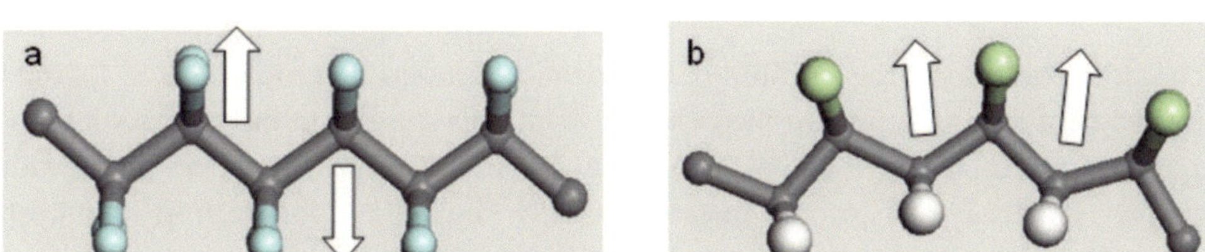

Figure 2. PTFE (a) and PVC (b) with arrow showing the net dipole moment.

The designing of dielectric material so as to achived the desired dielectric properties should take careful consideration of net polarity of the structure. This has been exemplified by the opposite effect in indiscriminately subsitituting fluorine atom into a polyimide chain resulting in an increase in otherwise low dielectric constant material.[11] There is no dipole polarization contribution for non-polar polymers as found in polar polymers. This different mode of mechanism lead to the resonance spectra in the case of electronic polarization which occur at frequency beyond 10^{12} Hz. At below this frequency, the relaxation spectra prevail relating to the behavior of dipole polarization. This observation can best be summarized as in the following Figure 3:

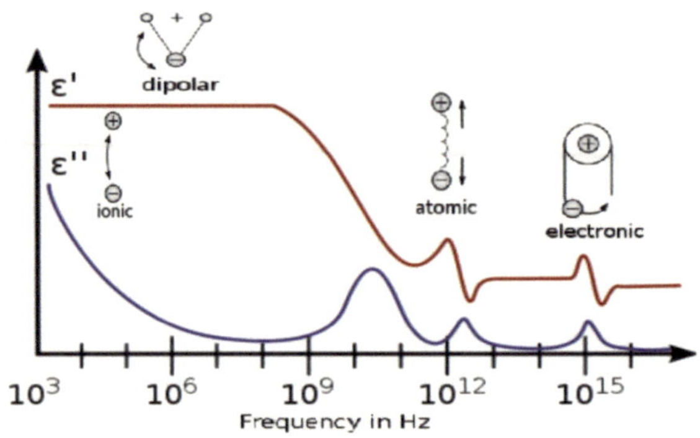

Figure 3. Dielectric constant and loss dispersion of dielectric materials against frequency (adapted from Wikipedia)

2.3. Relaxation and dielectric loss

Relative permittivity can be express in complex form as in Equation (4) below:

$$\varepsilon^* = \varepsilon' - j\varepsilon'' \tag{4}$$

It consist of the real part which is dielectric constant and the imaginary part which is the dielectric loss. The ratio between the dielectric loss with the dielectric constant is quantified as tan δ ie:

$$\tan \delta = \frac{\varepsilon''}{\varepsilon'}$$

Dielectric loss result from the inability of polarization process in a molecules to follow the rate of change of the oscillating applied electric field. This arise from the relaxation time (τ) in a polymer which is the time taken for the dipoles to return to its original random orientation. It does not occur instatntaneously but the polarization diminished exponentially. If the relaxation time is smaller or comparable to the rate of oscillating electric field, then there would be no or minimum loss. However when the rate of electric field oscillate well faster than the relaxation time, the polarization cannot follow the oscillating frequency resulting in the energy absorption and dissipated as heat. Dipole polarisability is frequency dependent and can be shown as in Equation (5)

$$\alpha_d = \frac{\alpha_o}{1 + j\omega t} \tag{5}$$

where α_d is the dipole polarisability and α_o is the low frequency (static) polarisability. It normally occur in the microwave region. Figure 3 above shows the variation in real dielectric constant with the imaginary dielectric loss. There is a sudden drop in dipole polarization region (< 10^{12} Hz) for dielectric constant ε accompanied with the maximum dielectric loss ε″. This maximum represent the complete failure for the dipole to follow the oscillating electric field beyond which the dipole remain freeze with no effective contribution to the dielectric constant. The mechanism for electronic and atomic polarization occur at higher frequency (shorter wavelength eg. infra-red region). This region involved excitation of electrons which is characterized by the quantized energy level hence is known as resonance behaviour. The dielectric constant display a maximum before a symmetrical drop about a certain frequency. These maximum and minimum represent the optimum polarization in phase with the oscillating frequency. The frequency at which the turning point occur is term the natural frequency ω_o. At this point the frequency of applied electric field is at resonant with the natural frequency hence there is a maximum absorption. Consequently this lead to maximum dielectric loss ε′.

2.4. Effect of temperature

Temperature affect dielectric properties. As the temperature is increased the intermolecular forces between polymer chains is broken which enhances thermal agitation. The polar group will be more free to orient allowing it to keep up with the changing electric field. At lower temperature, the segmental motion of the chain is practically freezed and this will reduce the dielectric constant. At sufficiently higher temperature, the dielectric constant is again reduced due to strong thermal motion which disturb the orientation of the dipoles. At this latter stage the polarization effectively contribute minimal dielectric constant. Beside the kinetic energy acquired, free space in the polymer matrix is of necessity so as to induce segmental movement. Throughout the measured frequency and temperature, electronic and atomic polarization are spontaneous. The dipole polarization, on the other hand, would significantly be affected during heat treatment by effectively reducing the relaxation time (τ) since the polymer chain τ would reduced as the temperature is increased hence the polymer segment would be better able to follow in phase with the oscillating electric field. Significant chain and segmental motions occur in polymers and they are identified as follows [39]:

i. α relaxation: Micro-Brownian motion of the whole chain. Formally this motion is designated as glass transition.
ii. β relaxation: Rotation of polar groups about C-C bond eg. CH_2Cl and $-COOC_2H_5$, conformational flip of cyclic unit.
iii. γ relaxation: libration of phenyl ring and limited C-H segmental chain movement.

The dielectric loss will show maxima at respective relaxation mechanisms as the temperature is increased. The loss in dielectric can be schematically represented as in the following Figure 4

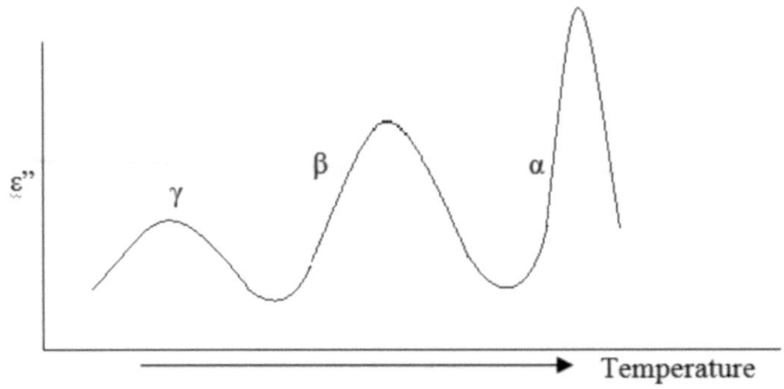

Figure 4. Schematic dielectric loss curve for polymer as temperature is increased.

The γ relaxation occur at lower temperature as it involved small entities of phenyl rings and C-H units whose motion are readily perturbed at low thermal energy. This is followed by β relaxation and finally α relaxation corresponding to the longer scale segmental motion. The broadness for each peaks signify dispersion in relaxation time as the result of different local environment of polarisable groups.

3. Structure-properties relationship.

3.1. Dielectric relaxation.

The earliest model of relaxation behavior is originally derived from Debye relaxation model [12] In this model, real and imaginary part of dielectric constant can be represented as in Figure 5

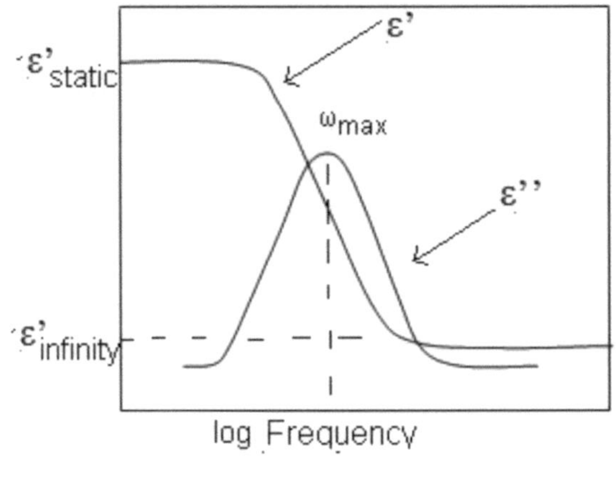

Figure 5. Debye dielectric dispersion curve.

where

$$\varepsilon' = \varepsilon_\infty + \frac{\varepsilon_{o-}\ \varepsilon_\infty}{1+\omega^2\tau^2}$$

$$\varepsilon'' = \frac{\varepsilon_{o-}\ \varepsilon_\infty}{1+\omega^2\tau^2}\omega\tau$$

This model relates the dielectric properties with the relaxation time. The relationship between ε′and ε″ can be formulated by eliminating the parameter ωτ to give Equation (6):

$$\left(\varepsilon' - \frac{\varepsilon_s + \varepsilon_\infty}{2}\right)^2 + \varepsilon''^2 = \left(\frac{\varepsilon_s - \varepsilon_\infty}{2}\right)^2 \quad (6)$$

This is a form of a semispherical plot which is popularly known as Cole-Cole plot. See Figure 6.

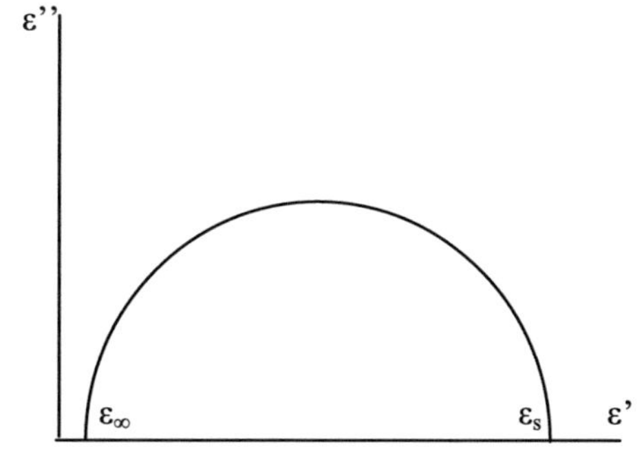

Figure 6. Cole-Cole Plot showing the relationship between dielectric constant and dielectric loss.

The plot shows that at dielectric constant of infinite frequency, ε_∞ and static dielectric constant, ε_s there will be no loss. Maximum loss occur at the midpoint between the two dielectric values. The larger the different between the static and infinite dielectric constant, the higher will be the loss. This model fit very well with polar small molecular liquids. However, polymeric materials are bigger in size, higher viscosity with entanglement between chains. This contribute to visco-elastic properties which requires some modifications to the original model. It can be noted that the above relationship involved only one specific relaxation time. This is contrary in polymeric system whose relaxation time is dependent on mobility of dipoles which behave differently in varying local environments. This result in distribution in relaxtion time. Modification include Cole and Cole semiemperical equation [13] Davidson and Cole [14] Williams and Watt [15] and Navriliak and Nagami [16]. The last modification lead to the new equation (7):

$$\varepsilon' = \varepsilon_\infty + \frac{\varepsilon_{o-}\ \varepsilon_\infty}{(1+\left(\omega^2\tau^2\right)^\alpha).^\beta} \quad (7)$$

where α and β is in the range 0 and 1. No physical meaning as yet is assignable to these parameters.[17] This modification result in a broader peak and smaller loss value with asymmetrical in features. The behaviour of dielectric constant and loss at variable frequencies and temperatures is exemplified in the following Figure 7 for polyvinylchloride.

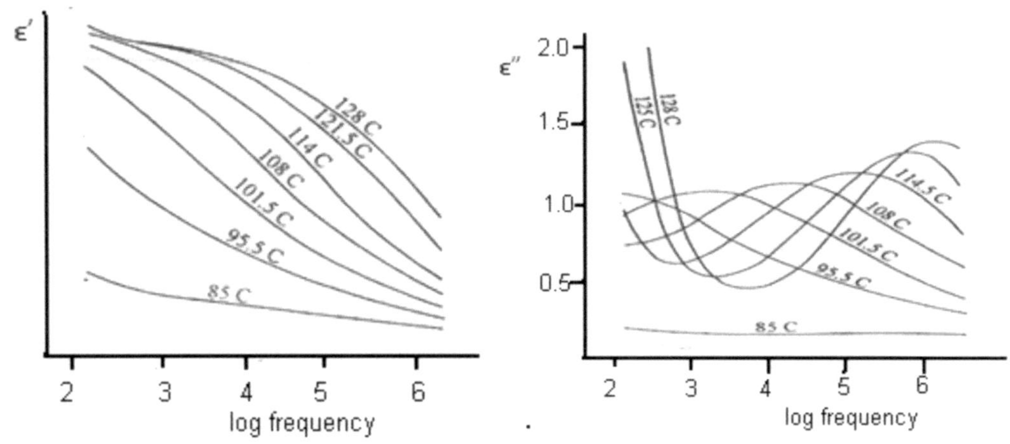

(a) (b)

Figure 7. Plot of dielectric constant (a) and dielectric loss with the change in frequency and temperature for polyvinylchloride. (From Ref 18)

Figure 7a shows the variation of ε' and ε" at the region of glass transition (85 ^{0}C) of polyvinylchloride. At the onset of glass transition the PVC showed a relatively low dielectric constant of 4.1 to 3.2 within the measured frequency range. With the increased in temperature, chain mobility begin to increase thus reducing the relaxation time. The dipole polarization of the polymer chain is better able to align in phase with the changing frequency and this account for the increase in dielectric constant as the temperature is increased. However this alignment with the applied oscillating field gradually failed as the frequency is increased. The optimum rate of decreased of dielectric constant occur at higher frequency as the temperature is increased. This correspond to the maximum dielectric loss in Fig 7b. Based on Cole-Cole plot, when there is a big difference in static and infinite dielectric constant, as under high thermal treatment, then the dielectric loss will be correspondingly large. It can be noted that at temperature 128 ^{o}C, there is a large dielectric loss occurring at higher frequency compared to those of lower temperature. Glass transition of polymer is a vital consideration that need to take into account during use of polymers as this affect the dielectric properties substantially. Substitution of fluorine into polyimide, for example, only affect the electronic polarization since PI is mostly used at temperature lower then its Tg (<260 ^{o}C). At this temperature, no effective polar orientation occurr which reduce any possibility of intrusion effect from this mechanism into the dielectric properties. The following Table 2 present the dielectric constant and loss of commercially used polymers.

Material	Dielectric constant (ε′)	Loss tangent (tan δ)	Frequency (Hz)
ABS (plastic)	2.0 – 3.5	0.005 – 0.0190	
Butyl rubber	2.35 2.35	0.001 0.0009	1 MHz 3 GHz
Gutta percha	2.6		
HDPE	1.0 – 5.0	0.00004 – 0.001	
Kapton (Type 100) (Type 200)	3.9 2.9		
Neoprene rubber	6.26 4.0	0.038 0.34	1 MHz 3 GHz
Nylon	3.2 - 5		
Polyamide	2.5 – 2.6		
Polycarbonate	2.8 – 3.4	0.00066 – 0.01	
Polypropylene	2.2		
Polystyrene	2.5 – 2.6	0.0001 0.00033	100 MHz 3 GHz
PVC	3		
Silicone (RTV)	3.6		
Teflon (PTFE)	2.0 – 2.1	0.0005 0.00028	100 Hz 3 GHz

Table 2. Dielectric parameters for some polymers at various frequencies.

3.2. Effect of cross-link between chains

Polymers are often cross-linked to improve their properties. The cross linking or curing process can be conveniently monitored based on relaxation time changes with the progress of reaction. This is exemplified during curing of diglycidylether bisphenol A (DGEBA) with diethyltetraamine (DETA)[19]. During the cross-linking process, the chains are covalently bonded to each other which induce chain rigidity.

Uncross-link chains

Cross-link network

Scheme 1. Affect of crosslink network on rigidity of polymer chains

This rigidity is proportional to cross-link density henceforth affecting the change in relaxation time. This can be illustrated in the following Figure 8:

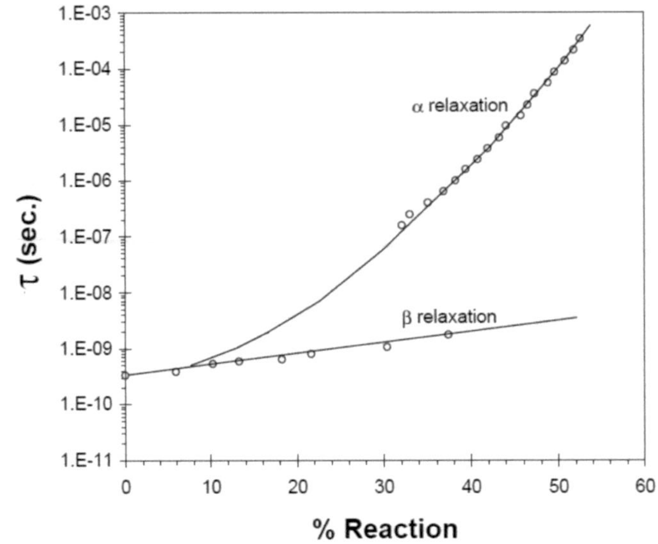

Figure 8. Effect of degree of reaction on the α and β relaxation time of DGEBA-DETA system

Figure 8 shows that both the α and β relaxation time increase with the increase in amount of cross-linking reaction. With the increase in cross-link density, the polymer chains are mostly bounded to each other much tighter hence inducing a longer time to return to their original equilibrium configuration. The rate of increase in α relaxation is higher as it approaches glassy state compared to β relaxation as the former relates to the segmental chain motion of larger in scale compared to the latter. The dielectric constant ε and loss is illustrated in the following Figure 9. In Figure 9a there is a significant drop in dielectric constant which correspond to the maximum frequency for dielectric loss in Figure 9b . This transition represent the frequency at which the dipole polarization is completely out of phase with the applied oscillating electric field. The maximum frequency ω_{max} of dielectric loss was extracted and applied into the equation $\tau = 1/\ \omega_{max}$ to yield the α relaxation time. It can be observed that as the level of curing is increased the maximum dielectric loss shift towards lower frequency while the change in dielectric constant at α_{static} with $\alpha_{infinity}$ become diminished. This behaviour represent the gradual transition from the rubbery state to glassy state of the polymer with the increase of cross-link density.

3.2.1. Polarizability and free volume

Polarizability and free volume are two important factors that influence the dielectric properties as formulated in the Clausius–Mossotti equation. These effects can be exemplified by introducing fluorine into a polymer chains.[20] Fluorination of polyimide film was performed through gaseous phase in a vacuum chamber. The impregnated fluorine content was determined using XPS analysis and the dielectric constant is shown as in following Table 3:

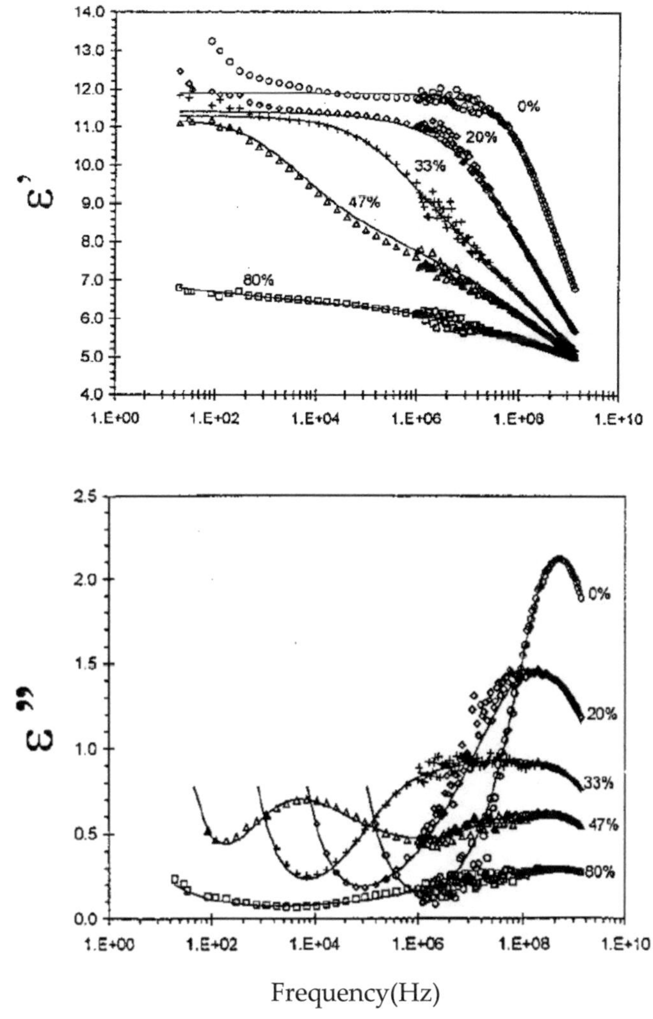

Figure 9. Effect of cross-link density on dielectric constant (ε′ above) and dielectric loss (ε″ below) for DGEBA-DETA system

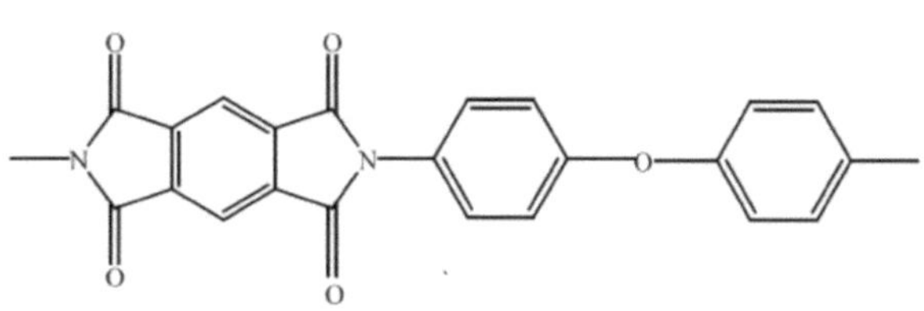

Scheme 2. Repeat unit of the polyimde (from Ref 20)

Sample	F_{1s}/C_{1s}(%)	Dielectric constant,ε	
		10^2(Hz)	10^6 (Hz)
F0	0	2.93	2.90
F2	57.8	2.64	2.60
F3	67.7	2.42	2.41
F4	78.6	2.28	2.27
F5	87.4	2.37	2.26

Table 3. The effect of Fluorine content on the dielectric constant of a polyimide

Similar result was obtained in a series of polyimides synthesised from starting monomers bearing varying percentage of fluorine content. [21] The decreased in dielectric constant as the fluorine content is increased can be explained as due to the low polarizability of fluorine. The electrons of fluorine being very tightly held and close to the nucleus. The polarizability of the fluorinated polyimides is decreased as the number of fluorine atoms is increased, due to the lower electronic polarizability of a C–F bond relative to that of a C–H bond that has been displaced. [22,23] The free volume concomitantly increases due to the relatively large volume of fluorine compared with hydrogen, which reduces the number of polarizable groups per unit volume.

The effect of free volume can be seen when introducing adamantane into a polyimide chain. [24] Adamantane is a bulky group which induce an increase in the free volume. The dielectric constant achieved was 2.7 at 1 KHz. This value is well below the commercial Kapton H film (25.4 μm) with a dielectric constant of approximately 3.5 at 1 kHz and 3.3 at 10 MHz. Besides, hydrophobicity was reduced thus preventing absorption of moisture. Low dielectric loss is important for a good capacitors and insulation. The strategy of introducing bulky substituents is further exemplified in a commercial Avatrel™ dielectric polymer made up of polynorbonene for passivation applications. It has a dielectric constant of 2.55, a loss tangent less than 0.002. These electrical properties held constant up to above 1 GHz. The bulky structures in these polymers are illustrated in the following Figure 10:

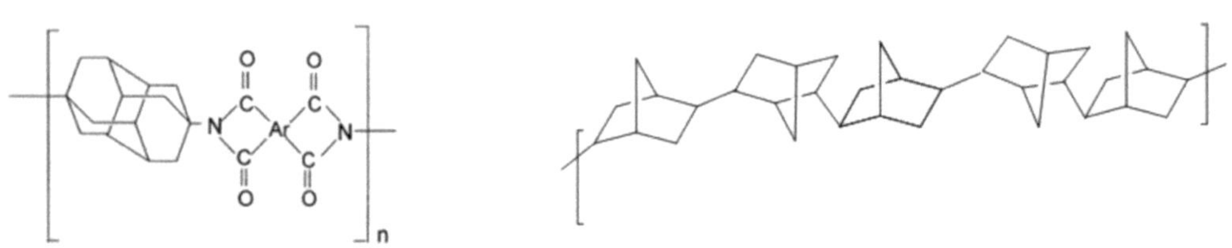

(a) (b)

Figure 10. Adamantane structure incorporated into polyimide chain (a, from Ref 12) and the generic structure for polynorbonene (b).

3.3. Dielectric breakdown

Electrical breakdown occurs when the dielectric strength which is the maximum electric field applicable on dielectric material is exceeded. It underwent catastrophic failure leading to short circuit or blown fuse. This occurs when at a given applied voltage the heat generated due to the losses is greater than the heat dissipated and if the voltage is applied long enough period then the dielectric is unable to reach a state of internal thermal equilibrium. The favourable condition for the occurrence of breakdown is large thickness of the dielectric, high temperature of both the dielectric and the surrounding, continous application of high voltage and large dielectric loss (high tan δ). The last factor is the most important to occur at high frequency. The high humidity in air can similarly affect dielectric breakdown through electrolytic process.

4. Designing of polymer dielectric materials

Based on the preceding discussions, designing of polymer dielectric materials can be made using several approaches. The following examples review two approaches undertaken of late namely free volume and copolymerization.

4.1. Free volume

Based on Maxwell-Garnet theory, the presence of second phase of lower dielectric constant in a composite will affect a significant decrease in dielectric constant.[25] This concept was applied in generating foam structure with the introduction of air-filled pores. At least two methods were utilised. One is to synthesised block copolymer of different thermal lability [26] and the other is performing solution etching of soluble component in a composite matrix. [27] The former method involved the use of block copolymer composed of high temperature and high Tg polymer and a second component of lower thermal property which can preferentially undergoes thermal decomposition. One of such a triblock polymer is shown below:

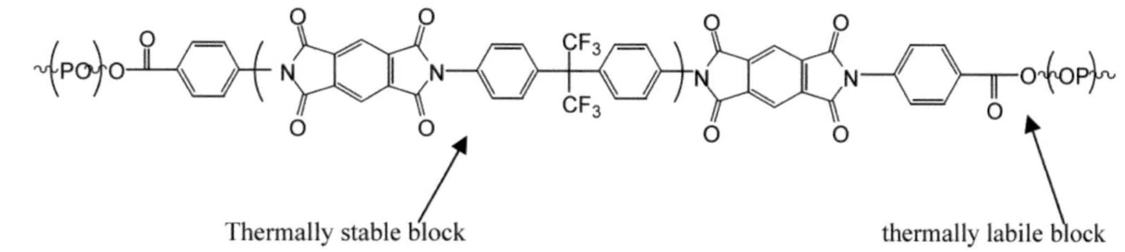

Scheme 3. Triblock polyimide structure illustrating the thermally labile and stable segments.

This triblock composed of thermally stable polyimide and thermally labile phosphate ester block. This copolymer is subjected to thermal treatment such that the temperature is sufficient to degrade the thermally labile block and leaving the thermally stable block intact. A small size scale of microphase saparation is then generated with spherical pore morphology, monodispersed in size and discontinuous. These nanopores are filled with air (ε = 1.0) which is responsible for the reduction in dielectric constant. Thermally labile oligomers include polymethylstyrene, polypropylene oxide and polymethylmethacrylate.

Nanofoam with dielectric constant of 2.3 was achievable with system made-up from PMDA/4BDAF/PPO triblock of void volume 16%. Figure 11 illustrate the relationship between the void content with the dielectric constant for PMDA/3FDA/PPO triblock system.[28]

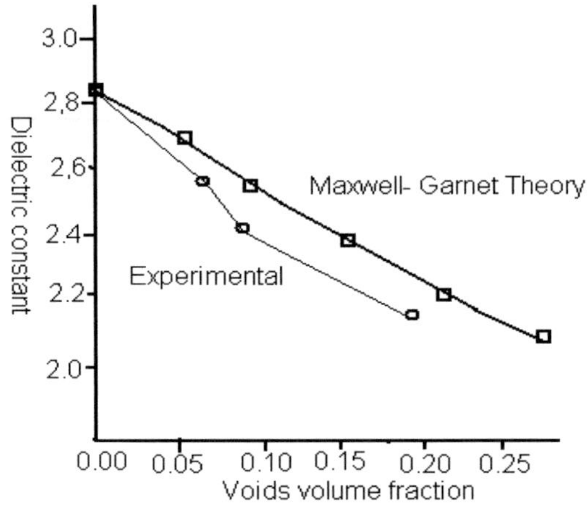

Figure 11. Relation between the dielectric constant with the void fractional volume in PMDA/3FDA/PPO triblock system.

In solution etching method, porosity were achieved by solution etching of soluble component in a nanocomposites leaving the chemically stable matrix intact. This was attempted using polyamic acid, a polyimide prepolymer, as the matrix while inorganic TEOS was incorporated through sol-gel method. Once the inorganic phase was homogeneously distributed in the polymer matrix, the composite was thermally cured followed by hydrofluoride etching. This will dissolved away the acid labile inorganic phase with the generation of nanosize closed cell pore of uniform density. The steps involved during its fabrication is illustrated as in the following Scheme (4):

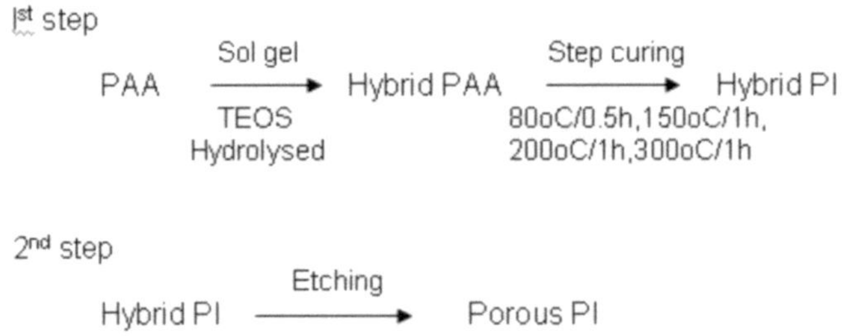

Scheme 4. Preparation of porous Polyimide using sol-gel method

The level of porosity is dependent on the TEOS content incorporated into the polymer matrix. Table 4 shows the dependence in dielectric constant on fluorine content and level of porosity based on TEOS content added during the materials fabrication

	F (Wt %)	0% TEO	10% TEOS	% TEOS
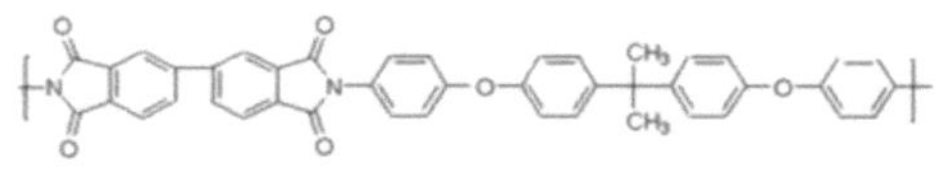 BAPP-BPDA	0	2.71	2.84	3.41
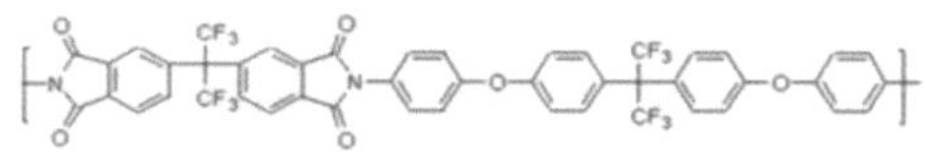 6FDA-BDAF	15	2.45	2.71	3.25
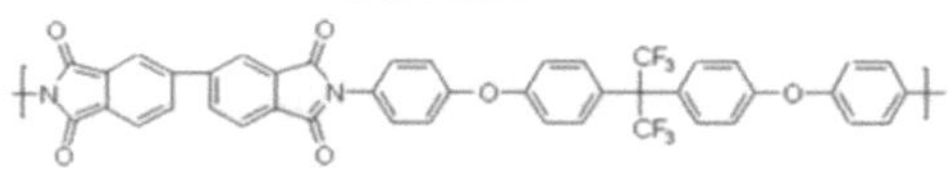 BDAF- BPDA	17	2.61	2.69	3.18
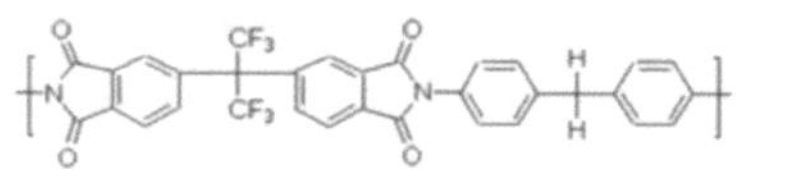 6FDA-MDA	33	2.50	2.62	2.98

Table 4. Dielectric constant of a series of polyimides at varying TEOS content

The results above display a general trend of decreasing dielectric constant as the TEOS concentration used during sol gel technique were increased. This was ascribed to an increased in void structures which reduced the dielectric property as the result of the presence of air. There was a linear decreased in dielectric constant as the weight percent of fluorine content in the structures were increased. Further the rate of decrease is almost constant between different TEOS content. Of the four synthesized polyimides, BAPP-BPDA showed the highest dielectric constant since this sample contains no fluorine. The SEM picture for the fracture surface morphology is shown in the following Figure 12.

(a) (b) (c)

Figure 12. SEM scan of fracture surface of pure (a) PI/SiO_2 10% (b) and PI/SiO_2 20 % porosity (c)

Simpsons *et al* [21] concluded that the presence of fluorine increases the free volume, lower electronic polarization and can either increase or decrease the dielectric constant depending on whether the substitution of the atoms are symmetric or asymmetric.

4.2. Copolymerisation

Copolymerisation of two or more polymers together is a strategy to produce a new materials of tailored dielectric properties. In copolymersation, two or more different monomer units were covalently bound thus producing a synergesic effect of respective constituents. Copolymerisation of polyimide with polysiloxane is popularly performed due to the complementary chemical and mechanical properties between the two. The polyimide has superior thermal and mechanical properties but too intractable to normal processing methods. For example the modulus of polyimides are in the range of 10^9 to 10^{12} Pa but their Tg are above 260 °C. On the other hand the polysiloxane is flexible and easily processable beside having a stable thermal degradation (> 400 °C). Copolymers of these materials produce an optimized dielectric material of practical application for several electronic packagings. Attempt was made with the following structures. [29]

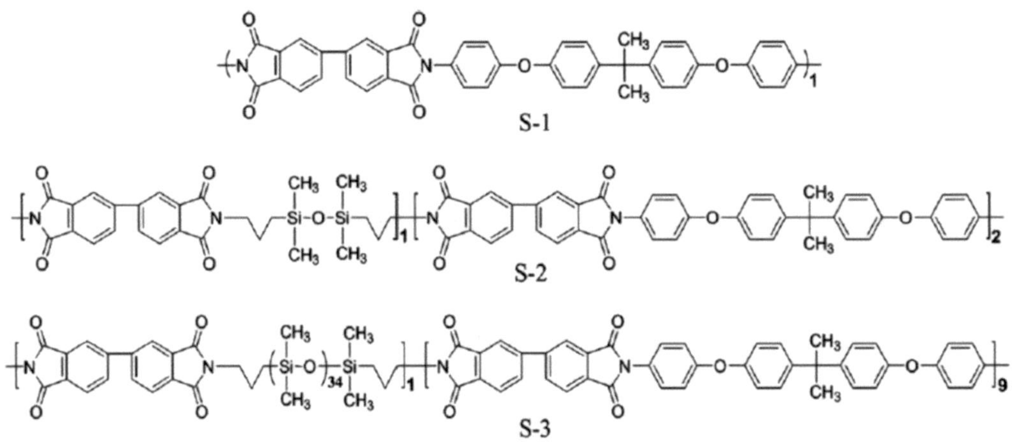

Scheme 5. Series of PI-polysiloxane copolymers

Their dielectric constant are shown in the following Table 5:

Sample	Dielectric constant at 1 kHz 293 °C	n (Si-O repeat unit)
S1	2.90	0
S2	2.57	1
S3	2.43	34

Table 5. Effect of silicone content in silicon-polyimide copolymers on dielectric constant.

The table above shows there is a decreasing trend in dielectric constant with the increase in siloxane units. Silicon is comparatively larger than a carbon atom and the Si - O bond is more flexible than the C - C bond. Thus, the bulky silicone units would be less mobile. Its presence affects the bulk movement of the whole polyimide network which reduces the efficiency of the dipole in reacting to polarity change during treatment with an alternating frequency. Furthermore, the molar polarization significantly decreases as the result of an increase in free volume. Several recent studies have demonstrated a similar trend of a decreasing dielectric constant, with an increasing siloxane content into polyimide structures [30,31]

5. Composites

The traditionally used inorganic material as a dielectric possesses several superior qualities such as excellent thermal, dielectric and magnetic properties. However they are brittle and consume high energy for processing. [32] On the other hand polymers are more flexible, strong resistivity and offer a tractable prosessibility. The disadvantages of polymeric materials are that they possesses lower thermal and dielectric properties. Combining the two materials in the form of nanocomposites offer an alternative in fabricating material of synergesic properties which displayed a tremendous improvement in dielectric properties with high flexibility and ease of processing. Their combination could readily geared towards miniaturization of electronic devices fabrication.

5.1. Polyimide-ceramic composites

Of late several attempts were made towards this strategy. Incorporation of alumina (Al_2O_3), barium titanate ($BaTiO_3$), titania (TiO_2) and zirconia (ZrO_2) into PI matrix were attempted.. [33,34] Several methods were employed in preparing these nanocomposites. It has been established that method of preparation affect the dielectric properties of these materials. A nanocomposite of PI/Al_2O_3 was prepared by mechanical stirring of prepolymer polyamic acid with the inorganic filler followed by thermal curing. [35] The nanocomposite showed an improved dielectric constant compared to a neat polymer material from about 3.0 to 3.4 at 1 MHz. This values increases correspondingly with the amount of filler loading. A further increase in dielectric constant was achieved when mixing was performed using ultrasonication. It has been shown from SEM result that this improvement was due to a better mixing during the latter treatment. Under these processes, the crystal structure of the inorganic fillers remains intact as shown by XRD data. The effect of good miscibility in improving the dielectric constant was proven when using a 3-Aminopropyltrimethoxysilane-treated (APS) ultrasonication. The APS served as an interface layer between the two immiscible organic PI with inorganic filler which reduced any agglomeration between the different phases. This is brought about possibly through the formation of hydrogen bond between the amine moeity of APS with the polar group of polyimide while the inorganic part of the methoxysilane of APS form secondary interaction with the inorganic fillers. Figure 13 reveals the SEM images of PI/ Al_2O_3 composites doped by the treated Al_2O_3 powder.

PA0 demonstrated a neat and clean morphology. The Al_2O_3 particles were homogeneously dispersed into PI matrix in all PA10, PA20 and PA30. The inset images revealed the average size of Al_2O_3 was around 2μm - 4μm. There was no obvious aggregation observed suggesting the improved compatibility between PI matrix and Al_2O_3 attributed to the APS coupling agent. The bahaviour of PI-nanocomposites for $BaTiO_3$, TiO_2 and ZrO_2 displayed similar trend with that of PI-Al_2O_3 nanocomposites. They can be summarized as in the following Figures 14:

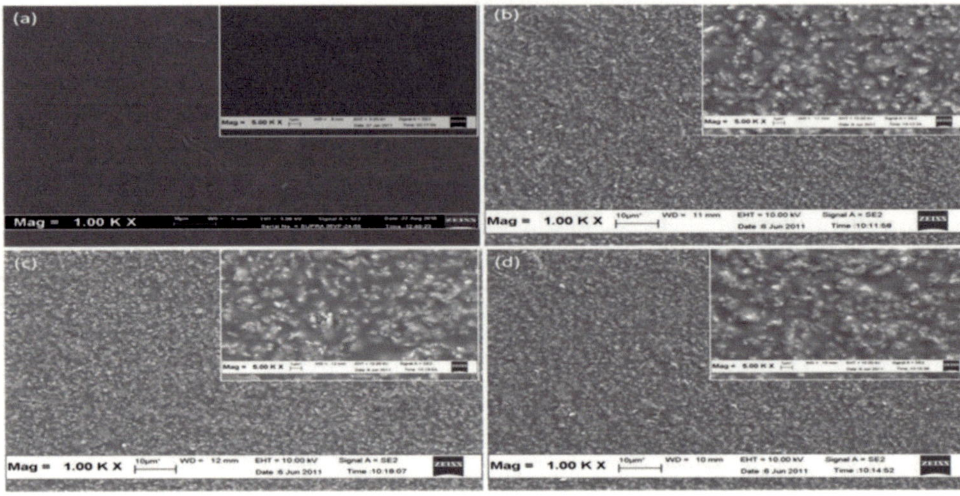

Figure 13. The morphology of PA0, PA10, PA20 and PA30 (a, b, c and d), respectively.

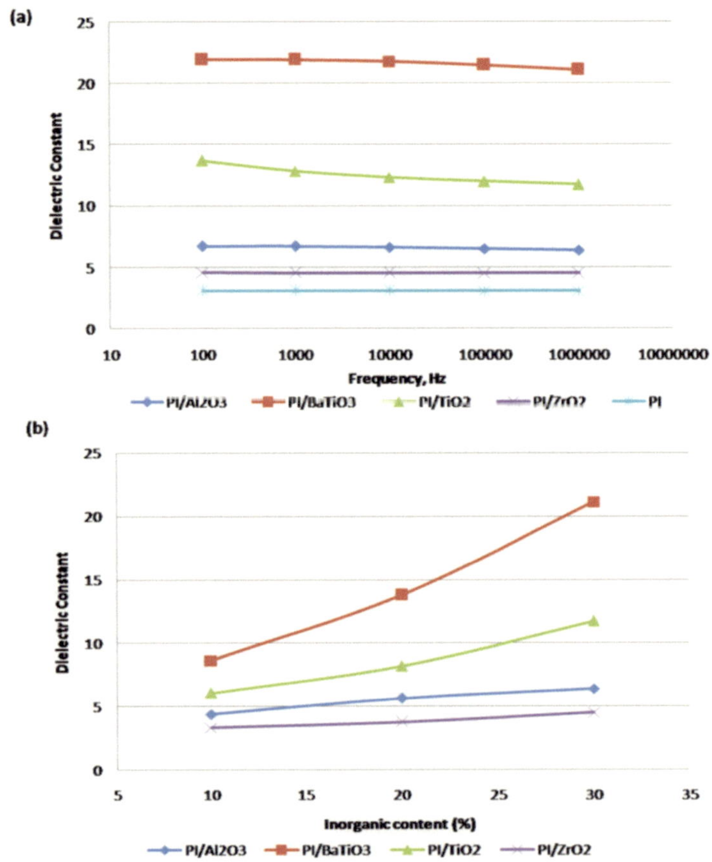

Figure 14. Dielectric constant of PI/inorganic (a) with 30 wt% inorganic content at varying frequency and (b) at 1MHz for several type of inorganic fillers.

All composite systems displayed a decreased in dielectric constant with the increase in frequencies. The dielectric constants increased as the inorganic filler content were increased. This can be attributed to the increase in polarizability group with the incorporation of the inorganic fillers which replace significant part of the PI in the matrix. As the result the polarizable units per unit volume and the space charge polarizability which occurred at the interfaces between PI matrix and inorganic particles were increased. Fig 13(a) shows the dielectric constant of PI/$BaTiO_3$ composite films demonstrating the highest value of dielectric constant followed by PI/TiO_2, PI/Al_2O_3, PI/ZrO_2 and neat PI films. Apparently this property is dependent on the dielectric constant of the respective fillers. $BaTiO_3$ was known to display highest value of dielectric constant [36] followed by TiO_2, ZrO_2 and Al_2O_3 in their neat form. $BaTiO_3$ possesed perovskite structure which is capable to polarize in the absence of electric field. This feature remains in the composite as the crystal structure remains intact as established in XRD data. The low dielectric constant for PI/ZrO_2 was attributed to the poor compatibility between phases resulted in the presence of voids and even led to cracks. The presence of voids naturally induce a low dielectric constant.

5.2. Composite models

Several models were proposed in predicting the dielectric constant of the composites which include Maxwell-Wagner model, Logarithmic Mixing Law and Bruggeman Model. [37] These models allow designing of composite materials based on respective dielectric constant of the polymer, inorganic filler, composition ratio as well as the filler sizes. The slight discrepancy of these models which do not fit to most composite systems are mainly due to inconsistency in treatment for the interphase interaction hence further modification is required. An interphase interaction factor, K, was introduced during fitting into this models.[38] A typical plot of composite dielectric constant with respect to the volume fraction of the fillers is illustrated in the following Figure 15:

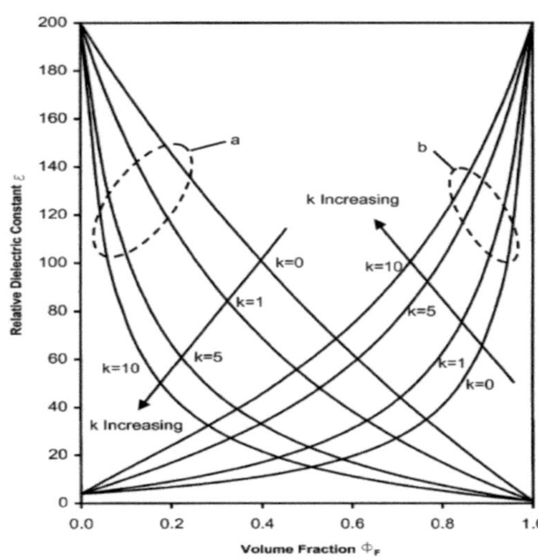

Figure 15. The prediction of the effective dielectric constant as a function of filler volume fraction for different K values. (a) The case of $\varepsilon_{polymer} > \varepsilon_{filler}$. (b) The case of $\varepsilon_{polymer} > \varepsilon_{filler}$ (Adapted from Ref 38)

At K = 0, there is no interaction between phases while a high K values showed a strong interaction. This interaction also dependent on the filler sizes. For a given volume fraction filler, a smaller particle size has a larger fraction of interphase volume in the region between the filler and the matrix granting more polarization to operate. Thus they lead to a relative increase in dielectric constant.

A major concern with polymer/ceramic composites is the heterogeneity in phase which lead to formation of cracks and voids. This effect is known as Maxwell-Wagner effect which reduce the dielectric constant. A more serious type of heterogeneity is that the composite comprised of conductive inorganic fillers which could lead to a mistaken interpretation of dipole polarization occurring at very low frequency region.

6. Conclusion

Polymers offer an alternative to the traditionally inorganic and ceramic material as dielectric amterials. This is due to their highly flexible, tractable processing, good chemical stability and readily tunable properties. The main drawback is they have lower thermal stability which limit their wider applications. Generally their dielectric constant is lower than non-polymeric materials. The mechanism which contribute to the dielectric properties are the interaction of electric field with electronic, atomic and dipole poalarization. These are dependent on polarizabity of constituents structure and the free volume as formulated in Clausius-Mossotti equation. The electronic and to some extent the atomic polarization are instantaneous throughout the measureable range of frequencies. However in dipole polarization there is relaxation time allowing an exponential decay of motion to return to equilibrium state. This different behavior contribute significantly to the values of dielectric constant and dielectric loss. These properties can be utilized to detect for any local or segmental motion during change in frequency and temperature treatment. Dielectric properties can be design by introducing polarizable groups into polymer chains, increasing free volume by inducing porosity as well as copolymerization. Increasing dielectric constant can be effectively made by producing nanocomposites with inorganic fillers possessing high dielectric constant.

Author details

Zulkifli Ahmad
University Sains Malaysia, Malaysia

Acknowledgement

The author wish to express a grateful acknowledgement to Universiti Sains Malaysia for financial support through Short Term and Research University (Individual) Grants awarded for the accomplishment of related works quoted in this paper.

7. References

[1] Milton H P, Electric Capacitor, US Patent 3311801 assigned to Intron International, NY, Mar 1967

[2] Alfred S C and Charles E W Jr, Electric Capacitor and method of Making the same, US Patent 3252830 assigned to General Electric Company, NY may 1966

[3] Yuxing R and Davud C L, Properties and Microstructures of Low-Temperature-Processable Ultralow-Dielectric Porous Polyimide Films, Journal of Electronic Materials, 2008, Vol. 37, No. 7.

[4] Tummala R R and Rymaszewski E J. Microelectronics Packaging Handbook. Van Nostrand Reinhold. New York, 1989. Chap. I; Numata S, Fujisaki K. Makino D and Kinjo N. Proceedings of the 2nd Technical Conference on Polyimides, Society of Plastic Engineers, Inc., October 1985, Ellenville, New York, p. 164.

[5] Ranjan V, Yu L , Nardelli M B and Bernholc J , Phase Equilibria in High Energy Density PVDF-Based Polymers, Phys. Rev. Lett. 2007, 99, 47801.

[6] Xie S H, Zhu B K, Li J B, Wei X Z and Xu Z K, Preparation and properties of polyimide/aluminum nitride composites. Polymer Testing, 2004, 23(7), 797-801.

[7] Liu L, Liang B, Wang, W. and Lei Q. Preparation of polyimide/inorganic nanoparticle hybrid films by sol-gel method. Journal of Composite Materials, 2006, 40(23), 2175-2183

[8] Li H, Liu G, Liu B, Chen W and Chen S, Dielectric properties of polyimide/Al_2O_3 hybrids synthesized by in-situ polymerization. Materials Letters, 2007, 61(7), 1507-1511.

[9] Indulkar C S and Thiruvengadam S, An Introduction to Electronic Engineering Materials, 2008, S Chand and Company, New Delhi, India.

[10] Blythe T and Bloor D, Electrical Proeprties of Polymers, 2nd Ed., (2005) Cambridge University Press.

[11] Hougham G, Tesoro G, Viehbeck A, Chapple-Sokol J D, Polarization Effects of Fluorine on the Relative Permittivity in Polyimides, *Macromolecules*, 1994, 27, 5964.

[12] Debye P, Polar molecules, (1929) Chemical Cataloque Company, reprinted in New York by Dover Publications.

[13] Cole R R and Cole K S, Dispersion and Absorption in Dielectrics I. Alternating Current Characteristics,1941, J Chem Phys , 9, 341.

[14] Davidson D W and Cole R H, Dielectric relaxation of glycerine, 1950, J. Chem Phys., 18, 1417

[15] Williams G and Watt D C, Non-symmetrical dielectric relaxation behaviour arising from a simple empirical decay function 1970, Trans Faraday Soc 66, 80

[16] Navriliak S and Havriliak S J, Dielectric and Mechanical Relaxation in Materials – Analysis, Interpretation and Application to Polymers, 1997, Munich, Hanser.

[17] Lukichev A A, Graphical method for the Debye-like relaxation spectra analysis, Journal of Non-Crystalline Solids, 2012, 358 447–453.

[18] Y Ishada, Colloid Z.,1960, 168, 29

[19] Oplicki M and Kenny J M, Makromol Chem. Makromol Symp, 1993, 68, 41

[20] Park S J, Cho K S and Kim S H, A study on dielectric characteristics of fluorinated polyimide thin film Journal of Colloid and Interface Science, 2004, 272, 384–390

[21] Simpson J O and St.Clair A K, Fundamental insight on developing low dielectric constant polyimides, Thin Solid Films,Volumes 1997, 308-309, 480-485

[22] Nansé G, Papirer E, Fioiux P, Mocuet F, Tressaud A, Carbon 1997, 35, 515

[23] Alegaonkar P S, Mandale A B, Sainkar S R, Bhoraskar V N, Nucl. Instrum. Methods Phys. Res. Sect. B Beam Interact. Mater. Atoms, 2002, 194, 281.

[24] Chern Y T and Huang C M, Synthesis and Characterisation of a new polyimides derived from 4,9-diaminodiamantane, Polymer, 1998, 39(25), 6643.

[25] Bergman D J and Stroud D, The physical properties of macroscopically inhomogeneous media, Solid State Phys. 1992, 46,148.

[26] Hedrick J, Labadie J, Russel TP, Hofer D, Wakharker V, High temperature polymer foam, 1993, Polymer, 34(22) 1747

[27] Nicholas Ang Soon Ming, Muhammad Bisyrul Hafi Othman, Hazizan Md Akil, Zulkifli Ahmad Dielectric Constant Dependence on Fluorine Content and Porosity of Polyimides, Journal of Applied Polymer Science, Volume, 2011, 121(6), 3192.

[28] Hedrick J L, Carter K K, Labadie J W, Miller R D, Volksen W, Hawker C J, Yoon D Y, Russell T P, McGrath J E, Briber R M, Nanoporous Polyimide, Adv Polym Sci., 1999, 40, 1 - 43

[29] Muhammad Bisyrul Hafi Othman, Mohamad Riduwan Ramli, Looi Yien Tyng, Zulkifli Ahmad, Hazizan Md. Akil, Dielectric constant and refractive index of poly (siloxane–imide) block copolymer, Materials and Design, 2011, 32, 3173–3182

[30] Tao L, Yang H, Liu J, Fan L, Yang S. Synthesis and characterization of highly optical transparent and low dielectric constant fluorinated polyimides. Polymer 2009;50:6009.

[31] Ghosh A, Banerjee S. Thermal, mechanical, and dielectric properties of novel fluorinated copoly (imide siloxane) s. J Appl Polym Sci 2008, 109, 2329–40.

[32] Wang S F, Wang T R, Cheng K C, Hsaio Y P, Characteristics of polyimide/ barium titanate composite films, Ceramics International 35 (2009) 265–268

[33] Bai, Y., Cheng, Z. Y., Bharti, V., Xu, H. S. & Zhang, Q. M. High-dielectric-constant ceramic-powder polymer composites. Applied Physics Letters, 2000, 76(25), 3804-3806

[34] Kuo, D. H., Chang, C. C., Su, T. Y., Wang, W. K. & Lin, B. Y. Dielectric behaviours of multi-doped $BaTiO_3$/epoxy composites. Journal of the European Ceramic Society, 2001, 21(9), p.1171-1177;

[35] Asliza Alias, Zulkifli Ahmad, Ahmad Badri Ismail, Preparation of polyimide/Al_2O_3 composite films as improved solid dielectrics Materials Science and Engineering: B, 2011, 176(10), 799-804

[36] Viswanathan B (2006), Structutre and properties of solid state materials, Oxford Alpha Science Internal LTd, p43 – 77

[37] Scaife B K P, Principles of Dielectric, Oxford Science Publication, USA198; NE Hill, Dielectric Properties and Molecular Behaviour, Van Nostrand-Rheinhold, England 1969

[38] Hung T V and Frank G S, Towards model based engineering of optoelectronic packaging materials: dielectric constant modeling., Microelectronics Journal, 2002, 33, 409 – 415

[39] R J Young, Introduction to Polymers, (1989) Chapman & Hall Ltd.

Chapter 2

Dielectric Materials for Compact Dielectric Resonator Antenna Applications

L. Huitema and T. Monediere

Additional information is available at the end of the chapter

http://dx.doi.org/10.5772/50612

1. Introduction

Dielectric resonators using high-permittivity materials were originally developed for microwave circuits, such as filters or oscillators as tuning element [1]. Indeed, in the late nineteen sixties, the development of low-loss ceramic materials opened the way for their use as high-Q elements [2-4]. Then, making use of dielectric materials to create the dielectric resonator antenna (DRA) illustrates the ingenuity of Professor S. A. Long [5], who was the first to propose such a procedure in the early nineteen eighties. Indeed, it introduced the use of a dielectric resonator as an antenna by exciting different modes using multiple feeding mechanisms. During the nineties, emphasis was placed on applying analytical or numerical techniques for determining input impedance, fields inside the resonator and Q-factor [6]. Kishk, Junker, Glisson, Luk, Leung, Mongia, Bhartia, Petosa and so on, have described a significant amount of DRAs' analyses and characterizations [7-18]. Petosa and al. proposed both in literatures and book [6,12] many of the recent advances on DRAs.

Current DRA literatures focus on compact designs to address portable wireless applications. Among them, new DRA shapes or hybrid antennas are developed to enhance the antenna impedance bandwidth [13-19] or for multiband antenna applications [20-22].

The first part will address a brief overview of the most common used DRA shapes and structures including both rectangular and cylindrical DRAs. The emphasis will be placed on better understanding what DRAs exactly are and how to develop such an antenna. This part will detail fundamental modes of DRAs, their resonant frequencies, fields inside the resonator and radiation patterns corresponding to these modes.

A second part will focus on the relevant dielectric material properties having a significant contribution to achieve better antenna performances. It will detail the kind of materials DRAs can use, which is closely linked to the targeted application.

Multiple techniques to miniaturize such an antenna will be presented in the third part, supported by concrete examples. At the same time, everyone will be able to appreciate that dielectric material properties have a major role to play in designing a DRA. It should be noted that the material choice is even more critical when the targeted challenge is the antenna size reduction.

Therefore, depending on the intended applications, this part will enable to find the best trade-off between the material choice and its shape.

Although some wideband or multiband DRA structures have been introduced in the third part, the fourth and last part will be dedicated to a new method to design a DRA. It will address engineering design data on hybrid modes creation to enhance the bandwidth or develop multiband antennas. This part will include many references to clearly explain this research method while highlighting their contribution to expand the use of DRA in new kind of mobile handheld devices (e.g. new tablets).

2. Overview on DRA studies

The design of a DRA in any geometry must satisfy various specifications including: the resonant frequency, the impedance bandwidth, the field distribution inside the resonator and also the radiated field. The intent of this part is to provide an understanding of fundamental operation of DRAs, emphasizing both design and implementation. Thus, to provide comprehensive research method, this part will start by presenting main findings of investigations on simple-shaped DRAs. Then, it will deal with the different DRA feeding methods. Finally, this part will focus on the study of two DRA shapes: cylindrical and rectangular.

2.1. Main DRAs characteristics

A non-exhaustive list of main simple-shaped DRAs characteristics is described below:

- The main dimension of a DRA is proportional to $\lambda_0 / \sqrt{\varepsilon_r.\mu_r}$ where λ_0 is the free-space wavelength at the resonant frequency, ε_r and μ_r are respectively the dielectric and the magnetic constant of the material. In a dielectric material case, $\mu_r = 1$ and the main dimension of a DRA is proportional to $\lambda_0 / \sqrt{\varepsilon_r}$.
- The radiation efficiency of the DRA is highly depending on the material losses. In case of a low-loss dielectric material, DRAs allow to achieve better efficiency than other kind of antennas because of minimal conductor losses associated with a DRA.
- For a given dielectric constant, both resonant frequency and radiated Q-factor are defined according to the resonator dimensions. That allows having a great flexibility and some degrees of freedom to design such an antenna.
- Another degree of freedom is the large spectrum of available dielectric materials. That allows doing the best trade-off between dimensions and impedance bandwidth according to the intended application.

- A number of modes can be excited within the DRA, many of them provide dipolar-like radiation characteristics.
- The most common targeted frequencies presented by the research literatures are ranging from 1GHz to 40 GHz.
- For a given DRA geometry, the radiation patterns can be made to change by exciting different resonant modes.
- A large number of DRA excitations are currently used, e.g. microstrip line, coaxial probe excitation, coplanar waveguide… The next subsection will deal with the most commonly used excitations.

2.2. Common DRAs feedings

Multiple feeding mechanisms are employed to excite different resonator modes. This subsection will summarise most widely-used excitations while giving many references in order designers to choose the most appropriate excitation.

2.2.1. Coaxial probe excitation

It can be located within the DRA or adjacent to it. Within the DRA, a good coupling can be achieved by aligning the probe along the electric field of the DRA mode as shown Figure 1.

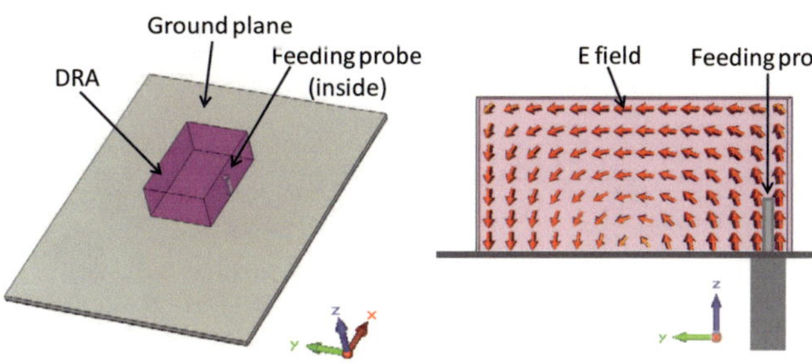

Figure 1. Coaxial probe coupling the E field

The adjacent position is currently used to couple the magnetic field of the DRA mode (Figure 2). In these both cases, the probe is exciting the TE_{111} fundamental mode of the rectangular DRA.

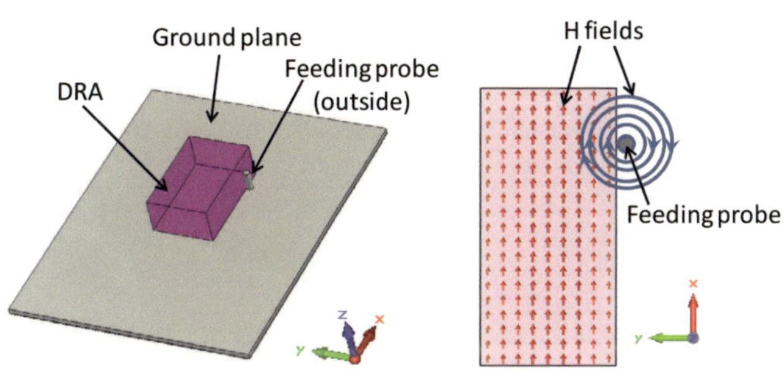

Figure 2. Coaxial probe coupling the H field

When the excitation probe is inside the resonator, particular attention has to be paid to the air gap between the probe excitation and the dielectric material. Indeed, an air gap results in a lower effective dielectric constant, which entails both a decrease in the Q-factor and a shift of the resonance frequencies [23-24]. The probe location allows choosing the intended excited mode and the coupling of the mode can be optimized by adjusting both length and height of the probe.

2.2.2. Microstrip feeding line and coplanar waveguide

The principle is similar to the probe excitation case. A microstrip line placed close to the DRA can couple the magnetic field of the DRA mode. However, this latter can affect the antenna polarisation and can thus increase the parasitic radiation. This could be reduced by placing the line under the resonator as shown Figure 3.

Another way is to replace the microstrip line by a coplanar waveguide, the Figure 3 is presenting a rectangular DRA excited by a coplanar waveguide.

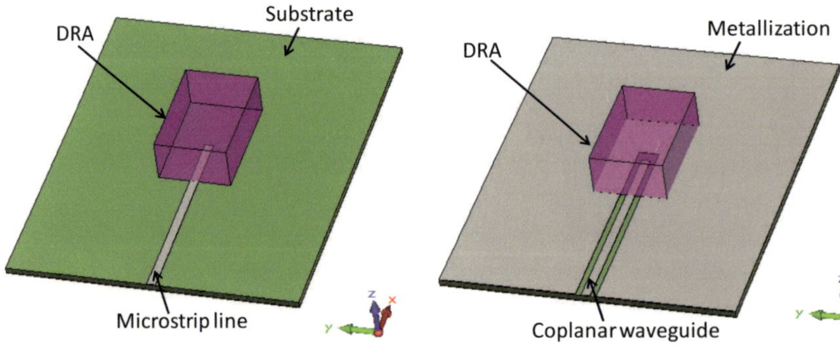

Figure 3. Microstrip feeding line and coplanar waveguide

In these both cases, the mode coupling can be optimized by changing the resonator position and/or its dielectric permittivity. For low dielectric permittivity materials (which allows obtaining a wide bandwidth), it is somewhat difficult to excite the mode. There are different solutions to obtain both miniaturization and good coupling, they will be explained in this chapter.

An important point is that these excitation methods are disturbing DRA modes by introducing electrical boundary conditions. This issue is all the more sensitive since the antenna is miniature. The last part of the chapter will show how to take advantage of this issue.

2.2.3. Aperture coupling

A common method of exciting a DRA is acting through an aperture in the ground plane. The Figure 4 shows an example of the excitation of the TE_{111} mode of a rectangular DRA with a rectangular slot. To achieve relevant coupling, the aperture has to be placed in a DRA strong magnetic area. Feeding the aperture with a microstrip line is a current approach [25-26].

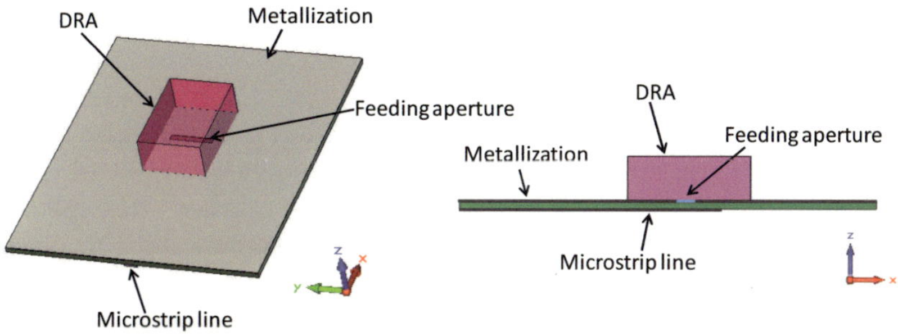

Figure 4. Aperture coupling the TE_{111} mode of the rectangular DRA

The main dimension of the aperture needs to be around $\lambda_g/2$, which is highly problematic at low frequencies.

On top of these multiple feeding methods, the choice of different DRAs shapes represents another degree of flexibility and versatility. The next subsection will deal with the cylindrical shape.

2.3. The cylindrical DRA

It offers great design flexibility, where both resonant frequency and Q-factor are depending on the ratio of radius/height. Various modes can be excited within the DRA and a significant amount of literature is devoted to their field configurations, resonant frequencies and radiation properties [27-28]. This part will present a complete and concrete study of a cylindrical DRA.

Like most realistic cases, the cylindrical DRA presented Figure 5, is mounted on a finite ground plane. Because dielectric material properties will be studied in very great depth in the third part, the dielectric permittivity ε_r of the DRA is fixed and chosen equal to 30. It is also characterized by its height d and its radius a. To keep the chapter concise while remaining comprehensive, only relevant results and equations will be given.

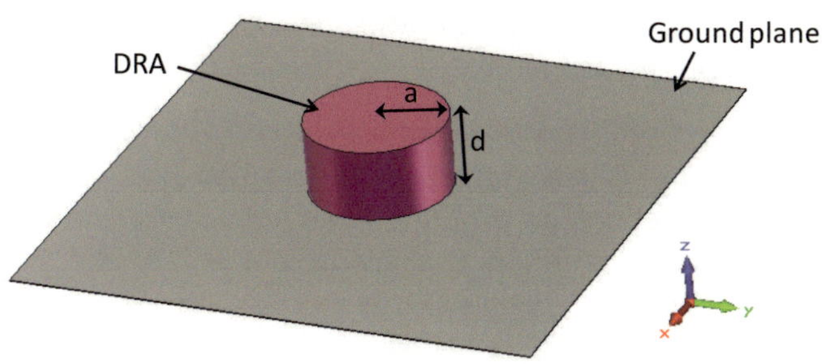

Figure 5. Cylindrical DRA

First of all, the study can begin with a modal analysis. This allows determining the most appropriate excitation method. Modes of cylindrical DRA can be divided into three types: TE, TM and hybrid modes, i.e. EH or HE [9-10,30]. The latter have a dependence on azimuth ϕ, while TE and TM modes have no dependence on azimuth. To identify fields' variations according ϕ (azimuth), r (radial) and z (axial) directions, subscripts respectively noticed n, p and m are following the mode notation. Because TE and TM modes have no azimuthal variation, n=0 for these modes. Finally, all cylindrical DRA modes can be defined such as: $TE_{0pm+\delta}$, $TM_{0pm+\delta}$, $HE_{npm+\delta}$ and $EH_{npm+\delta}$. The δ value is ranging between 0 and 1, it approaches 1 for high ε_r. It should be noted that n, m and p are natural numbers. The modal analysis of a DRA can be deducted either with analytical calculations or thanks to electromagnetic simulators like CST Microwave Studio (CST MS).

2.3.1. Modal analysis

- Analytical equations derive from the analytical calculations of a cylindrical dielectric resonator by assuming perfect magnetic and/or electric walls on resonator faces. The perfect magnetic wall boundary condition was demonstrated to be valid for high ε_r values [30], it remains accurate for lower values as well.

Fields equations inside the DRA, resonant frequencies and Q factor are detailed in [7,9] and [12]. Resonant frequencies of all modes are provided hereafter:

$$\begin{pmatrix} f_{TMnpm} \\ f_{TEnpm} \end{pmatrix} = \frac{c}{2\pi a\sqrt{\varepsilon_r \mu_r}} \sqrt{\begin{pmatrix} X'_{np} \\ X_{np} \end{pmatrix}^2 + \left(\frac{(2m+1)\pi.a}{2d}\right)^2} \tag{1}$$

Where X_{np} and X'_{np} are Bessel's solutions, $(n,m,p) \in \mathrm{N}^3$, a and d are the radius and the height of the dielectric resonator.

In the case presented Figure 5, the fundamental excited mode is the $HE_{11\delta}$ and its resonant frequency equals:

$$f_{110} = \frac{3.10^8}{2\pi\sqrt{30}} \sqrt{\left(\frac{X'_{11}}{0.04}\right)^2 + \left(\frac{\pi}{2\times 0.045}\right)^2} = 503.6MHz \tag{2}$$

This method requires the resolution of both Maxwell and propagation equations. It was therefore reserved for simple-shaped DRA.

- The "Eigenmode solver" of CST MS allows viewing 3D fields of each mode and having their resonant frequencies. When considering the studied example and defining perfect magnetic boundary conditions on all DRA walls, except for the DRA bottom where perfect electric boundary condition is considered (due to the ground plane), the software gives 504 MHz for the resonant frequency value of the $HE_{11\delta}$ mode. It is also possible to see both H and E fields of this mode, they are presented Figure 6. This valuable information allows designer to choose the most appropriate excitation of the considered mode.

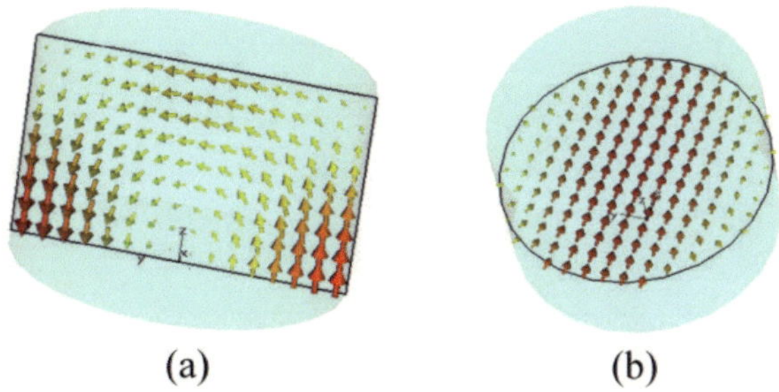

Figure 6. E field (a) and H field (b) of the $HE_{11\delta}$ mode

In light of this above, the best way to excite the $HE_{11\delta}$ mode is to integrate a coaxial probe along the E field inside the DRA.

Now that the modal analysis was explained and the excitation determined, we can go straight to the electromagnetic study of the DRA.

2.3.2. *Electromagnetic study*

Since input impedance and also S_{11} parameter cannot be calculated with the magnetic wall model, their study is solely possible with an electromagnetic simulator or of course can be experimentally done. Moreover, electromagnetic simulators facilitate accurate and efficient antenna analysis by providing the complete electric and magnetic fields inside and outside the antenna taking into account the finite ground plane. The impedance, the S_{11} parameter, the power radiated and the far field radiation pattern are determined everywhere and at any frequency in a single analysis thanks to the Finite Integration Temporal method.

The Figure 7 presents the cylindrical DRA excited by a coaxial probe.

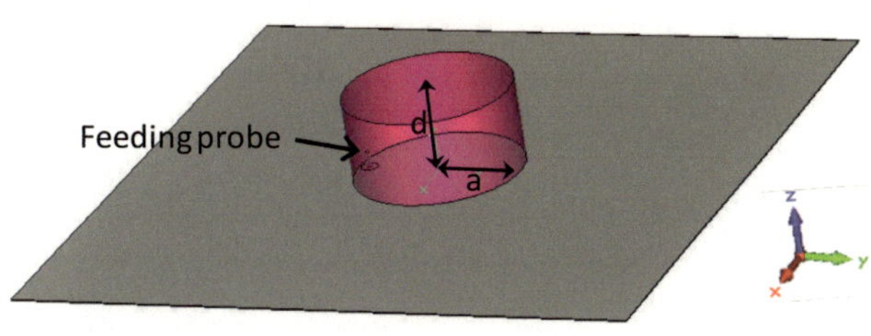

Figure 7. Excitation of the $HE_{11\delta}$ mode with a coaxial probe

Using the CST MS software, the input impedance is presented Figure 8. It shows that the fundamental mode is excited at the resonant frequency fairly corresponding to the predicted value by the modal analysis of 504 MHz. The minor shift between resonant frequencies deducted with modal analysis and electromagnetic study is due to magnetic wall model used during the modal analysis, which is not absolutely accurate.

An important data for antenna designers is the S_{11} parameter (Figure 8). It is directly deducted from the input impedance parameter.

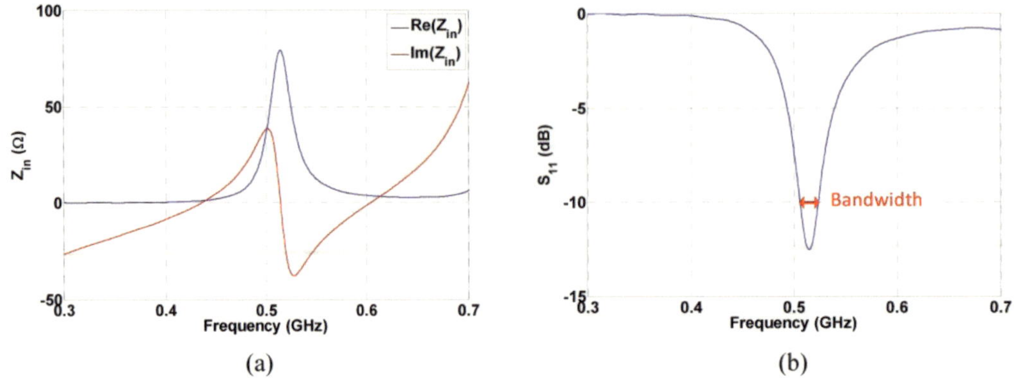

Figure 8. Input impedance (a) and S_{11} parameter (b) of the DRA

It allows knowing the matching frequency as well as the impedance bandwidth, basically corresponding to the working frequencies of the antenna. It this case, the matching frequency equals 510 MHz with 3.6% of bandwidth at -10 dB.

Another important issue is the radiation pattern. It can be expressed in spherical coordinates by using equivalent magnetic surface currents [7]. This can only be done until the DRA is mounted on an infinite ground plane. Using the electromagnetic software is another way of accessing to the 3D radiation pattern. This method is more accurate because it is taking into account the realistic structure (i.e. the finite ground plane). The Figure 9 shows the 3D radiation pattern of the cylindrical DRA using the simulator.

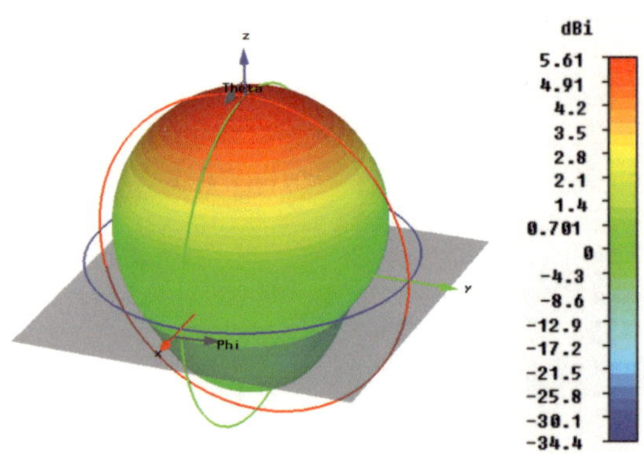

Figure 9. 3D radiation pattern

The simulated 3D radiation pattern provides antenna information such as the radiation efficiency, which is defined as the ratio between the radiated to accepted (input) power of the antenna.

The $HE_{11\delta}$ mode of the cylindrical DRA radiates like a short horizontal magnetic dipole. Concerning other modes, the $TM_{01\delta}$ radiates like a short electric monopole, while the $TE_{01\delta}$ mode radiates like a short magnetic monopole.

2.4. The rectangular DRA

The rectangular DRA has one degree of freedom more than the cylindrical DRA. Indeed, it is characterized by three independent lengths, i.e. its length *a*, its width *b* and its height *d*. Thus, there is great design flexibility, since the large choice of both dielectric materials and different lengths ratios.

Usually the dielectric waveguide model is used to analyze the rectangular DRA [8-10]. In this approach, the top surface and two sidewalls of the DRA are assumed to be perfect magnetic walls, whereas the two other sidewalls are imperfect magnetic walls. Since the considered case is a realistic one (Figure 10), the DRA is mounted on a ground plane, thus, an electric wall is assumed for the bottom surface.

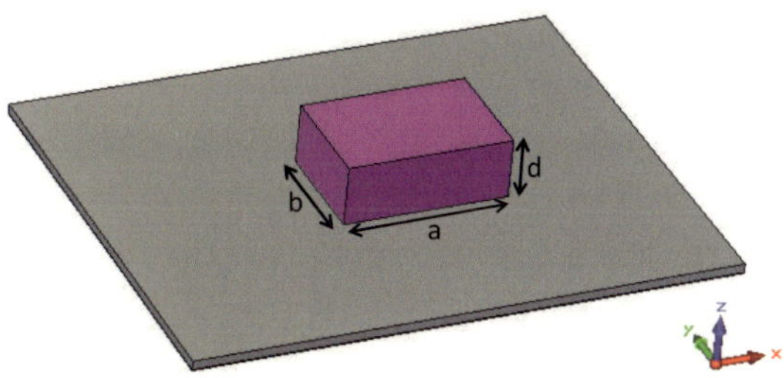

Figure 10. Rectangular DRA

The modes in an isolated rectangular dielectric resonator can be divided in two categories: TE and TM modes, but in case of the DRA mounted on a ground plane, only TE modes are typically excited. The fundamental mode is the TE_{111}. As the three dimensions of the DRA are independent, the TE modes can be along the three directions: x, y and z. By referring to the Cartesian coordinate system presented Figure 10, if the dimensions of DR are such as a>b>d, the modes in the order of increasing resonant frequency are TE^{z}_{111}, TE^{y}_{111} and TE^{x}_{111}. The analysis of all the modes is similar. The example of the TE^{z}_{111} mode is discussed in [8], the field components inside the resonator and resonant frequencies are analytically presented.

As for the cylindrical case, CST MS can be used to see both E and H fields, as presented Figure 11.

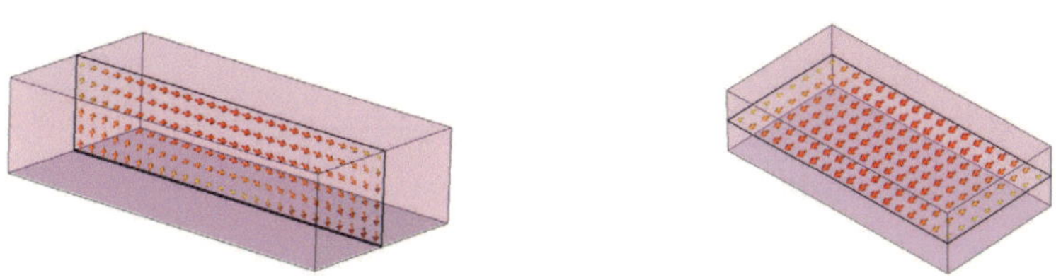

(a) (b)

Figure 11. E field (a) and H field (b) of the TE_{111} mode

The resonant frequencies definition is reminded hereafter:

$$f_0 = \frac{c}{2\pi\sqrt{\varepsilon_r \mu_r}}\sqrt{k_x^2 + k_y^2 + k_z^2} \tag{3}$$

It is found by solving the following transcendental equation:

$$k_z \tan\left(\frac{k_z a}{2}\right) = \sqrt{(\varepsilon_r - 1)k_0^2 - k_z^2} \tag{4}$$

where $k_x = \frac{\pi}{a}$, $k_y = \frac{\pi}{b}$, $k_0 = \frac{\omega_0}{v} = \frac{2\pi f_0 \sqrt{\varepsilon_r \mu_r}}{c}$ and $k_x^2 + k_y^2 + k_z^2 = k_0^2$

Values of resonant frequencies predicted by using this model are close to the measured ones for moderate to high value of ε_r. A frequency shift appears for low ε_r but it remains a good approximation method. If more accuracy is required, the electromagnetic study with CST MS (for example) presented in the cylindrical DRA case will have to be undertaken. Moreover, it allows taking into account feeding mechanism and ground plane dimensions.

Now DRA research method has been initiated, presenting resonant frequencies, fields configuration and feeding mechanisms, the next part will focus on the relevant dielectric material properties having significant influences on antenna performances.

3. Analysis on the dielectric material choice

To supply satisfactory answers about the effects of dielectric material properties, this section will present a careful and extensive investigation into relevant cases. Indeed, properties of the dielectric material have an influence on antenna characteristics, i.e. impedance bandwidth, Q factor, resonant frequency and radiation efficiency. Thus, this part will allow the reader to correctly select a dielectric material for a targeted application.

The cylindrical DRA example presented Figure 5 with a radius a=40mm and a height h=45mm will be pursue here. However, all results included in this part are generally applicable to most shapes of DRAs. A first sub-section will detail the influence of the dielectric permittivity and a second one will be interested in the impact of the dielectric loss tangent on the DRA performances.

Dielectric material properties having an impact on antenna characteristics are dielectric permittivity values and loss tangents.

3.1. Influence of the dielectric permittivity

To show the real impact of dielectric permittivity values, this sub-section will deal with a loss less dielectric material. Analytical studies show that the Q factor of the $HE_{11\delta}$ mode is defined as

$$Q = 0.01007.\varepsilon_r.\frac{a}{h}.\left(1 + 100.\exp(-2.05(\frac{a}{2h} - \frac{1}{80}\left(\frac{a}{h}\right)^2\right) \tag{5}$$

It is plotted as a function of a/h for different values of ε_r in the Figure 12.

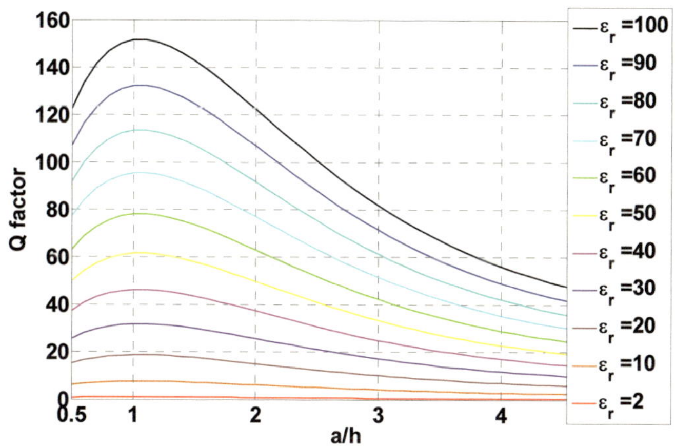

Figure 12. Q factor according the a/h values

Q factor is increasing with ε_r and reaching a maximum for a/h=1.05. This Q factor can be used to estimate the fractional bandwidth of an antenna using:

$$BW = \frac{\Delta f}{f_0} = \frac{s-1}{Q\sqrt{s}} \tag{6}$$

Where Δf is the absolute bandwidth, f_0 is the resonant frequency and s the maximum acceptable voltage standing wave ratio (VSWR).

The Q factor equation is deriving from the cylindrical dielectric resonator model approach by assuming perfect magnetic and/or electric walls on resonator faces. These equations are not absolutely accurate but they offer a good starting point for the design of cylindrical DRAs.

Let's consider the electromagnetic study presented in the first section with the cylindrical DRA example (Figure 7). Previously, the dielectric permittivity was fixed and equaled 30, it is now a variable. The Figure 13 plots both resonant frequencies and impedance bandwidths according to the dielectric permittivity ε_r. Because the coupling of the mode is depending on both length and height of the probe, this latter has been optimized for each ε_r value. The Figure 13 is thus the result of a large number of simulations.

As expected (see equation 3), the resonant frequency decreases when the dielectric permittivity increases. Moreover, this Figure shows that the bandwidth is the widest for ε_r=10. Fields are less confined for a low dielectric permittivity DRA, it is thus more difficult to couple the mode inside the resonator. Indeed, for higher dielectric values (ε_r>10), strong coupling is achieved, however, the maximum amount of coupling is significantly reduced if the dielectric permittivity of the DRA is lowered. That is why the bandwidth is low for ε_r

under 10. For a dielectric permittivity over 10, the Q factor is increasing and therefore the impedance bandwidth is decreasing (see equation 5).

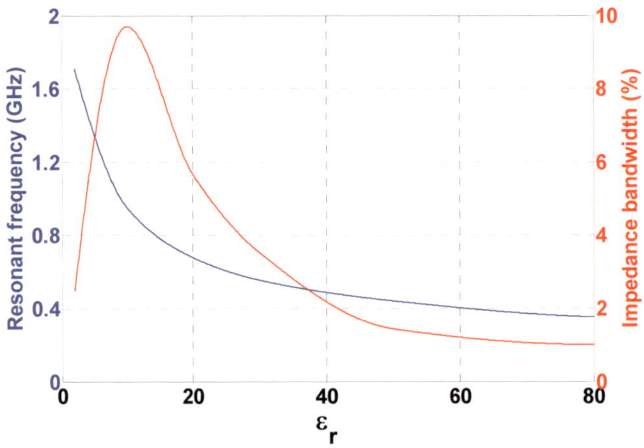

Figure 13. Resonant frequency according the dielectric permittivity values ε^r

Now that the influence of the dielectric permittivity has been shown, we can consider in the next sub-section more realistic cases by studying the impact of losses on the antenna characteristics.

3.2. Influence of dielectric material losses

Dielectric material losses directly impact the impedance bandwidth and antenna radiation efficiency. Their influence depends on the dielectric permittivity of the material.

Because analytical considerations do not take into account losses, only electromagnetic approaches are available to complete this study. To be more readable and relevant, charts will be provided to present and take into account all effects of losses.

The chosen example is still the same: the cylindrical DRA mounted on a ground plane with a radius and a height respectively equal to 40 mm and 45 mm and excited on its fundamental mode. The study of losses is done for different dielectric permittivity values.

The impedance bandwidth is studied as a first step. The Figure 14 presents the impedance bandwidth according to both dielectric permittivity and tangent loss of the dielectric material.

Several information have to be noted:

The most important loss tangents are, the widest the impedance bandwidth is.
The impedance bandwidth is the widest when for ε_r=10, whatever the losses.

The radiation efficiency is now investigated. Same simulations are done and the Figure 15 presents the antenna radiation efficiency according to both dielectric permittivity and tangent loss of the dielectric material.

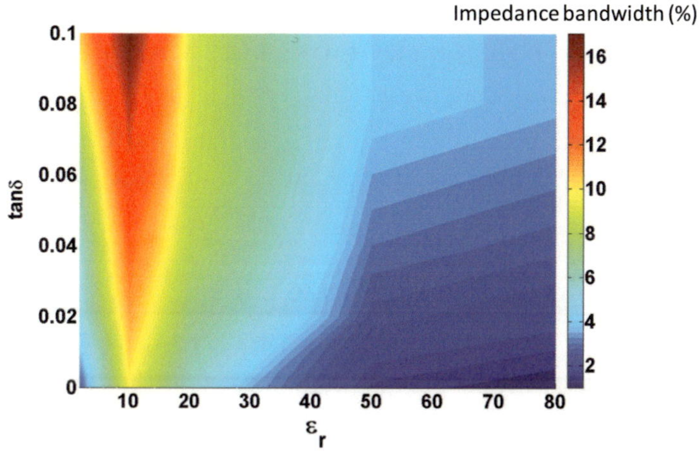

Figure 14. Impedance bandwidth according to the dielectric permittivity and loss tangent

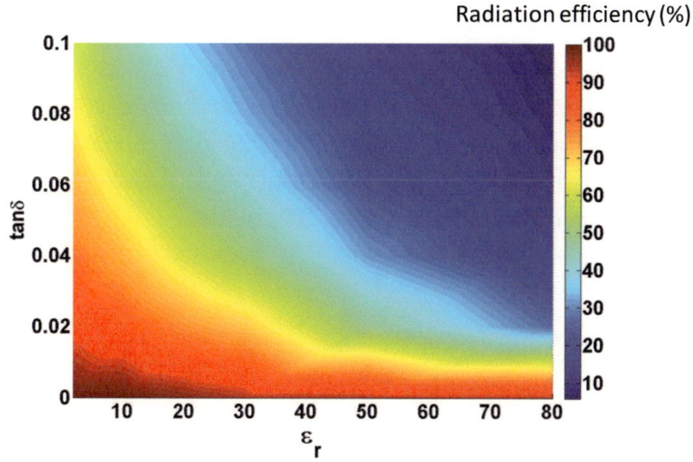

Figure 15. Radiation efficiency according to the dielectric permittivity and loss tangent

Other information can be deduced from this new graph:

- Antenna radiation efficiency is all the more affected by the losses as the dielectric permittivity increase.
- For low dielectric permittivity, even in case of a high losses material, the radiation efficiency remains higher than 50%.

To conclude this part, a DRA designer has to choose the dielectric material according to the application for which he is aiming. If he targets a wide bandwidth application, he could choose an alumina ceramic (ε_r ~10). Depending on the radiation efficiency he aims, the chosen ceramic would have more or less losses.

Now, if he targets an ultra-miniature DRA, it will be in his interest to choose a dielectric material with a higher dielectric permittivity. In this case, the impedance bandwidth will be affected, even more if the losses are high.

The dielectric material choice is one of the most important degree of freedom in the DRA design. It is necessary to highlight the best tradeoff, keeping in mind the targeted application.

Thus, using high dielectric permittivity values is one method for achieving a compact design, but it is not the only one. The next section will deal with different techniques to miniaturize a DRA.

4. Overview of techniques to miniaturize DRAs

This part examines techniques to design compact DRAs. Targeted applications are mobile handsets or wireless tablet. There are several techniques to make DRAs more compact. By adding metal plates, inserting a high permittivity layer (multisegment DRA) or removing portions of the DRA, a significant size reduction can be achieved.

4.1. Addition of a metallic plate on a DRA face

The rectangular DRA shape has been studied in the first part. The perfect metallic wall implies that electric fields are normal to this conductor, while magnetic fields are tangential. E and H fields presented Figure 11 assume that a metallic plate can be inserted in the middle of the DRA according to the y-component. The principle is detailed and explained by the Figure 16. It also shows the E and H fields of the TE_{111} mode.

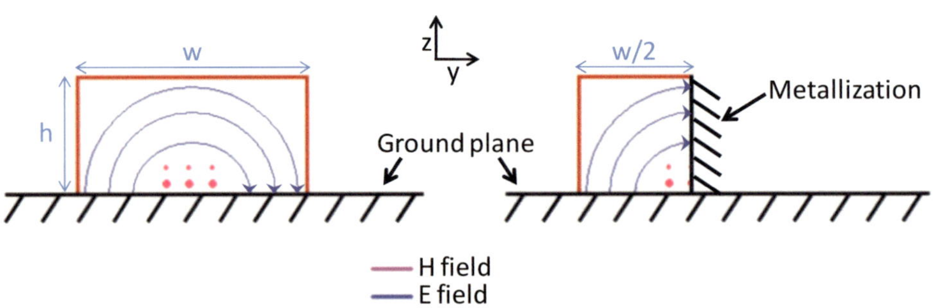

Figure 16. Integration of a metallic plate

By applying the image theory, it is possible to insert a metal plate in the y=w/2 plane. The Table 1 extracted from [12] shows the influence of the metallic plate insertion on resonant frequency and impedance bandwidth.

ε_r	w (cm)	d (cm)	h (cm)	Metallization	f_0(GHz)	Bandwidth
12	2.75	2.75	2.95	No	1.98	10%
12	2.75	2.75	2.95	Yes	1.24	5.6%

Table 1. Influence of the metallic plate insertion on both resonant frequency and impedance bandwidth

Thus, the metal plate insertion allows dividing by two the DRA size, while reducing the resonant frequency. However, as pointed by the Table 1, the metallic plate insertion involves also the decrease of the impedance bandwidth.

4.2. Multisegment DRA

Another way to decrease the DRA size is to insert different substrate layers as illustrated Figure 17.

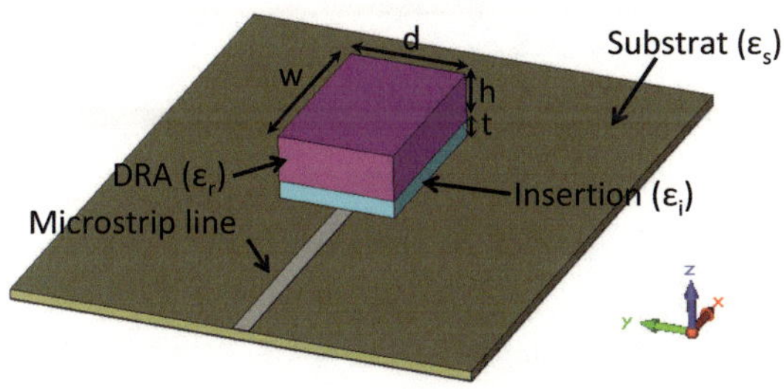

Figure 17. Multisegment DRA

It allows achieving strong coupling when the first insertion has a relatively high dielectric permittivity. This technique is detailed in [12] and [31]. The Table 2 summarizes a parametrical study done in [31] for one layer inserted (Figure 17) with w=7.875 mm, d=2 mm, h=3.175 and ε_r=10. It is mounted on a 0.762 mm height substrate of permittivity ε_s=3. The TE_{111} mode of the DRA is excited with a 50Ω microstrop line.

t (mm)	ε_i	Measured f_0(GHz)	Bandwidth
0	-	15.2	21%
0.25	20	14.7	18%
0.635	20	14.5	18%
1	20	13.9	16%
0.25	40	14.7	20%
0.635	40	13.7	13%
1	40	12.9	5%
0.25	100	14.7	16%
0.635	100	13.1	7%
1	100	10.8	5%

Table 2. A parametrical study done in [31] for one layer inserted

Thus, a thin layer insertion allows improving the coupling of modes inside the DRA while decreasing the resonant frequency thanks to the decrease of the effective dielectric permittivity of the DRA. As the previous technique, the downside is the decrease of the impedance bandwidth.

4.3. Circular sector DRAs

To clearly explain this miniaturization technique, we need to take up the cylindrical DRA example with the equation 1 of the resonant frequencies. In [32], DRA size and resonance frequencies significant reductions have been demonstrated by using cylindrical sector DRA which is shown in Figure 18.

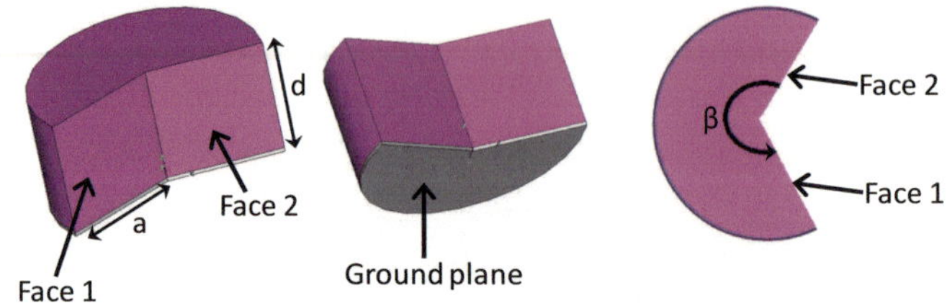

Figure 18. Circular sector DRA

As shows the Figure 18, a cylindrical sector DRA shape consists of a cylindrical DRA of radius *a* and height *d* mounted on a metallic ground plane, with a sector of dielectric material removed. β is the angle between the face 1 and 2, which can be metalized or left open. Thus, a cylindrical sector DRA is formed when $\beta < 2\pi$. For such a DRA, considering the equation 1, the n subscript can be substitute by the ν subscript, which is a positive real number that depends on the boundary conditions on the sector faces as well as the sector angle β. In this case, first excited modes are writing such as $HE_{\nu pm+\delta}$ and the corresponding resonant frequencies as defined by the following equation:

$$f_{\upsilon pm} = \frac{c}{2\pi\sqrt{\varepsilon_r\mu_r}}\sqrt{\left(\frac{X'_{\upsilon p}}{a}\right)^2 + \left(\frac{(2m+1)\pi}{2d}\right)^2} = \frac{c}{2\pi a\sqrt{\varepsilon_r\mu_r}}\sqrt{\left(X'_{\upsilon p}\right)^2 + \left(\frac{(2m+1)\pi.a}{2d}\right)^2} \quad (7)$$

It should be noted that for the $\beta = 2\pi$ case, ν=n.

An important point has to be highlighted: For a given cylindrical sector DRA (radius, height, permittivity and permeability), the resonant frequency is only depending on the $X'_{\nu P}$ value. The lower the $X'_{\nu P}$ value will be, the lower the resonant frequency will be. $X'_{\nu P}$ values are summarized in the Table 3.

	ν=0	ν=1/4	ν=1/3	ν=1/2	ν=2/3	ν=1	ν=2
p=1	3.832	0.769	0.910	1.166	1.401	1.841	3.054
p=2	7.016	4.225	4.353	4.604	4.851	5.331	6.706

Table 3. $X'_{\nu P}$ values according to the ν and p values

By applying the good boundary conditions on each faces, fundamental modes of the different shapes presented in Table 4 can be determined. This table is reminding the $X'_{\nu P}$ values with the corresponding resonant frequencies for a DRA such as a=40mm and d=45mm with a dielectric permittivity equals to 10.

Following these results, resonance frequencies depend on whether parts of some faces are coated with a metal or not. A good trade-off between antenna size and low resonant frequency is to choose the third line of the Table 4, when one face is coated by metal. Finally, this study has demonstrated that the metallization of some DRA faces allows creating new resonant modes, which have resonant frequencies lower than the fundamental modes inside a classical cylindrical DRA. It should be noted that these results have been done with cavity resonator model, i.e. the outer surfaces of the cavity are approximated by perfect magnetic walls. They thus show an approximate analysis of the fields inside the resonator. However, this model provides reasonable accuracy for prediction of resonant frequencies.

Mode	ν	p	$X'_{\nu P}$	$f_{\nu pm}$	Shape
$HE_{21\delta}$	2	1	3.054	1.27 GHz	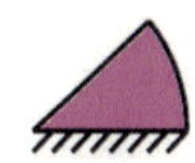
$HE_{11\delta}$	1	1	1.841	872 MHz	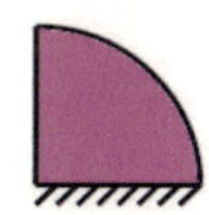
$HE_{1/2\,1\,\delta}$	1/2	1	1.166	687 MHz	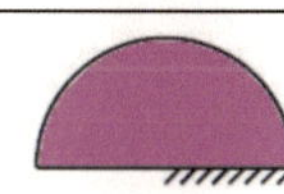
$HE_{1/3\,1\,\delta}$	1/3	1	0.910	629 MHz	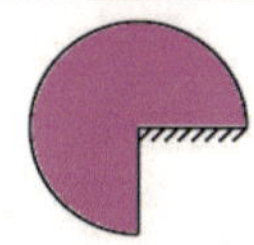
$HE_{1/4\,1\,\delta}$	1/4	1	0.769	602 MHz	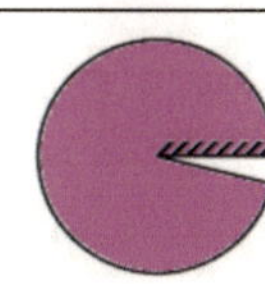

Table 4. Excited modes with the corresponding DRA shapes

5. New approaches for wireless applications

Antenna design for mobile communications is often problematic by the necessity to implement multiple and/or ultra wideband applications on the same small terminal.

Some of currently antennas integrated in portable wireless systems have a planar structure based on microstrip patches or PIFAs [33-35]. These kinds of antennas present a low efficiency, especially when small, because of metallic losses. DRAs do not suffer from such losses, which makes them a good alternative for these more conventional antennas. That is why, in recent years, much attention has been given to DRA miniaturization [16,36] in order to integrate them inside mobile handheld.

This part is divided in two sub-sections. The first one will focus on the bandwidth enhancement of a DRA for ultra wideband applications, and the second part will aim multiband applications. The common thread is the miniaturization of DRAs while obtaining good performances. For that, new hybrid modes will be studied with the application of partial electric boundary on sides of DRAs.

5.1. For ultra wideband application

A tendency and well known ultra wideband application is the Digital Video Broadcasting-Handheld (DVB-H). Indeed, the allocated frequency range is divided in two sub-bands, i.e. [470 MHz – 790 MHz] and [790 MHz – 862 MHz]. Considering the entire DVB-H frequency range, it presents 58% of bandwidth and a few kind of antenna can cover it instantaneously while being miniature and having good performances.

As presented in previous parts, for a single-mode excitation, the DRAs bandwidth doesn't exceed 15%. Recently, different shapes and stacked resonators have been proposed to enhance the bandwidth. These techniques are generally difficult to implement without increasing the size of the antenna. A novel DRA design method is proposed in this part with detailed parametric studies [37]. After seeing the proposed DRA and its miniaturization technique, the study will focus on the bandwidth enhancement. In order to design and optimize the proposed antenna, both eigenmode solver and the Finite Integration Temporal method of CST Microwave Studio are used. The measurement and simulation results will be shown, followed by a discussion.

Previous parts have detailed the miniaturization technique by using a circular sector DRAs. It has been shown that resonance frequencies depend on whether parts of some faces are coated with a metal or not. According to the equation 7 and in order to easily integrate the antenna inside a terminal, *d* has to be lower than 25 mm. In this case, the lowest mathematical resonance frequency which can be obtained is 949MHz. The best compromise between the size and the desired frequency band is to use the $HE_{\frac{1}{2}1\delta}$ mode (Table 4). In order to entirely cover the DVB-H band, the non-coated face of the selected shape is transformed to a cubic part. The final designed DR shape is shown Figure 19.

(a) (b)

Figure 19. Proposed DR shape (a) and metallization of two different sides (b)

As mentioned previously, coating some faces with a metal allows resonance frequencies decreasing. A further parametric study is performed in order to determine how to coat some faces with a metal and how to choose the resonator position on the ground plane. In the Figure 19, *a* and *b* are defined as the metallization lengths of two different metallic sides.

Figure 20 represents the variation of the first and the second resonance frequencies mode according to *a* and *b* lengths.

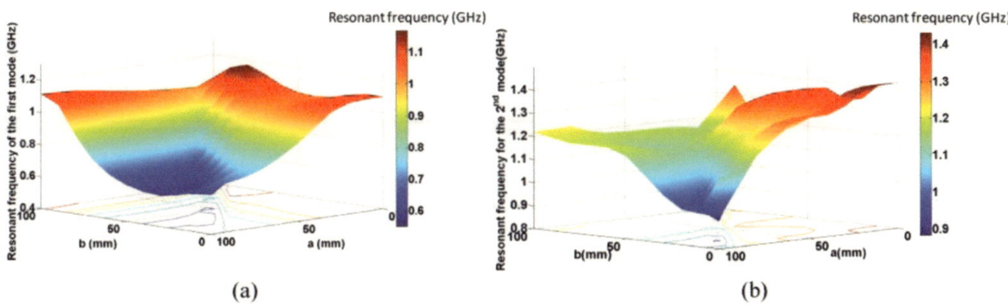

Figure 20. Variation of the resonance frequency of the 1st (a) and 2nd (b) modes according to a and b lengths

The resonance frequencies decrease significantly when the two metallic sides have the same potential and each one covers half of the face. So, the optimum metallization lengths are *a*=50mm and *b*=50mm and the two lowest resonance frequencies are 568MHz and 1GHz. Through the above results, the chosen structure looks like the Figure 19 with *a*=*b*=50mm. Now, it is necessary first of all to glance over the E-field distribution inside the DR. Figure 21 shows the E-field inside the DR for the first and the second mode respectively at 568 MHz and 1 GHz. The first mode is derivative from the $HE_{\frac{1}{2}1\delta}$ mode.

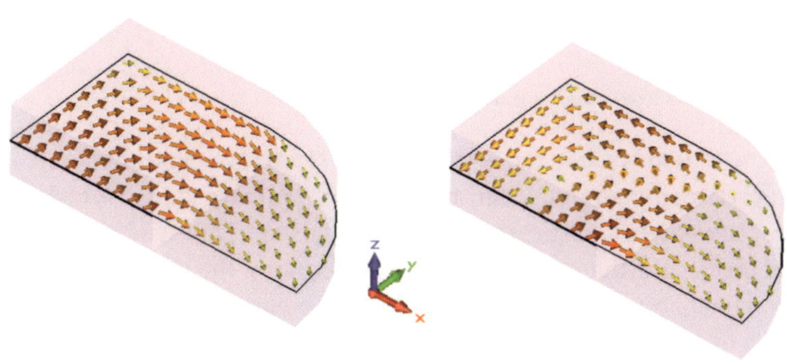

Figure 21. E fields of the two first modes

In the following, a ground plane is inserted on the lower metallic face as shown Figure 22. The ground plane size is 230mm x 130mm, chosen to correspond to a standard DVB-H handheld receiver. The DR of $\lambda_0/7x\lambda_0/13x\lambda_0/28$ dimensions at 470 MHz is mounted on such ground plane.

Considering the E fields distribution, a probe is chosen to feed the DRA. It is placed on the lateral metallic side in order to simultaneously excite the two first modes (Figure 22).

To reach an ultra wideband, the probe position, diameter and length are tuned. The optimal probe diameter and length values are 1.5mm and 49mm.

Following this optimization, input impedance and return loss are shown in the Figure 23.

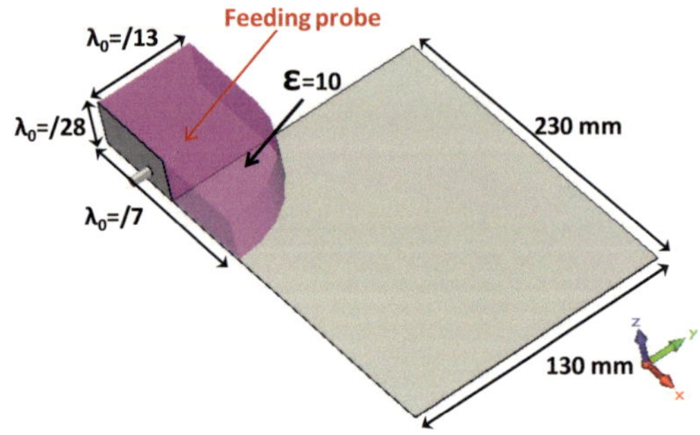

Figure 22. Antenna design fed by a coaxial probe

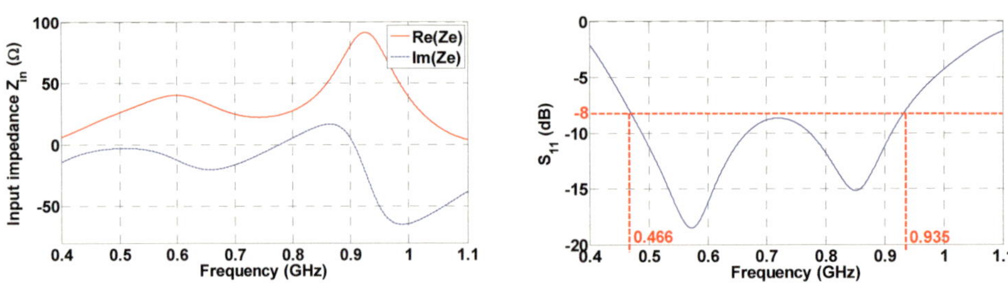

(a) (b)

Figure 23. Input impedance (a) and S_{11} parameter (b) of the considered DRA

They show that two modes exist at 0.6 MHz and 0.93 MHz, excited by the coaxial probe. As expected, a difference appears between resonance frequency values which are obtained by the eigenmode solver and electromagnetic simulation. This discrepancy is due to the presence of the probe and the ground plane. Furthermore, the boundary conditions applied during the modal analysis are perfect electric or magnetic conditions on the DR walls contrarily to the electromagnetic simulation.

The simulated coefficient reflection shows 70% of bandwidth for a -8 dB impedance bandwidth definition over the frequency range [466 MHz – 935 MHz]. The -8dB impedance bandwidth definition is sufficient to achieve a good efficiency. Thus, the antenna satisfies the DVB-H system specifications mentioned previously. To demonstrate the antenna performances, the antenna has been realized with a ceramic material. It is fed by a probe and is half-mounted on a ground plane. The manufactured antenna is shown Figure 24.

The input impedance and the coefficient reflection are shown Figure 25.

There is a good agreement between simulations and measurements. It should be underlined that these simulations series take into account that the probe is inserted in a 2 mm diameter hole instead of a 1.5 mm hole due to mechanical constraint, therefore an air gap appears around the probe. This air gap results in lowering the DRA effective dielectric constant, which is turn lower the Q-factor accompanied by a shift in the resonance frequency. Furthermore, after material characterizations, the ceramic's dielectric constant turns out equal to 9.5 at 500 MHz. So, the -8dB impedance matching bandwidth is included in [540 MHz – 1.05 GHz].

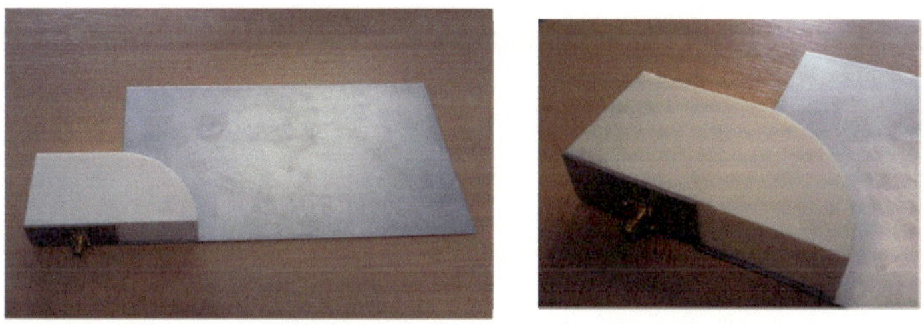

Figure 24. Antenna prototype

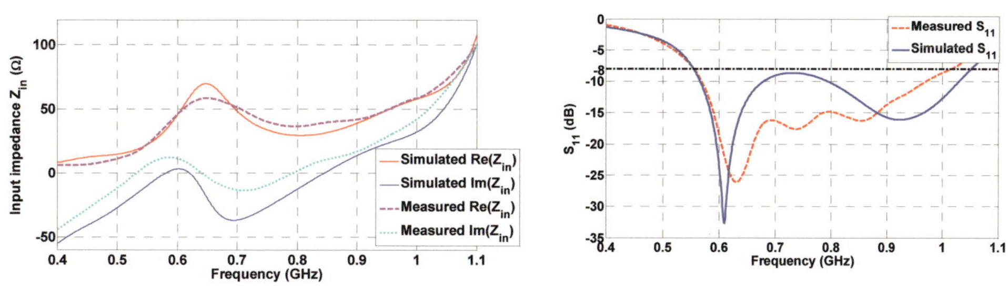

(a) (b)

Figure 25. Measured and simulated input impedances (a) and S_{11} parameters (b)

Radiation patterns of the antenna were characterized in an anechoic chamber. Simulated and measured radiation patterns at 620MHz and 870MHz are shown Table 5.

It can be noticed that the *xz*-plane and *yz*-plane radiation patterns indicate a correct omnidirectional radiation patterns. The antenna operates at two different modes inside the operating band, so the antenna doesn't offer the same radiation pattern at the two resonance frequencies, particularly in the *xy*-plane. The simulated and measured total efficiencies of the proposed antenna are illustrated Figure 26.

This efficiency disagreement is probably due to the coaxial feeding cable, the discrepancy between the simulated and measured reflection coefficient and the dielectric loss tangent.

In spite of this 10% difference, there is a good agreement between measured and simulated efficiencies. The measured efficiency remains higher than 75% above 600MHz.

This part has proposed a novel dielectric resonator antenna. Parametric studies have been realized to decrease resonance frequencies and to increase impedance matching bandwidth. The results have shown that by shaping the dielectric structure and coating some faces with a metal, resonance frequencies have been reduced for a fixed structure size. Good performances have been obtained and the proposed antenna can be used for DVB-H and/or GSM900 applications with a wide bandwidth and a good efficiency.

	ϕ=0°	ϕ=90°	θ=90°
f=620 MHz	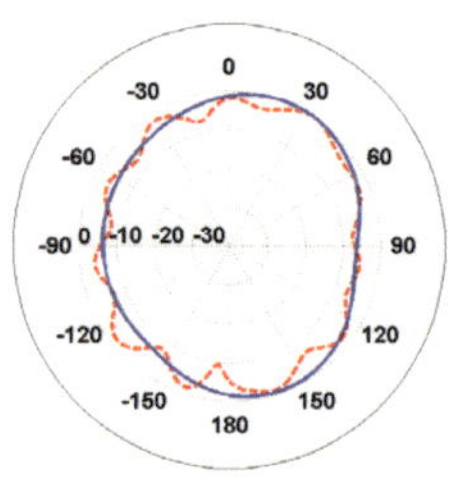	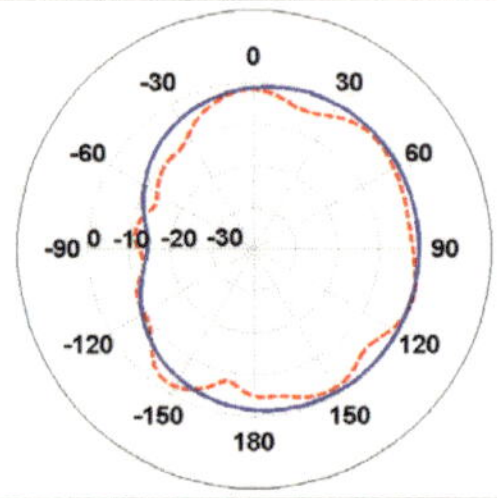	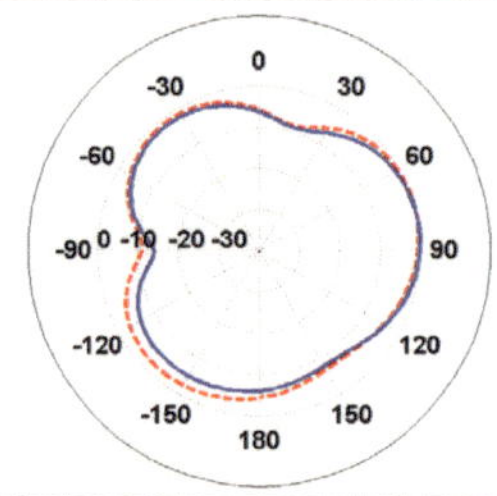
f=870 MHz	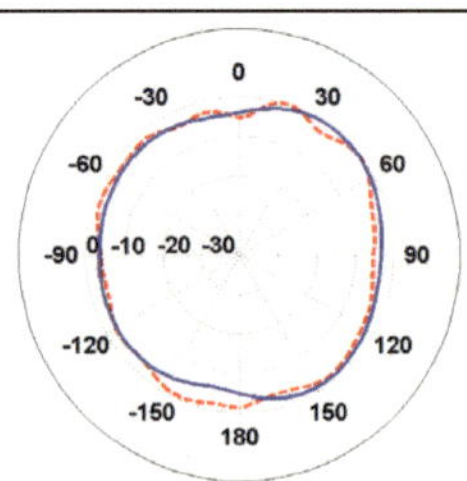	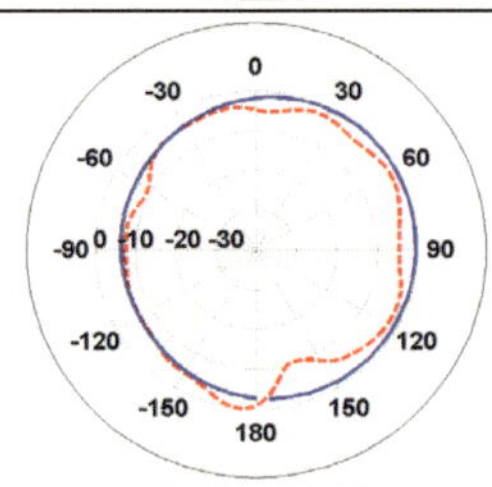	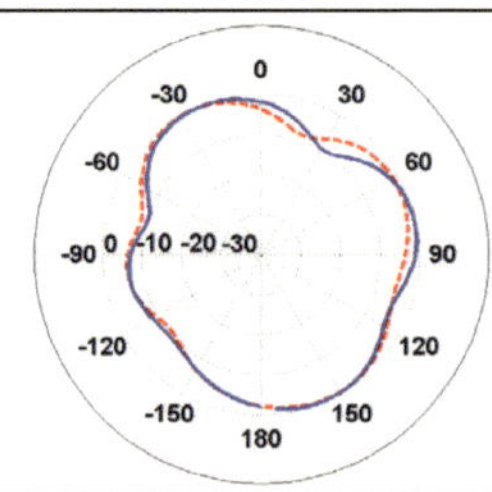

Table 5. Simulated (blue line) and measured (red line) radiation patterns at 620 MHz and 870 MHz

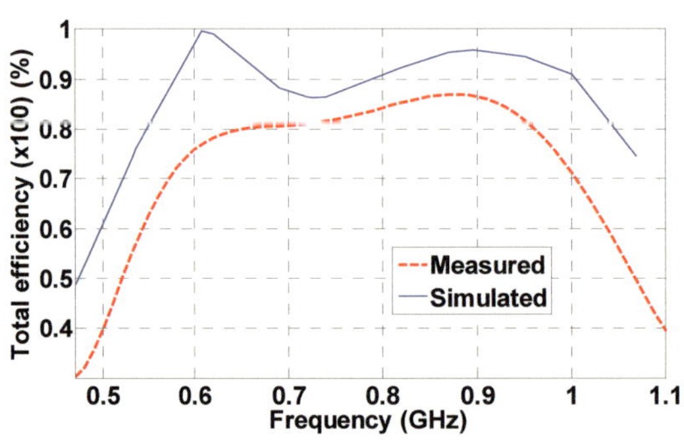

Figure 26. Measured and simulated total efficiencies of the considered DRA

5.2. For multiband application

In the last decade, the huge demand for mobile and portable communication systems has led to an increased need for more compact antenna designs. This aspect is even more critical when several wireless technologies have to be integrated on the same mobile wireless communicator. All the new services and the increased user density are driving the antenna design toward multi-band operation.

Recently, many studies have been devoted to multi-band antennas [38-41], some of them dealing with DRAs [42-44]. A dielectric resonator indeed supports more than one resonant mode at two close frequencies, which allows them to meet the requirements of different applications with a unique device. Some studies furthermore use both the dielectric resonator and the feeding mechanism as radiator elements [43-45]. This explains why the DRAs present a major advantage for multi-standard devices, when compared to other kinds of antennas.

The objective of this part is to show the integration of a small antenna in a multi-band mobile handheld device, working on the nine channels of the second sub-band of the DVB-H, i.e. [790 MHz – 862 MHz], the WiFi band at 2.4 GHz and the WiMax band at 3.5 GHz. Additionally, in order to improve the quality and the reliability of the wireless links, i.e. obtain pattern diversity, two antennas will be integrated in the same device. Setting maximum limits, we have decided to integrate two orthogonally aligned antennas in the allocated space of 30 mm x 41 mm ($\lambda_0/13$ x $\lambda_0/9$ at 800 MHz), on a 230 mm x 130 mm ground plane. Each DRA will therefore have to be very compact to be able to fit in such a limited area, and will operate around 850 MHz, 2.4 GHz and 3.5 GHz, thus covering the nine channels of the DVB-H band, the WiMax and the WiFi band.

Firstly, only one radiating element is described. It will be done thanks to a modal analysis [46] of the dielectric resonator. The DRA design and the choice of the dielectric permittivity will be discussed. As for the previous sub-section, both the Finite Integration Temporal method and the eigenmode solver of CST Microwave Studio were used to carry out this work.

Getting three resonant frequencies for a single dielectric resonator is aiming. Firstly, the resonator has to be integrated in a handheld receiver, which means that it will be placed on a FR4 substrate (ε_r=4.9). Secondly, to be integrated in a handheld device, the allocated space for the antenna system must not exceed 41mm x 30 mm x 4mm as shown Figure 27.

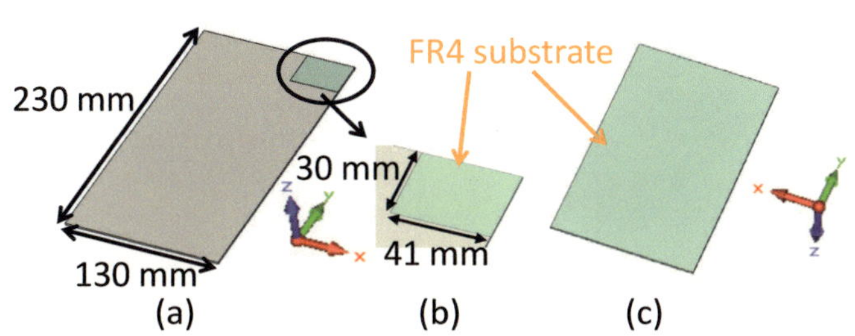

Figure 27. Top view of the defined PCB card (a), allocated size for the antenna system (b) and bottom view of the defined PCB card (c)

The dimensions and dielectric permittivity of each resonator need to be chosen according to these constraints, to ensure the integration of the antenna in the final device.

As a result, the dimensions of one resonator were chosen to be 25 mm x 10 mm x 4 mm, with a very high dielectric constant ε_r of 37. The resulting geometry is depicted Figure 28.

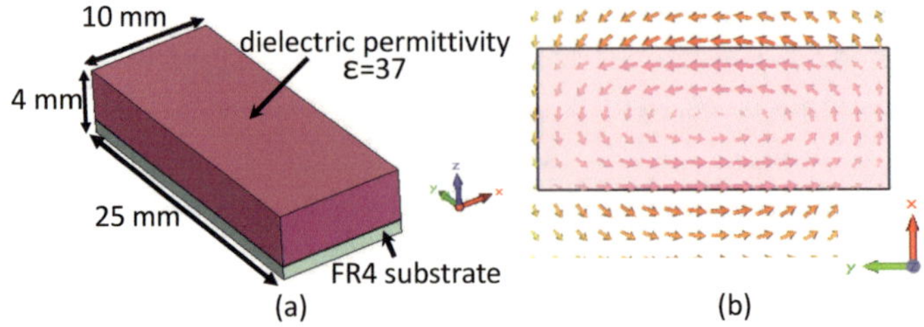

Figure 28. Dimensions and properties of one resonator (a) and E-field distribution of the first natural mode for the resonator in (b) in the z=2mm plane

The first natural mode of this resonator is the $TE_{11\delta}$ (Figure 28) which resonates at 3.99 GHz. This resonance frequency remains too high for the intended applications. It has however been shown [37], that this resonance frequency depends on the metallization of the DRA's faces. It is therefore necessary to envision the feeding mechanism of the resonator, before performing the modal analysis. Indeed, if the antenna is not fed by proximity coupling, the excitation cannot be ignored during the modal analysis. Furthermore, it can be used to adjust the resonance frequency of the resonator, especially in the case of an electrically small DRA. In this study, the feeding will indeed play a preeminent role. The chosen excitation is a line printed on the FR4 substrate and positioned under the dielectric resonator as shown Figure 29. The E-field distribution of the first mode (Figure 29) is completely different from the one obtained without this line. Indeed, as stated before, the E-field and H-field distributions inside the DRA depend on the boundary conditions on its faces. The feeding line introduces partially perfect electric conducting conditions, which disturb the field distribution inside the resonator when compared with the $TE_{11\delta}$ mode. As a result, the resonance frequency of each mode will vary in accordance with the length and width of the feeding line (defined in the Figure 29).

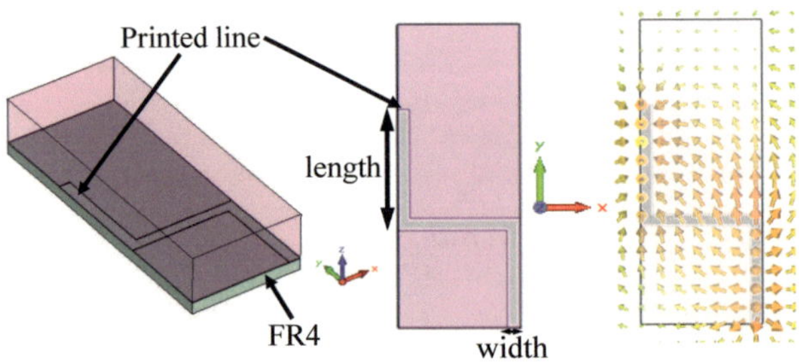

Figure 29. Design of the DRA fed by the printed line and E-field distribution of the first natural mode for the resonator fed by the line

Considering the previous study, the resonator design method is as follows.

- First of all, in spite of the high dielectric permittivity, there is no mode around 800 MHz. In order to allow the antenna to operate on the nine channels of the DVB-H band, the printed line will be designed to resonate around 800 MHz. It will therefore behave like a printed monopole loaded by a dielectric. The length of this line, which will from now be referred as the "monopole", will be set to obtain the first resonance frequency around 800 MHz.

- This being done, the second and third band will be covered by the resonances of the dielectric resonator disturbed by the presence of the monopole. As previously explained, the length, width and shape of this monopole entail a modification of the DRA resonance frequencies. The shape and width of the monopole will therefore be optimized to obtain the desired resonance frequencies, while its length remains set by the first resonance.
- An important point concerns the radiation Q-factor. The high dielectric permittivity involves a high Q-factor [7]. With such an ε_r, the radiation Q-factor is also important, making it difficult to obtain a wide impedance bandwidth for a given mode. So, to have a suitable impedance bandwidth, the antenna will have to be matched between two peaks of the real part of input impedance. By this way, the resonances won't have to be close to the operating bands. The goal being to match the DRA over the WiFi and WiMax bands, the monopole (its shape and width) has to be optimized to obtain resonance frequencies around 2 GHz, 2.8 GHz and 4 GHz.
- A modal analysis has been performed to show the variations of the resonance frequencies of the first three modes, according to the shape and width of the monopole. Figure 30 shows the resonance frequency of the first mode according to the monopole geometry. With the first resonance frequency set to match the antenna at 2.4 GHz and the length of the monopole set to have the first resonance at 800 MHz, the graph of the modal analysis allows an easy determination of the monopole width, which is 1 mm.

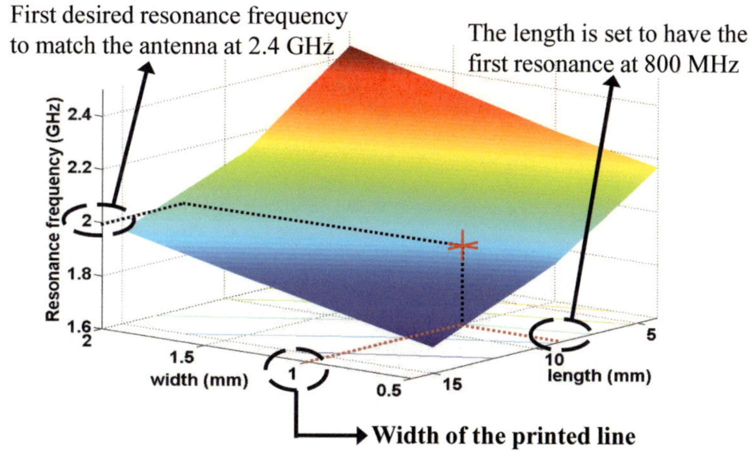

Figure 30. : Resonance frequency of the first dielectric resonator mode according to the width and the length of the monopole previously defined

Same studies have been performed for different shapes of the monopole. All these studies allowed the shape, length and width of the monopole to be set, which led to the final design of the dielectric resonator. Table 6 shows the values of the resonance frequencies for the first three modes inside the resonator, which were obtained through the modal analysis.

Mode	Resonance frequency
First mode	1.969 GHz
Second mode	2.773 GHz
Third mode	4.135 GHz

Table 6. Values of Resonance Frequency for the Three First Modes

- Electromagnetic study of the DRA

In order to validate the previous modal study, the dielectric resonator, placed in the area dedicated to the antenna (Figure 27) and fed by a 50Ω discrete port, has been simulated with the FIT method using CST Microwave Studio. It must be noticed that a discrete port is modeled by a lumped element, consisting of a current source with a 50 Ω inner impedance that excites and absorbs power.

The Figure 31 shows the simulated input impedance of the dielectric resonator with its feed. The resonance frequencies are in agreement with the modal analysis. The radiation Q-factor is important, and the input impedance variations confirm that the antenna matching is easier between two resonances. The reflection coefficient is shown Figure 31. The antenna is matched on all of the desired bands, i.e. the nine channels of DVB-H going from 790 MHz to 862 MHz, the WiFi band at 2.4 GHz and the WiMax band at 3.5 GHz. It can be noted that the matching over the first band is obtained due to the resonance of the λ/4 monopole.

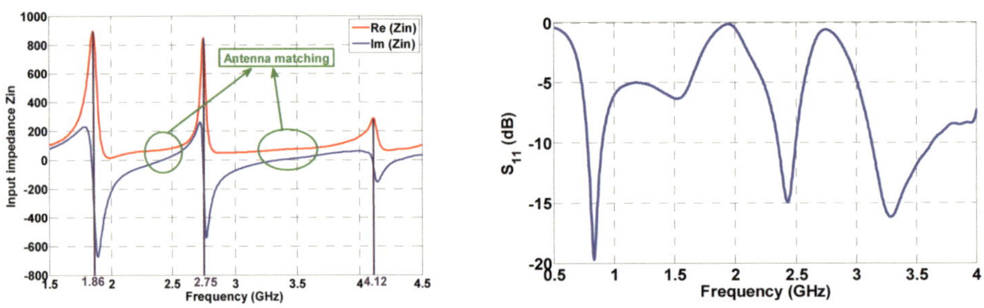

(a) (b)

Figure 31. Input impedance (a) and S_{11} parameter of the fed DRA

The dielectric resonator must now be integrated in its context, i. e. on a 230 mm x 130 mm ground plane, chosen to correspond to a standard DVB-H handheld receiver. As explained before, another specification was to obtain a reconfigurable radiation pattern. Thus, two instances of the previously studied resonator are orthogonally integrated on the ground plane.

- Final structure

Based on the previous parametric study, the final structure has been designed as shown Figure 32. In order to obtain pattern diversity, two dielectric resonators are orthogonally disposed in a 30 mm x 41 mm area, both fed by a printed line acting as a monopole. Each line is fed by a 50Ω coaxial cable. They are studied on a 230 mm x 130 mm ground plane, as defined by the specifications. It will be shown in a following section that the antenna matching is not affected by the ground plane dimensions. In order to ascertain the performances of this antenna, a prototype has been fabricated as shown Figure 32.

The resonators are manufactured with a ceramic material with a dielectric permittivity of 37 and a loss tangent tanδ=0.005 on the 0.5-10 GHz band. During the simulation and measurements, each resonator has been excited by a printed line fed by a coaxial cable.

During the manufacturing process of the antenna, a special care has to be given to minimize the air gap between the excitation and the resonator. In the case of this prototype, the resonators have been pressed onto the PCB to avoid this air gap.

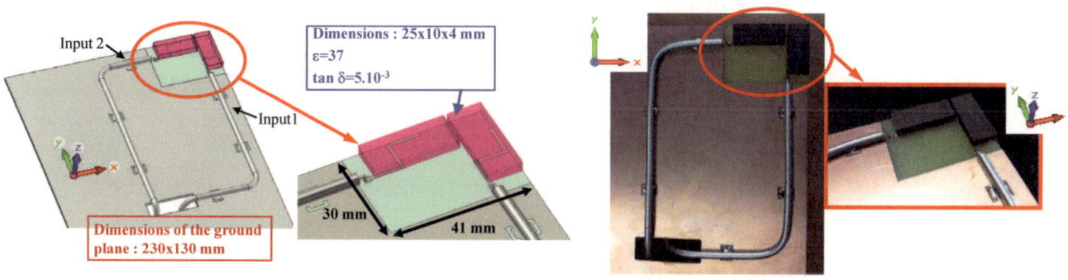

(a) (b)

Figure 32. Final simulated design (a) and the corresponding prototype (b)

- Measured and simulated performances of the antenna

The comparison between the simulated and measured results is now studied. The first ones have been obtained using the transient solver of CST Microwave Studio, while the measurements have been performed inside an anechoic chamber. Both the simulated S11 and S22 parameters are compared with the measured ones Figure 33.

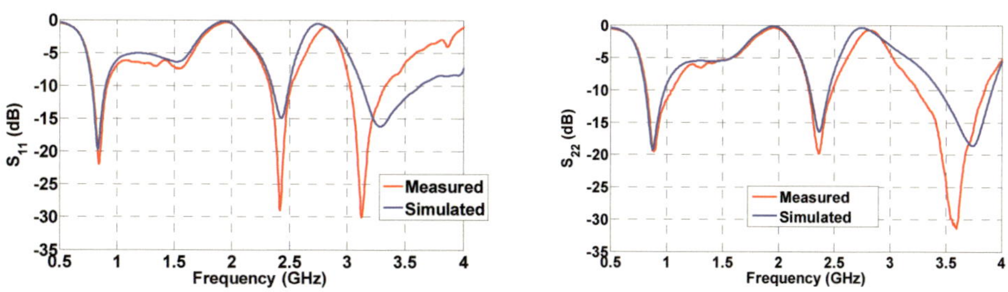

(a) (b)

Figure 33. Measured and simulated S_{11} (a) and S_{22} (b) parameters of the DRA

The measurements and simulations are in very good agreement. Moreover, the antenna is matched over all the desired bands and for both inputs.

Radiation patterns: The radiation patterns have been measured inside an anechoic chamber. Table 7 shows the 3D simulated radiation patterns for both inputs at 830 MHz, 2.4 GHz and 3.5 GHz.

It can be seen that the radiation pattern at a given frequency will depend on the excited port. While promising, this result is not sufficient to conclude that the radiation pattern is reconfigured. More details are explained in [47] (this requires the characterization of the whole system in a reverberation chamber, in order to determine the correlation coefficient).

Thus, this study started with modal analyses, which allowed the shape and dimensions of the antenna's excitation to be defined with a dual objective. Indeed, this line had first to behave like a monopole and cover the nine channels of the DVB-H band (going from 790 MHz to 862 MHz). Secondly, it had to excite the dielectric resonator and set its resonance frequencies so as to match the antenna on the WiFi and WiMax bands.

After performing these preliminary studies, two instances of the conceived dielectric resonator have been orthogonally integrated on a 230 mm x 130 mm ground plane, which is consistent with a tablet. Finally, the antenna system, which only occupies a 30 mm x 41 mm area, is matched on the three desired bands, i.e. the nine channels of the DVB-H band, the WiFi band and the WiMax band with pattern diversity.

This part has presented the design method, the realization and the measurement of a two compact DRAs, one for ultra wideband application and the second for multiband applications.

	f=830 MHz	f=2.4 GHz	f=3.5 GHz
1st input	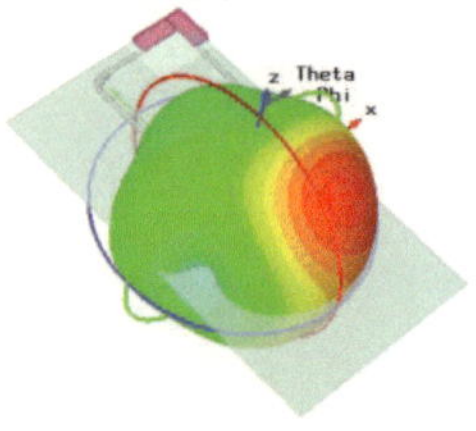	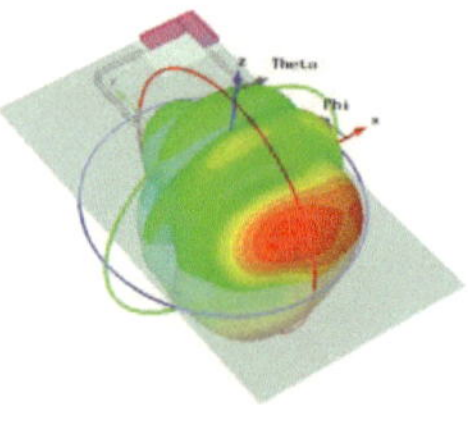	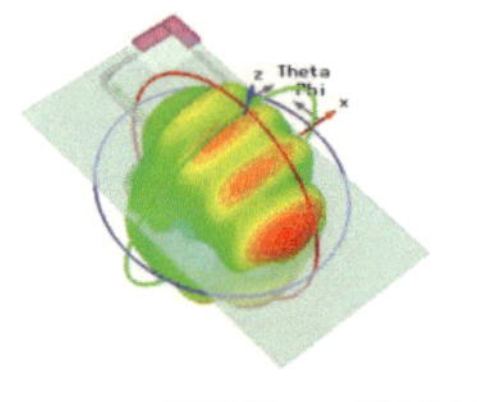
2nd input	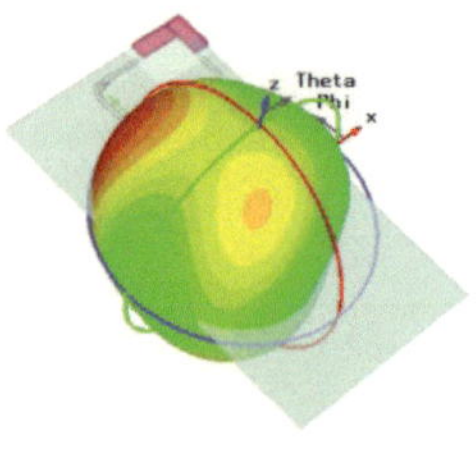	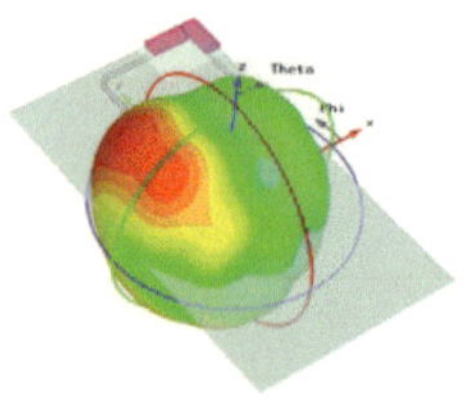	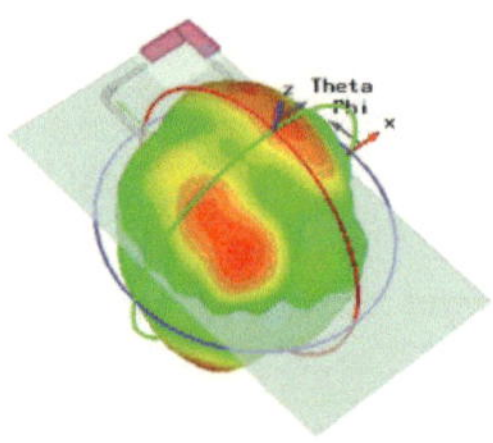

Table 7. 3D Radiation patterns at 830 MHz, 2.4 GHz and 3.5 GHz for the two inputs

6. Conclusion

To conclude, an affordable chapter has been presented allowing the reader to find an overview of main DRA shapes, properties and approaches while appreciating the influence and the impact of the dielectric material properties. Indeed, a broad spectrum of dielectric materials can be used depending on the intended application. In addition to the advantages common to all DRAs described at the beginning of this chapter, a dedicated part has focused on other advantages of compact DRAs, which are desirable for many emerging wireless and mobile communication systems. Finally, a specific part had presented relevant data for postgraduate researchers, antenna design engineers in general and particularly the ones engaged in the innovative design of mobile and wireless systems by focusing on the hybrid modes creation to enhance the bandwidth or develop multiband antennas.

Author details

L. Huitema and T. Monediere
University of Limoges, Xlim Laboratory, France

7. References

[1] R. D. Richtmyer, "Dielectric Resonator", J. Appl. Phys., vol. 10, pp. 391-398, Jun. 1939

[2] D. Kajfez and P. Guillon, Eds., Dielectric Resonators. Norwood, MA: Artech House, 1986

[3] Cohn, S.B., "Microwave Bandpass Filters Containing High-Q Dielectric Resonators," Microwave Theory and Techniques, IEEE Transactions on, vol.16, no.4, pp. 218- 227, Apr 1968

[4] S. J. Fiedziuszko, "Microwave Dielectric Resonators", Microwave Journal, vol. 29, September 1986, pp 189-200

[5] S.A. Long, M.W. McAllister and L.C. Shen, "The Resonant Dielectric Cavity Antenna", IEEE Transactions on Antennas and Propagation, Vol. 31, n°3, March 1983, pp. 406-412

[6] A. Petosa, A. Ittipiboon, Y.M.M. Antar and D. Roscoe, "Recent Advances in Dielectric Resonator Antenna Technology", IEEE Antennas and Propagation Magazine, Vol. 40, n°3, 06/1998, pp. 35-48

[7] K.M Luk and K.W Leung, "Dielectric Resonator Antennas", Electronic & Electrical Engineering Research Studies

[8] R.K Mongia and A. Ittipiboon, "Theoretical And Experimental Investigations on Rectangular Dielectric Resonator Antenna", IEEE Transactions on Antennas and Propagation, Vol. 45, n°9, September 1997, pp. 1348-1356

[9] D. Drossos, Z. Wu and L.E. Davis, "Theoretical and experimental investigation of cylindrical Dielectric Resonator Antennas", Microwave and Optical Technology Letters, Vol. 13, No. 3, pp. 119-123, October 1996

[10] R. K. Mongia and P. Bhartia, "Dielectric Resonator Antennas – A review and General Design Relations for resonant Frequency and Bandwidth", International Journal of

Microwave and Millimeter-wave Computer-Aided Engineering, Vol. 4, No. 3, pp. 230-247, Mar. 1994

[11] A.A. Kishk, B. Ahn and D. Kajfez, "Broadband stacked dielectric resonator antennas", IEE Electronics Letters, Vol. 25, n°18, Aug. 1989, pp. 1232-1233

[12] A. Petosa, "Dielectric Resonator Antenna Handbook", Artech House, Boston/London, 2007

[13] R. Chair, A. A. Kishk, K. F. Lee, "Wideband Stair-Shaped Dielectric Resonator Antennas," IET Microwaves, Antennas & Propagation, Vol. 1, Issue 2, pp. 299-305, April 2007

[14] Wei Huang and Ahmed Kishk,"Compact Wideband Multi-Layer Cylindrical Dielectric Resonator Antenna," IET Microwave Antenna and Propagation, Vol. 1, no. 5, pp. 998-1005, October 2007

[15] R. Chair, A. A. Kishk, K. F. Lee, "Low Profile Wideband Embedded Dielectric Resonator Antenna," IET Microwaves, Antennas & Propagation, Vol. 1, Issue 2, pp. 294 - 298, April 2007

[16] Y. Gao, B. L. Ooi, W. B. Ewe, A. P. Popov, "A compact wideband hybrid dielectric resonator antenna", IEEE Microw.Wirel. Compon. Lett., 2006, 16, (4), pp. 227–229

[17] K. A. A. Wei Huang, "Use of electric and magnetic conductors to reduce the DRA size", Int. Workshop on Antenna Technology: Small and Smart Antennas Metamaterials and Applications, IWAT '07, 2007

[18] J.M Ide, S.P Kingsley, S.G O'Keefe, S.A Saario, "A novel wide band antenna for WLAN applications," Antennas and Propagation Society International Symposium, 2005 IEEE , vol.4A, no., pp. 243- 246 vol. 4A, 3-8 July 2005

[19] L. Huitema, M. Koubeissi, C. Decroze, T. Monediere, "Handheld Dielectric Resonator Antenna for Ultra Wideband Applications", 2010 IEEE International Workshop on Antenna Technology: iWAT2010: « Small Antennas and Novel Metamaterials» March 1–3, 2010, pp. 1-4, Portugal

[20] L. Huitema, M. Koubeissi, C. Decroze, T. Monediere, "Compact and multiband dielectric resonator antenna with reconfigurable radiation pattern," Antennas and Propagation (EuCAP), 2010 Proceedings of the Fourth European Conference on , vol., no., pp.1-4, 12-16 April 2010

[21] A. Sangiovanni, J. Y. Dauvignac and Ch. Pichot, "Stacked dielectric resonator antenna for multifrequency operation", Microw. and Opt. Techn. Lett., vol. 18, pp. 303-306, July 1998

[22] Z. Fan and Y.M.M. Antar, "Slot-Coupled DR Antenna for Dual-Frequency Operation," IEEE Trans. Antennas and Propagation, Vol. 45, No. 2, Feb. 97,pp. 306-308

[23] G.P. Junker, A.A. Kishk, A.W. Glisson and D. Kajfez, "Effect of an air gap around the coaxial probe exciting a cylindrical dielectric resonator antennas", Electronics Letters, Vol. 30, No. 3, pp. 177-178, 3rd February 1994

[24] G.P. Junker, A.A. Kishk, A.W. Glisson and D. Kajfez, "Effect of air gap on cylindrical dielectric resonator antennas operating in TM_{01} mode", Electronics Letters, Vol. 30, No. 2, pp. 97-98, 20th January 1994

[25] Kwok-Wa Leung; Kwai-Man Luk; Lai, K.Y.A.; Deyun Lin; , "Theory and experiment of an aperture-coupled hemispherical dielectric resonator antenna," Antennas and Propagation, IEEE Transactions on , vol.43, no.11, pp.1192-1198, Nov 1995

[26] A.A. KISHK, A. ITTIPIBOON, Y. ANTAR, M. CUHACI "Slot Excitation of the dielectric disk radiator", IEEE Transactions on Antennas and propagation, vol. 43, N°2, pp.198-201, Feb 1993

[27] A.W. Glisson, D. Kajfez and J. James, "Evaluation of modes in dielectric resonators using a surface integral equation formulation," IEEE Trans. Microwave Theory Tech., Vol. MTT-31, pp. 1023-1029, 1983

[28] D. Kajfez, A. W. Glisson and J. James, "Computed modal field distributions for isolated dielectric resonators," IEEE Trans. Microwave Theory Tech., Vol.MTT-32, pp. 1609-1616, 1984

[29] Y. Kobayashi and S. Tanaka, "Resonant Modes of a dielectric rod resonator short circuited at both ends by parallel conducting plates", IEEE Trans. Microwave Theory and Tech., Vol. 28, pp. 1077-1085, Oct. 1980

[30] Van Bladel, J., "On the Resonances of a Dielectric Resonator of Very High Permittivity," Microwave Theory and Techniques, IEEE Transactions on , vol.23, no.2, pp. 199- 208, Feb 1975

[31] Petosa, A.; Simons, N.; Siushansian, R.; Ittipiboon, A.; Cuhaci, M.; , "Design and analysis of multisegment dielectric resonator antennas ," Antennas and Propagation, IEEE Transactions on , vol.48, no.5, pp.738-742, May 2000

[32] M.T.K Tam and R.D Murch, "Compact circular sector and annular sector dielectric resonator antennas", IEEE Transactions on antennas and Propagation, Vol. 47, n° 5, 1999, pp. 837-842

[33] C. W. Ling, C. Y. Lee, C. Y. Tang, and S. J. Chung *"Analysis and Application of an On-Package* Planar Inverted-F Antenna", IEEE Transactions on Antennas and Propagation, Vol. 55, n°6, June 2007, pp. 1774 - 1780.

[34] M. J. Ammann and L. E. Doyle, "A loaded inverted-F antenna for mobile handset," Microwave Opt. Technol. Lett., vol. 28, pp. 226–228, 2001.

[35] M. Ali andG. J. Hayes, "Analysis of integrated inverted-F antennas for Bluetooth application," in Proc. IEEE-APS Conf. Antennas and Propagation for Wireless Communications, Waltham, MA, 2000, pp. 21–24.

[36] K. A. A. Wei Huang, "Use of electric and magnetic conductors to reduce the DRA size", Int. Workshop on Antenna Technology: Small and Smart Antennas Metamaterials and Applications, IWAT '07, 2007

[37] L. Huitema, M. Koubeissi, C. Decroze, T. Monediere, "Ultrawideband Dielectric Resonator Antenna for DVB-H and GSM Applications" IEEE Antennas and Wireless Propagation letter, vol. 8, pp. 1021-1027, 2009

[38] Y.-Y. Wang and S.-J. Chung, "A new dual-band antenna for WLAN applications," in Proc. IEEE AP-S Int. Symp., Jun. 20–25, 2004, vol. 3, pp. 2611–2614.

[39] Jan, J. Y. and L. C. Tseng, "Small planar monopole antenna with a shorted parasitic inverted-L wire for wireless communications in the 2.4-, 5.2-, and 5.8-GHz bands," IEEE Trans. Antennas Propag., Vol. 52, 1903-1905, 2004

[40] Z. D. Liu andP . S. Hall, "Dual-band antenna for handheld portable telephones," Electron. Lett., vol. 32, pp. 609–610, 1996.

[41] J. Y. Jan andL. C. Tseng, "Planar monopole antennas for 2.4/5.2 GHz dual-band application," in Proc. IEEE-APS Int. Symp. Dig., Columbus, OH, 2003, pp. 158–161

[42] K. Hady, A. A. Kishk and D. Kajfez, "Dual-Band Compact DRA With Circular and Monopole-Like Linear Polarizations as a Concept for GPS and WLAN Applications", IEEE Trans. on ant. and prpoag., Vol. 57, No. 9, 2591-2598, September 2009.

[43] Rao, Q.; Denidni, T.A.; Sebak, A.R.; Johnston, R.H.; , "Compact Independent Dual-Band Hybrid Resonator Antenna With Multifunctional Beams," Antennas and Wireless Propagation Letters, IEEE , vol.5, no.1, pp.239-242, Dec. 2006

[44] Rotaru, M.; Sykulski, J.K.; , "Numerical investigation on compact multimode dielectric resonator antennas of very high permittivity," Science, Measurement & Technology, IET , vol.3, no.3, pp.217-228, May 2009

[45] Qinjiang Rao; Denidni, T.A.; Sebak, A.R.; , "A hybrid resonator antenna suitable for wireless communication applications at 1.9 and 2.45 GHz," Antennas and Wireless Propagation Letters, IEEE , vol.4, no., pp. 341- 343, 2005

[46] R.A Kranenburg, S.A Long, "Microstrip transmission line excitation of dielectric resonator antennas," Electronics Letters , vol.24, no.18, pp.1156-1157, 1 Sep 1988

[47] L. Huitema, M. Koubeissi, M. Mouhamadou, E. Arnaud, C. Decroze And T. Monediere, "Compact and Multiband Dielectric Resonator Antenna with Pattern Diversity for Multi Standard Mobile Handheld Devices", IEEE Transaction on antennas and propagation, vol. 59, pp.4201-4208, 2011

Chapter 3

Low Dielectric Materials for Microelectronics

He Seung Lee, Albert. S. Lee, Kyung-Youl Baek and Seung Sang Hwang

Additional information is available at the end of the chapter

http://dx.doi.org/10.5772/51499

1. Introduction

Over the past half century, low dielectric materials have been intensively researched by ceramic and polymer scientists. However, these materials possess a vast myriad of electrical, thermal, chemical, and mechanical properties that are just as crucial as the name that classifies them. Therefore, in many cases, the applications of low dielectric constant materials are dictated by these other properties, and the choice of low dielectric material may have a tremendous effect on a device's performance and lifetime.

In the field of microelectronics, many of the early low dielectric materials have been satisfactory in covering the required properties. But as the microelectronics industry continuously boomed through the 21st century, more and more advanced processes and materials have been in demand. Since the invention of microprocessor, the number of active devices on a chip has been exponentially increasing, approximately doubling every year, famously forecast by Gordon Moore in 1965. All of this is driven by the need for optimal electrical and functional performance.

Figure 1 shows the shrinking of the device dimensions over signal delay value. And while the total capacitance can be traded for resistance and vice versa by changing the geometry of the wire cross-section, the RC will always increase for future nodes. In other words, in order to enhance performance, decreasing the device size, as well as decreasing the interconnecting wire distance, gate and interconnect signals delay is the main challenge for ceramic and polymer scientists to overcome. In another approach to solve this RC delay problem, researchers have already changed the aluminum line to Cu line, which has lower resistance. But due to limitations in metal lines being applicable for use, research of low dielectric materials are continually being pursued today. The main challenge for researchers in the microelectronic industry is not to develop materials with the lowest dielectric constant, but to find materials that satisfy all of the electrical, thermal, chemical, and mechanical properties required for optimal device performance.

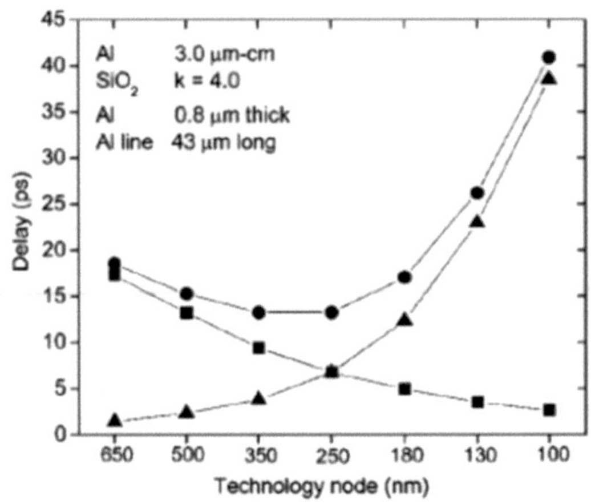

ITRS Technology Trend Targets (including PIDS 2011 roadmap flash and DRAM trend driver proposals)

	Year of Production	'09	'10	'11	'12	'13	'14	'15	'16	'17	'18	'19	'20	'21	'22	'23	'24	'25
2010 ORTC	Flash 1/2 Pitch (nm) (un-contacted Poly)(f) [A]	32	38	28	25	23	20	18	15.9	14.2	12.6	11.3	10.0	8.9	8.0	7.1	6.3	N/A
2010 PIDS Projection based on survey data	Flash 1/2 Pitch (nm) (un-contacted Poly)(f) [B]	N/A	26	24	22	20	19	18	16	14	13	12	11	9	8	8	8	N/A
2010 WAS	DRAM 1/2 Pitch (nm) (contacted) [C]	52	45	40	36	32	28	25	22.5	20.0	17.9	15.9	14.2	12.6	11.3	10.0	8.9	N/A
2010 PIDS Projection based on survey data	DRAM 1/2 Pitch (nm) (contacted) [D]	N/A	42	36	31	28	25	24.0	21.0	18.0	16.0	14.0	13.0	12.0	10.0	9.0	8.0	N/A
	MPU/ASIC Metal 1 (M1) 1/2 Pitch (nm)	54	45	38	32	27	24	21	18.9	16.9	15.0	13.4	11.9	10.6	9.5	8.4	7.5	N/A
	MPU Printed Gate Length (GLph)(nm)	47	41	35	31	28	25	22	19.8	17.7	15.7	14.0	12.5	11.1	9.9	8.8	7.9	N/A
	MPU Physical Gate Length (GLph)(nm)	29	27	24	22	20	18	17	15.3	14.0	12.8	11.7	10.7	9.7	8.9	8.1	7.4	N/A
	ASIC/Low Operating Power Printed Gate Length (nm)	54	47	41	35	31	25	22	19.8	17.7	15.7	14.0	12.5	11.1	9.9	8.8	7.9	N/A
	ASIC/Low Operating Power Physical Gate Length (nm)	32	29	27	24	22	18	17	15.3	14.0	12.8	11.7	10.7	9.7	8.9	8.1	7.4	N/A
	ASIC/Low Standby Power Physical Gate Length (nm)	38	32	29	27	22	18	17	15.3	14.0	12.8	11.7	10.7	9.7	8.9	8.1	7.4	N/A
	MPU Etch Ratio Glpr/Glph (nm)	1.6409	1.5296	1.4588	1.4237	1.3895	1.3561	1.3235	1.2917	1.2607	1.2304	1.2008	1.1729	1.1435	1.1163	1.0895	1.0633	N/A

Figure 1. Calculated gate and interconnect dely as a function of technology node according to the National Technology Roadmap for Semiconductores(NTRS) in 1997 (top): ■ gate delay; ▲ interconnect delay (Al and SiO_2); ● sum of delays (Al and SiO_2) and ITRS technology trend targets (bottom)

2. Definition of dielectric constant

Dielectric constant k (also called relative permittivity εr) is the ratio of the permittivity of a substance to that of free space. A material containing polar components, such as polar chemical bonds, which are presented as electric dipoles in Figure 2, has an elevated dielectric constant, in which the electrical dipoles align under an external electric field. This alignment of dipoles adds to the electric field. As a result, a capacitor with a dielectric medium of higher k will hold more electric charge at the same applied voltage or, in other words, its capacitance will be higher. The dipole formation is a result of electronic polarization (displacement of electrons), distortion polarization (displacement of ions), or orientation polarization (displacement of molecules) in an alternating electric field. These phenomena have characteristic dependencies on the frequency of the alternating electric field, giving rise to a change in the real and imaginary part of the dielectric constant between the microwave, ultraviolet, and optical frequency range.

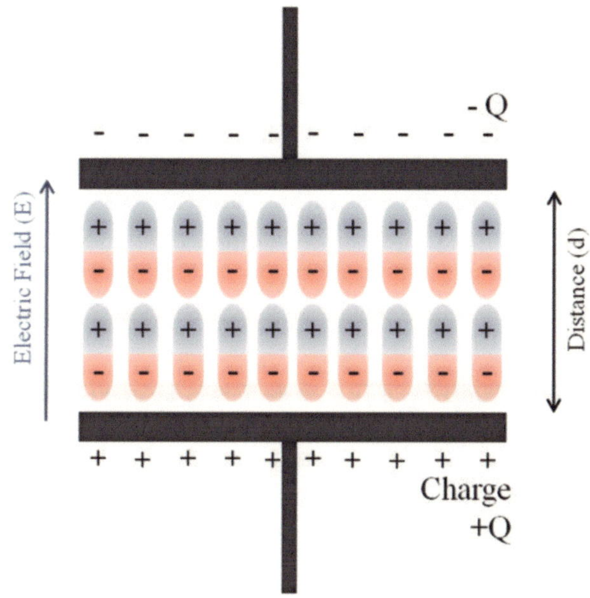

Figure 2. Schematic illustration of a capacitor.

3. Property requirements of low dielectric material

Dielectric materials must meet stringent material property requirements for successful integration into the interconnect structures. These requirements are based on electrical properties, thermal stability, thermomechanical and thermal stress properties, and chemical stability. The desired electrical properties can be outlined as low dielectric constant, low dielectric loss and leakage current, and high breakdown voltage. As RC delay and crosstalk are primarily determined by the dielectric constant, in a typical CVD SiO_2 film, the dielectric constant is around 4. And although many polymeric materials satisfy these electrical criteria, the dimensional stability, thermal and chemical stability, mechanical strength, and thermal conductivity of polymers are inferior to those of SiO_2.

Electrical	Chemical	Mechanical	Thermal
Dielectirc constant Anisotropy Low dissipation Low leakage current Low charge trapping High electric-field strength High reliability	Chemical resistance Etch selectivity Low moisture uptake Low solubility in H_2O Low gas permeability High purity No metal corrosion Long storage life Enviromentally safe	Thickness uniformly Good adhesion Low stress High hardness Low shrinkage Crack resistance High tensile modulus	High thermal stability Low coefficient of thermal expansion Low thermal weight loss High thermal conductivity

Table 1. Property Requirements of Low-k Dielectrics

In the fabrication of the multilevel structures, as many as 10 to 15 temperature treatments are repeated at elevated temperatures exceeding 400-425°C. This inherent processing of inter-dielectric (ILD) materials makes thermal stability a key prerequisite of low dielectric materials in microelectronics. Not only is the thermal stability in terms of degradation key, but the insensitivity to thermal history may be just as important. For example, changes in the crystallinity and/or crystalline phases during these thermal cycles may cause changes in the electrical and/or mechanical properties, making the material dependent on its thermal history. Other problems seldom seen in thermal processing include outgassing of volatile solvents and/or decomposition products which may cause poisoning, delamination, blistering, or cracking in the ILD.

Another thermomechanical concern of ILD materials is its coefficient of thermal expansion (CTE). The extensive thermal cycling of microelectronics may also cause stresses in the interconnect structure if there is a CTE mismatch between the ILD material and the metal or substrate. These stresses invariably cause delamination if adhesion is poor. And while adhesion promoters may be added to enhance wetting and chemical bonding at the interface between the ILD and substrate, this is mostly undesired from manufacturing point of view, as it adds unnecessary processing steps. Also, if the adhesion promoter thermally degrades, it may lead to adhesion failures or create a leakage path.

Adhesion is determined by chemical bonding at the metal/ILD interface and the mechanical interaction between the metal and ILD. Thus, ideal ILDs should have good mechanical properties such as a large Young's modulus (E), tensile strength, and elongation-at-break. And although it is uncertain what constitutes sufficient mechanical strength for successful integration into a manufacturable process, the elongation-at-break should be as large as possible to sustain the deformation and impart crack resistance, even at elevated temperatures. Also, a high modulus retention at elevated temperatures, E(T), is required for the ILD to maintain its structural integrity and dimensional stability during subsequent processing steps. Related to E(T) is the glass transition temperature, Tg. Since exceeding the Tg causes a large decrease in the modulus and yield stress in amorphous, non-crosslinked polymers, a Tg greater or equal to the highest processing temperature is desired. For example, residual compressive stresses in capping layers can cause buckling and delamination of the capping films due to the compliance of an ILD above its Tg [1,2]. Buckling has also been observed in capping layers deposited below the ILD's Tg if the capping film is highly compressive [3].

Other processing concerns include chemical resistance to the solvents and etchants commonly used during chip fabrication, chemical interaction with the metal lines causing corrosion, and moisture uptake. Moisture is a primary concern because even trace amounts can have a detrimental impact on the dielectric constant. The ILDs should also be free of trace metal contaminants, have long shelf-lives, and, preferably, not require refrigeration. Metal contamination, which can compromise the device and provide a leakage path between lines, is often a problem for polymers synthesized using metal catalysts. Other processing requirements include the ability to pattern and etch the film, etch selectivity to resists, good thickness uniformity, gap-fill in submicron trenches, and planarization.

The long-term reliability of chips fabricated using low-k materials must also be evaluated. Electromigration and stress voiding are primary failure mechanisms in integrated circuits [4-6] and these are reliability concerns when replacing SiO_2 with an alternative ILD that has thermal and mechanical properties inferior to those of SiO_2.

4. Design of low dielectric material

There are two strategies for designing a low dielectric material: decreasing dipole strength or the number of dipoles (Fig. 4) or a combination of both. In the first strategy, materials with chemical bonds of lower polarizability than Si-O or lower density would be used. Today, the microelectronics industry has already moved to certain low-k materials, where some silica Si-O bonds have been replaced with less polar Si-F or Si-C bonds. A more elementary reduction of the polarizability can be attained by utilizing all nonpolar bonds, such as C-C or C-H, as in the case of organic polymers.

Bond	Polarizability ($Å^3$)	Average bond energy (Kcal/mole)
C - C	0.531	83
C - F	0.555	116
C - O	0.584	84
C - H	0.652	99
O - H	0.706	102
C=O	1.020	176
C=C	1.643	146
C≡C	2.036	200
C≡N	2.239	213

[a] Reference [7]·
[b] Reference [8]

Table 2. Electronic polarizability[a] and bond enthalpieds[b]

The second strategy involves decreasing the number of dipoles within the ILD material by effectively decreasing the density of a material. This can be achieved by increasing the free volume through rearranging the material structure or introducing porosity. Porosity can be constitutive or subtractive. Constitutive porosity refers to the self-organization of a material. After manufacturing, such a material is porous without any additional treatment. Constitutive porosity is relatively low (usually less than 15%) and pore sizes are ~ 1 nm in diameter. According to International Union of Pure and Applied Chemistry (IUPAC) classification[9], pores less than 2 nm are denoted 'micropores'. Subtractive porosity involves selective removal of part of the material. This can be achieved via an artificially added ingredient (e.g. a thermally degradable substance called a 'porogen', which is removed by annealing to leave behind pores) or by selective etching (e.g. Si-O bonds in SiOCH materials removed by HF).

5. Utilization of low dielectric materials in microelectronics

A particularly difficult challenge for low dielectric materials development has been to obtain the combination of low dielectric constant and good thermal and mechanical stability. Generally, the types of chemical structures that imbue structural stability are those having strong individual bonds and a high density of such bonds. However, the strongest bonds often are the most polarizable, and increasing the bond density gives a similar increase in polarization. For example, the rigidity and thermal stability of SiO_2 is in part due to the dense (2.2–2.4 g/cc) chemical network. Unfortunately, the high bond and material density in SiO2 lead to a large atomic polarizability, and therefore a high dielectric constant. Organic polymeric materials often have a lower dielectric constant due to the lower material density (<1.0 g/cc) and lower individual bond polarizabilities.

In this part, the relationship between molecular structure and low dielectric properties is discussed with consideration of factors such chemical bond, density, and polarizability.

5.1. Linear structure

Linear structured materials have been actively researched for various microelectronic applications. In the early stages of microelectronics development, IBM implemented a polyimide-based material in microchips based on its good thermal, mechanical, chemical, and electrical properties. However, as required properties have become stricter because of narrowing interconnect line distance, polyimide-based materials have been unable to satisfy device performance with the main reason due to its high water absorption. Despite its superior properties, it became apparent that a linear polymeric structure was unfeasible for application as more high performance devices were being demanded.

However, linear polymeric structures have given polymer scientists invaluable clues into the possible molecular content of low dielectric materials. According to the definition of a dielectric, the material density has a direct relationship with respect to its dielectric constant. Linear polymers occupy a free volume, derived from large steric hindrance compared to single small molecules. For this reason, linear structured materials such as organic polymers, polyethylene and polypropylene show quite low density (0.8~0.9), and thus low dielectric value (2.1~2.6). Unfortunately, these organic polymers suffer from critical disadvantages such as thermal instability such as low glass transition temperature and low degradation temperature.

Therefore, many scientists turned to polymeric materials having an aromatic moiety. This chemical structure showed enhanced thermal properties and was expected to have a low density due its rigid molecular structure. The high polarizability of these materials due to their relatively high dipole moment was expected to compensate for the inherently large free volume. Some of the various aromatic, linear polymers are outlined below.

5.1.1. Polyimides (PIs)

Excellent thermomechanical properties can be obtained by incorporating a very stiff polymer. The classic example of a stiff polymer chain is aromatic polyimides, which have a rigid backbone due to the many aryl and imide rings along the chain. These structural characteristics give rise to excellent mechanical and thermal properties in the form high modulus (8–10 GPa) and high Tg (350 to 400°C) [10]. However, the rigid chain structure causes the PI chains to align preferentially parallel to the substrate, especially when deposited as thin films, which results in anisotropic properties [11-18]. For example, while the out-of-plane k value of BPDA-PDA is 3.1, the more important in-plane value is >3.5 [14].

The thermomechanical properties are likewise anisotropic. For instance, the CTE of thin films of rigid PIs is often <10 ppm/°C in the plane of the film, but can be more than ten times larger in the out-of-plane direction [14]. Another drawback to PIs is that they absorb water effectively owing to the carbonyl groups, which raises the dielectric constant further. The release of this water during processing can cause blistering of overlying layers [19].

Some of the drawbacks mentioned above can be ameliorated by tailoring the chemical structure of the PI. The k value and water adsorption can be lowered by incorporating fluorine into the material, while the anisotropy can be reduced by introducing single bonds between rings, making the chain less rigid. For example, PMDA-TFMOB-6FDA-PDA, which utilizes both of these design strategies, has an out-of-plane k=2.64 [20] and absorbs less moisture than unfluorinated PIs such as BPDA-PDA [10]. However, the in-plane k value is still >3.0, and the water uptake, although reduced, is significant enough to cause blistering in overlying layers during integration [19].

5.1.2. Poly (aryl ethers) (PAE)

The utilization of spin-on PAE materials results from attempts to balance the dielectric and thermomechanical properties. The aryl rings in these materials provide better thermomechanical properties than do PIs, but the flexible aryl linkages allow bending of the chains, which results in a more isotropic material than is obtained for PIs.

Additionally, the lack of polar groups, such as carbonyl, results in a lower k value and lower water uptake than the PIs. Fluorinated versions of PAEs had a k value of 2.4 [21]. However, because of concerns about fluorine corrosion, the fluorine was removed from later versions of the material. The nonfluorinated PAEs typically have a k of 2.8–2.9, whereas typical values for the modulus and CTE are 2.0 GPa and 50-60 ppm/°C, respectively. Resistance to thermal decomposition can be quite good for PAEs as weight losses of only <2% over 8 h at 425°C have been reported. One drawback of uncrosslinked PAEs is that they have a relatively low Tg of <275°C, which is lower than many of the thermal treatment temperatures of microelectronic devices.

5.1.3. Polynorbornene

Polynorbornene [22] is a pure hydrocarbon polymer without any polar or polarizable groups. Known for their high thermal stability among organic polymers (Tg ~365°C) and low dielectric constant [23] (~2.2), polynorbornenes are soluble in common organic solvents despite its rigid backbond due to the randomly coiled nature of the polymer chains and lack of polar interactions.

This combination of properties makes polynorbornene an interesting candidate for ILD/IMD use. However, polynorbornenes exhibit insufficient adhesion to substrates with polar surfaces such as Si, oxides or metals and its rigid backbone results in a rather brittle material. To overcome these shortcomings, a copolymer with alkoxysilyl and aikyi side groups at the norbornane rings in the backbone of polynorbornene was developed by BFGoodrich (Avatre[). [23,24] The alkoxysilyl groups enhance adhesion to surfaces with hydroxyl groups and increase the relative mobility of the polymer chains, and hence the elongation at break of thin films [23]. However, the dielectric constant increases from 2.2 of the unsubstitituted polymer to 2.67 with an aikoxysilyl content of 20% [23]. Copolymers from alkoxysilyl norbornene and alkylnorbornene derivatives show improved elongation-at-break and reduced dielectric constant [23] of E = 2.56. The glass transition temperatures and thermal stabilities of these materials are reduced compared to non-functionalized polynorbornene [25,26]

5.1.4. Polytetrafluoroethylene (PTFE)

Proposals to use fluorinated organic materials like PTFE are aimed toward minimizing the dielectric constant using the bonds of lowest polarizability. PTFE, which consists of singly bonded carbon chains saturated with fluorine atoms, has one of the lowest k values (<1.9) of any nonporous material, and is normally deposited by spin-on films [27]. One drawback of PTFE is that the flexible and uncrosslinked chain structure limits the thermomechanical stability of the material. For example, one PTFE material evaluated in our laboratory was found to have a low yield stress (12 MPa), low elastic modulus (0.5 GPa), low softening temperature (<250°C), and high thermal expansion coefficient (CTE) (>100 ppm/°C). Together these factors can cause buckling or wrinkling of the film during process integration. A second issue for PTFE, which is a concern for all fluorine-containing materials, is the potential release of fluorine atoms that can cause corrosion of metals or other reliability problems in the interconnect structure.

5.1.5. Polysilsesquioxane

While research of linear, rigid, organic polymers have centered on lowering the density and enhancing thermal and mechanical properties [28,29], many of these materials tend to have elevated dielectric constants and decreased processability because the main frame of these polymers are composed of aromatics, double, and triple bonds, which can be readily polarized or have weak thermal stability.

Polysilsesquioxanes $(RSiO_{3/2})_n$ comprise a class of polymers that exhibit unique physical properties different from those of purely organic or inorganic compounds [30]. The various structures of polysilsesquioxanes, including ladder-like polysilsesquioxanes with double stranded backbones, cage-type polyhedral oligomeric silsesquioxanes (POSS), and sol-gel processed random branched structures have shown many inherent advantageous properties such as high thermal stability, low dielectric constant, good mechanical properties, and chemical resistance. Such properties have made polysilsesquioxanes a material of interest for polymer and ceramic scientists in the microelectronics industry.

In particular, polymethylsilsesquioxanes are of particular interest, as these materials exhibit low loss at high temperatures exceeding 500°C. However, to approach the favorable properties of polysilsesquioxanes in a reproducible manner, a regular structure with high molecular weight, such as ladder-structures, is to be favorable. This rigid ladder backbone with high molecular weight would support lower dielectric constants stemming from increase of inter-molecular space and high mechanical strength, as well as minimizing shrinkage during the ILD process. This material had 6.3 GPa of modulus and 2.7 of dielectric constant without curing process. [31]

5.2. Branched structures

In polymer chemistry, polymer branching induces a lower profile of material density without significant changes of in chemical properties. Because of this, many researchers have investigated polymer geometries such as graft and hyper branched structure for application as low dielectric material.

Branched polymers are advantageous in that through chemical modification of the side or end groups can give unique functionality that could not be realized by composites of two or three different materials. Also, the chemical bonding between the two components may offset one or more deficient property of the singular component.

5.2.1. Graft Polymers

Graft copolymers are a special type of branched copolymer in which the side chains are structurally distinct from the main chain. The Figure (3) depicts a special case where the main chain and side chains are composed of distinct homopolymers.

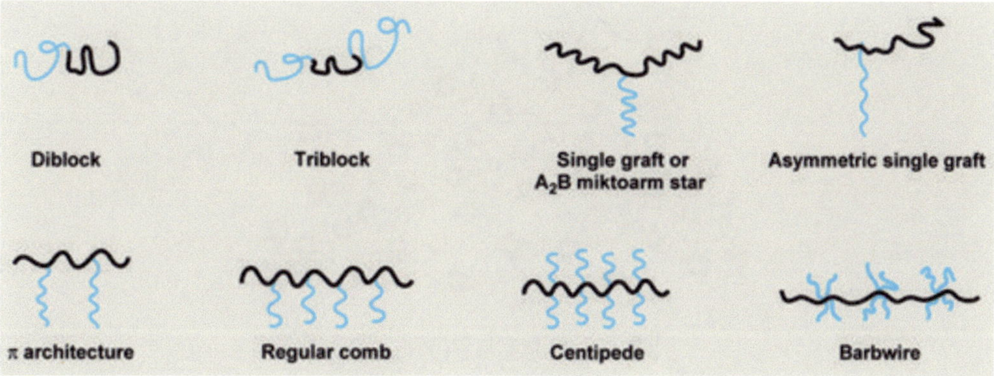

Figure 3. Special cases of grafted polymer

Kung-Hwa wei et al. reported a polyimide grafted polyhedral oligomeric silsesquioxane (POSS). They presented the dielectric constants and densities of the POSS/polyimide nanocomposites (figure 4). The dielectric constants of the POSS/polyimide nanocomposites decreased as the amount of POSS increased. The maximum reduction in the dielectric constant of POSS/polyimide nanocomposites was found to be about 29%, compared to 16 mol% POSS/polyimide to pure polyimide (*k*) 2.32 vs 3.26. However, these graft polymers exhibited a slightly lower glass transition temperature about 40°C and increased thermal expansion efficient (CTE) from 31.9 ppm/K to 57.1 ppm/K after the grafting of POSS. [32]

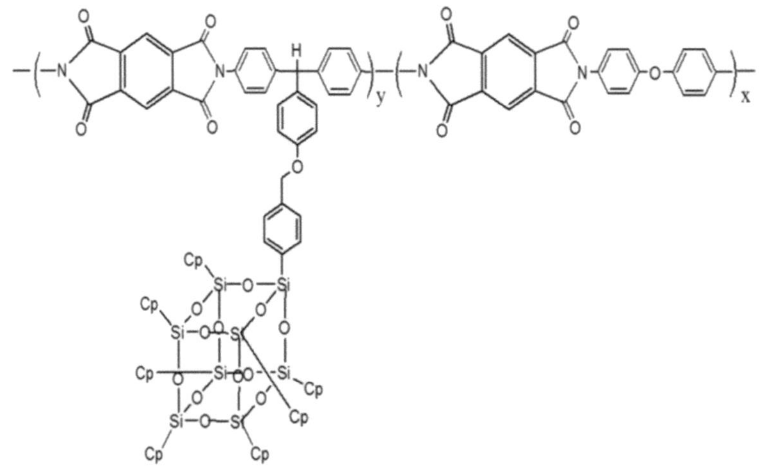

Figure 4. POSS/polyimide nanocomposites by grafted method

Another researcher studied about polyimide with grafted POSS structure. By introducing a polymerizable methyl methacryl functional groups to POSS and subsequent free-radical graft polymerization to an ozone treated polyimide, POSS grafted polyimide structures were obtained. Copolymers with dielectric constants approaching 2.2 could be achieved in the PI-*g*-PMA-POSS film containing 23.5 mol % MA-POSS. In this approach, POSS content could be easily tuned by the grafting ratio of MA-POSS.[33]

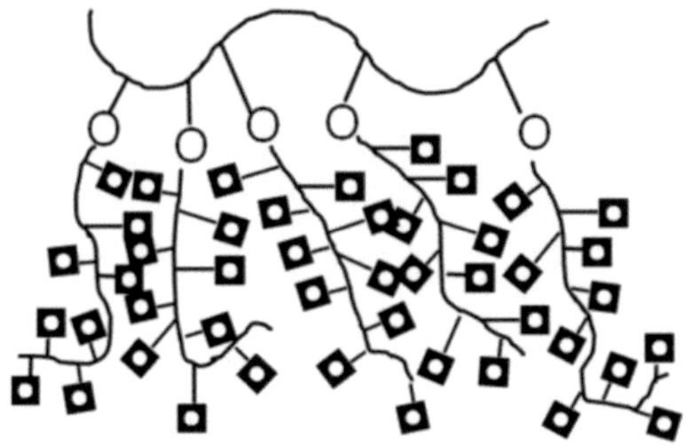

Figure 5. Synthesis of PI-*g*-PMA-POSS for low dielectrics

5.2.2. Hyperbranched Polymer

Hyperbranched polymers are densely branched structures with large number of reactive groups. They are polymerized from monomers with mixed reactivities, commonly denoted A2B or A3B monomers, thus giving branched structures with exponential growth, in both end-group functionalities and molecular weights.

One property often mentioned of hyperbranched polymers is the non-Newtonian relationship between viscosities and molecular weight, where hyperbranched polymers showed *low viscosities* at *high molecular weights.* For coating applications, this should be highly interesting in terms of microelectronics, where they may be used as an aid in critical patterning applications for **back-end**-of-line (BEOL) inter-level dielectric (**ILD**) materials.

Jitendra et al. showed that dense hyperbranched carbosiloxane (HBCSO) thin films have better mechanical properties than traditional organosilicates.[34] These materials are obtained by sol-gel processing of methane-bridged hyperbranched polycarbosilanes (HBPCSs), with the incorporated methane bridges being reminiscent of the systems described above (Figure 6). For example, Young's moduli of 17-22 GPa are obtained for films with dielectric constants ranging from 2.6 to 3.1. These materials have excellent electrical properties, breakdown voltages higher than 5 MV/ cm, and leakage currents $<10^{-8}$ A/cm2 measured at 2 MV.

It was also shown that the HBPCS structure is of considerable importance in determining the properties of the thin films generated after sol-gel processing.

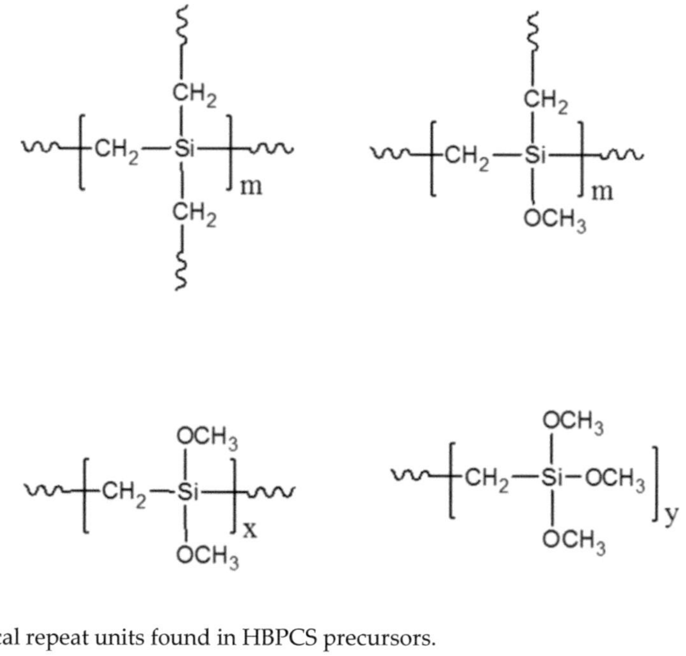

Figure 6. Chemical repeat units found in HBPCS precursors.

5.3. Network structure of low dielectric materials

In defining the different types of network structure materials for low dielectric constant applications in microelectronics, two classifications may exist. One is organic networks based on elemental carbon including amorphous carbon (diamond-like-carbon(DLC)) [35,36] and interpenetrating polymer network (IPN). The other is inorganic networks based on silicon oxide bonds such as amorphous SiO_2 and mesoporous crystalline silicon oxide.

Intuitively, network structures have excellent thermal stabilities and mechanical properties, and chemical resistance, but have relatively high density which is a factor in elevated dielectric constants. Because of this, dielectric constant and mechanical properties should be carefully controlled and careful consideration of its crystalline structure, as crystalline structures tend to be easily polarized.

5.3.1. Organic Network

When compared with linear structured polymers, network polymers have significant advantages in thermal stability with increasing glass transition temperature and complying CTEs. Dielectric materials for microelectronics needed to have high Tg temperature up to 400°C and endurance of repeated thermal cycling, creating the thermal mismatches which can lead to flow, delamination, adhesive failure, etc. Examples or organic network materials are discussed below.

5.3.1.1. Diamond-like carbon [DLC]

Amorphous diamond-like carbon [DLC], which can be prepared by chemical vapor deposition [CVD] method, [35,36] are metastable materials composed of sp2, sp3, and even sp1 hybridized carbon atoms with hydrogen concentrations, *CH*, ranging from 1% to 50%, with the composition being primarily determined by the nature of the precursor and the corresponding deposition conditions.

These sp^3 bonds can occur not only with crystals - in other words, in solids with long-range order - but also in amorphous solids where the atoms are in a random arrangement. In this case, there will only be bonding between a few individual atoms and not in a long-range order extending over a large number of atoms. The bond types have a considerable influence on the material properties of amorphous carbon films. If the sp^2 type is predominant the film will be softer, if the sp^3 type is predominant the film will be harder.

Under the right conditions, it is possible to deposit DLC films with compressive stress, spanning values from 200-800 MPa, and dielectric constants approaching 2.7.[36, 37]

5.3.1.2. SiLK and BCB resins

A very promising class of network polymers has been developed by Dow under the name 'Silk'. The formulations presumably consist of a mixture of monomeric and/or oligomeric aromatic starting compounds, which contain ortho-bisethinyl or -phenylethinyl groups [38]. The materials exhibit k values on the order of 2.6-2.7, with decomposition temperatures in excess of 500°C, no softening up to 490°C, good gap fill properties down to below 0.1 um, a maximum water uptake of 0.25%, and a coefficient of thermal expansion 4s of 66 ppm/K.

Benzocyclobutene (BCB) resins were developed by Dow in the 1980's [39], with a silicon-containing derivative for microelectronics applications, and are commercially available under the name 'Cyclotene'. BCB resins with imide structures can be extremely tough, and the dielectric constant of cured films from this monomer is 2.6-2.7, with thermal stability up to 375°C and water adsorption of only 0.2% [40,42].

5.3.2. *Inorganic network*

Inorganic networks mainly consist of ceramics or amorphous silica. In various materials for microelectronics, the silicon oxides play a major role due to its low polarizability, superior thermal and mechanical properties. In addition, tunable microspores can be made to reduce the dielectric constant through control of microstructure under special conditions.

5.3.2.1. Ordered Mesoporous Materials

Microporous zeolite thin films were first investigated by Yan and co-workers. [43] These films offer good thermal stability (i.e. no pore collapse or unidirectional shrinkage) and inter-particle mechanical strength. They can be prepared via a simple spin-on method [43,44] or by in situ growth. [44,45] With simple spin-on methods a dispersion of small zeolite particles are prepared and cast onto a surface. The porosity originates from the interparticle porosity within the zeolite nanoparticles and the intra-particle porosity owing to the packing of the near-spherical nanoparticles in thin film format.

5.3.2.2. Network polysilsesquioxane

Some of the most promising materials for dielectric materials are poly(silsesquioxanes). Most common are polymethylsilsesquioxane (MSQ), e.g. Accuspin T-18 from Allied Signal [46], or poly(hydridosiLsesquioxane) (HSQ), e.g. FOx from Dow Corning [47]. Synthesis of these silsesquioxanes(MSQ and HSQ) have traditionally been through the sol-gel method, as its utility in being able to obtain highly cross-linked structures through acidic and or basic conditions has been well documented [48,49,50]. Dielectric constant values of around 2.6 can be achieved for HSQ and MSQ. But while MSQ exhibits this dielectric constant after curing at temperatures up to 450°C, HSQ must be cured at temperatures lower than 210°C [51]. Curing of HSQ at temperatures of 250°C or above results in dielectric constant around 3 or even higher 32°[51,52]. Gap fill and planarization properties are also acceptable and because of their chemical structure, which is closely related to SiO_2, polymethylsilsesquioxanes are also compatible with existing lithographic procedures.

Efforts to further decrease the dielectric constant without decreasing mechanical strength, POSS skeletons have been introduced in MSQ. To suppress the phase separation, incompletely condensed methyl functionalized POSS precursors have been used to form chemical bonds with oligomeric sol precursors. These incompletely condensed POSS moieties functioned as coupling agents while expanding the free volume of the final sol after curing which was accomplished to 4 GPa of modulus and 2.3 of dielectric constant. [53](figure 7).

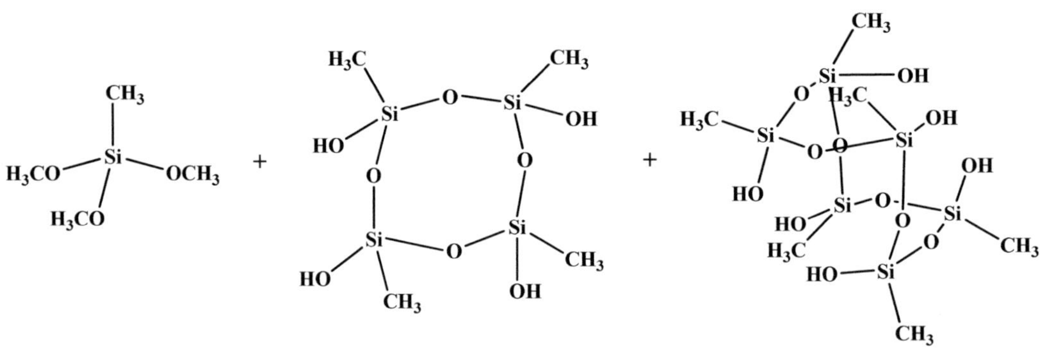

Figure 7. Introduction of POSS moiety by sol-gel method

5.4. Porous Network polymer by subtraction of porogen

Numerous methods of introducing subtractive porosity into spin-on deposited materials exist. Conventional methods of substractive porosity involve the addition of molecular or supramolecular particles called 'porogens' with tailored thermal stability to the dielectric precursor[54]. The stability of these particles is such that they are not affected by the coating drying step, and they are removed by pyrolysis during final film sintering or cure at temperatures typically in the range from 300 to 400 °C. An example of a material for which the pore size and porosity, or the pore size and porogen load can be controlled independently. [55] However, it should be noted that the use of porogens should only be applied to dense materials having a k less than 2.5 and modulus over 5 GPa in order for the final material to satisfy the required mechanical property.

In organic materials, the SilK matrix has been the only known material to provide the thermal and mechanical properties at temperature up to 500°C for use in combination with porogens. C.E. Mohler et al. [56] reported on porous SiLK dielectric film properties such as pore volume, porosity, size distribution, and showed a 2.2 dielectric constant at 30% load of porogens.

In comparison with organic porous dielectric material, inorganic porous dielectric materials have been more rigorously investigated because of their superior mechanical properties. Representative studies have used polymethylsilsesquioxane (PMSQ) as matrix for the addition of various porogens such as the block copolymers, poly(styrerene-block-acrylic acid) [57], macromolecules of cyclodextrin [58], poly(caprolactone [59], and calix[4]arene [60].

Many of these studies with porogens have reported materials that have excellent mechanical and electrical properties, but lack in other practical aspects for application in microelectronics. When porogens are introduced into a matrix, critical problems may occur, such as thermal degradation products acting as a poison or contaminant within the matrix or interfacial adhesion problems. Therefore, use of porogens has yet to remain a difficult process for practical applications in microelectronics.

6. Conclusions

The search for materials with low dielectric constant in the microelectronics industry has and will continue feverishly into the future as the demand of faster processing speeds increases. Reduction of the dielectric constant of a material can be accomplished by selecting chemical bonds with low polarizability and introducing porosity. Integration of such materials into microelectronic circuits, however, poses a number of challenges, as the materials must meet strict requirements in terms of properties and reliability. The introduction of low k materials in microelectronics research and development is a good example of how industrial needs drive new fundamental and applied research topics in science. Examples include pore structure characterization, deposition of thin films on porous substrates, mechanical properties of porous films, and conduction mechanisms in these materials. The substantial efforts made by materials and IC researchers to integrate the low-k films and continue historical device performance improvements have contributed to, and are still leading to, innovative fundamental and applied science.

Author details

He Seung Lee, Albert. S. Lee, Kyung-Youl Baek and Seung Sang Hwang
Center for Materials Architecturing, Korea Institute of Science Technology, Seoul, Korea

Acknowledgement

This work was financially supported by a grant from the Fundamental R&D Program for Core Technology of Materials funded by the Ministry of Knowledge Economy, Republic of Korea and Partially by a grant from Center for materials architecturing of Korea Institute of Science and Technology (KIST)

7. References

[1] Ray, GW. 1998. Mater. Res. Soc. Symp. Proc.511:199
[2] Fox, R, Pellerin JR. 1997. Res. Rep. Austin TX: SEMATECH

[3] Hummel, JP. 1995. In Advanced Multilevel Metallization Materials Properties Issuesfor Copper Integration, ed. CS Schuckert,6:547.Wilmington, DE: DuPont Symp.Polyimides in Microelectronics
[4] Ho, PS; Kwok, T. 1989. Rep. Prog. Phys.52:301
[5] Hu, C-K.; Rodbell, KP; Sullivan, TD.; Lee, KY.; Bouldin, DP.; 1995. IBM J. Res. Dev. 39:465
[6] Wilson, SR.; Tracy, CJ.; eds. 1993. Handbook of Multilevel Metallization for Integrated Circuits. Park Ridge, NJ: Noyes
[7] Miller, KJ.; Hollinger, HB.; Grebowicz, J.; Wunderlich, B.; 1990. Macromolecules 23:3855
[8] Pine, SH. 1987. Organic Chemistry. New York: McGraw-Hill. 5th ed.
[9] Rouquerol, J., et al., Pure Appl. Chem. (1994) 66, 1739
[10] Auman, BC. 1995. Mater. Res. Soc. Symp.Proc. San Francisco. 381:19
[11] Kang, Y-S. 1994. Microstructure and strengthening mechanisms in aluminum thin films on polyimide film. Ms thesis. Univ. Texas, Austin
[12] Lee, J-K. 1998. Structure-property correlation of polyimide thin films on line structure. PhD thesis. Univ. Texas, Austin
[13] Molis, SE. 1989. In Polyimides: Materials,Chemistry and Characterization, ed. Amsterdam: Elsevier
[14] Ree, M.; Chen, KJ.; Kirby, DP. 1992. J. Appl.Phys. 72:2014
[15] Chen, ST.; Wagner, HH. 1993. J. Electron.Mater. 22:797
[16] Lin, L.; Bastrup, SA. 1994. J. Appl. PolymerSci. E 54:553
[17] Boese, D.; Lee, H.;Yoon, DY.; Rabolt, JF.; 1992 J. Polymer Sci. Polymer Phys. 30:1321
[18] Hardaker, SS.; Moghazy, S.; Cha, CY.;Samuels, RJ. 1993. J. Polymer Sci. Polymer Phys. 31:1951
[19] Wetzel, JT.; Lii, YT.; Filipiak, SM.; Nguyen, BY.;Travis, EO.;, et al. 1995. Mater. Res. Soc.Symp. Proc. San Francisco. 381:217
[20] DeMaggio, GB.; Frieze, WE.; Gidley, DW.;Zhu, M.; Hristov, HA.; Yee, AF. 1997. Phys.Rev. Lett. 78:1524
[21] Hendricks, NH.;, Lau, KSY.; Smith, AR.; Wan, WB. 1995. Mater. Res. Soc. Symp. San Proc. Francisco. 381:59
[22] Heitz, W. 1995. Pure Appl.Chem. 67:1951
[23] Grove, N.R. et al 1997Mater Res Soc Symp Proc 476:3
[24] Grove, N.R. et al. 1997. Proceedings of the 6th International Conference on Multichip Modules, Institute of Electrical and Etecronics Engineers, New York
[25] Treiche, L.H. et al. 1999. Low dielectric constant materiaLs for interlayer dielectrics. In: Nalwa, H.S. (Ed.) Low-k and high-k materials, Academic Press,Boston
[26] Grove, N. et al. 1997. Proceedings of the 6th International Conference on Polyimides and other Low-k Dielectrics, McAfee, N J
[27] Rosenmayer, CT.; Bartz, JW.; Hammes, J. 1997. Mater. Res. Soc. Symp. Proc. San Francisco. 476:231
[28] Voit, B. 2000. Journal of Polymer Science Part a-Polymer Chemistry, 38:2505

[29] Mathews, A.S.; Kim, I. and Ha, C.S. 2007. Macromolecular Research, 15:114
[30] Baney, R. H.; Itoh, M.; Sakakibara, A. and Suzukit, T. 1995. Chem. Rev. 95:1409
[31] Hwang, S.S., et al. Macromolecular Research, in press
[32] Leu C-M; Chang, Y-T; and Wei,K-H, 2003. Macromolecules, 36: 9122
[33] Chen,Y.; Chen, L.; Nie, H.; Kang, E. T. 2006. Journal of Applied Polymer Science, 99:2226
[34] Rathore, J. S.; Interrante, L. V.; Dubois, G. 2008. AdV. Funct. Mater. 18: 4022.
[35] Grill, A. In Dielectric Films for AdVanced Microelectronics; Baklanov, M., Maex, K., Green, M., Eds.; Wiley: New York, 2007.
[36] Grill, A.; Patel, V.; Saenger, K. L.; Jahnes, C.; Cohen, S. A.; Schrott, A. G.; Edelstein, D. C.; Paraszczak, J. R. 1997. Mater. Res. Soc. Symp. Proc., 443:155.
[37] Grill, A. 2001. Diamond Relat. Mater. 10: 234.
[38] Babb, D. et el Wodd Patent WO 97/10193, The Dow Chemical Company,March 20, 1997
[39] Kirchhoff, R.A. and Bruza, K.J. 1994. Adv Polym Sci 117:1
[40] Yang G.R. et al 1997. Mater Res Soc Symp Proc 476 :161
[41] Gutmann, R.J. et al 1995.Mater Res Soc Symp Proc 381:177
[42] MilLs, M.E. et al 1997. Microelectron Engng 33: 327
[43] Hunt, H. K.; Lew, C. M.; Sun, M.; Yan, Y. and Davis, M. E. 2010. Micro. And Meso. Mater., 128:12
[44] Wang, Z.; Wang, H.; Mitra, A.; Huang, L. and Yan, Y. 2001. Adv. Mater., 13:746 .
[45] Li, S.; Demmelmaier, C.; Itkis, M.; Liu, Z.; Haddon, R. C. and Yan, Y. 2003. Chem.Mater., 15:2687.
[46] Hacker, N.P. et al 1997. Mater Res Soc Symp Proc 476:25
[47] Bremmer, J.N. et al 1997. Mater Res Soc Symp Proc 476:37
[48] Lee, D. et al., 2008.Macromolecular Research, 16:353-359
[49] Rankin, S.E. et al., 2000. Macromolecules, 33:7639
[50] Mackenzie, J.D. 1988. Journal of Non-Crystalline Solids, 100:162
[51] Kim, S.M. et al 1998. Mater Res Soc Symp Proc 511:39
[52] Tobben, D. et al 1997. Mater Res Soc Symp Proc 443:195
[53] Lee, A. S.; Lee,H.S.; Hwang, S. S. preparing to publish
[54] Lu, Y. F. G.; Cao, Z.; Kale, R. P.; Prabakar, S.; Lopez, G. P. and Brinker, C. J. 1999. Chem. Mater. 11:1223
[55] Baklanov , M. R. et al., Proceedings of the Advanced Metallization Conference (Materials Research Society, Pittsburgh, PA, 2002)
[56] Mohler, E. B.; Landes, G.; Meyers, F.; Kern, B. J.; Ouellette, K. B. et al. 2003. AIP Conf. Proc. 683:562
[57] Chang, Y.; Chen, C-Y.; Chen, W-C. 2004. Journal of Polymer Science: Part B: Polymer Physics, 42:4466
[58] Lyu, Y-Y.; Yim, J-H.; Byun, Y.; Kim, J. M.; Jeon, J-K.; 2006. Thin Solid Films 496:526
[59] Hyeon-Lee, J.; Lyu, Y. Y.; Lee, M. S.; Hahn, J-H.; Rhee, J. H.; Mah, S. K.; Yim, J-H.; Kim, S. Y. 2004. Macromol. Mater. Eng., 289:164

[60] Vallery, R. S.; Liu, M.; Gidley, D. W.; Yim, J-H. 2011. Microporous and Mesoporous Materials 143:419

Section 2

Magnetodielectric Materials

Chapter 4

Magnetodielectric Materials – Use in Inductive Heating Process

D.A. Hoble and M.A. Silaghi

Additional information is available at the end of the chapter

http://dx.doi.org/10.5772/2781

1. Introduction

In the induction heating processes, the main problem is to increase process efficiency, which can be achieved through the intervention of different parts of the installation [9,10,11]

A method used to increase electrical energy conversion efficiency, which is referred to in the literature and that also was considered in the study discussed in this paper is to use magnetic flux concentrator[1,2,6]. If we analyze the structure of hypothetical wound inductors, located close to a work piece, as shown in Figure 1, we can see that for this structure is equivalent electric circuit in Figure 2.

Ohm's Law applied to the magnetic circuit is:

$$NI = R_m \times \Phi_i \tag{1}$$

The magnetic reluctance equivalent of the system consists of two parallel reluctance: magnetic reluctance of the work piece R_{mp} and the air gap magnetic reluctance between spiral inductors and piece of heated, R_{ma}, in series with the magnetic reluctance of outside environment, Rme, of the inductor coils

If the magnetic reluctance of the piece, R_{mp}, depends on the material characteristics, and the air gap reluctance R_{ma} can not be reduced below a value that depends on technological conditions of the heating process, remains the method of intervention to reduce the equivalent reluctance of the system, improving the environment reluctance outside coils inductors, R_{me}.

This is actually the role of magnetic field concentrator using in inductive heating processes.

To achieve these magnetic concentrators in practice it is using a variety of materials and could therefore be useful brief review of their focusing on magnetodielectric materials made and used in the study approached the work.

Requirements of magnetic materials in induction heating applications can be very severe in many cases[1,8]. They must operate within a broad category of frequencies, to possess permeability and high saturation flux densities, have stable mechanical properties and resistance to high temperatures caused by heat loss due to magnetic concentrators and heat transferred from the heated parts.

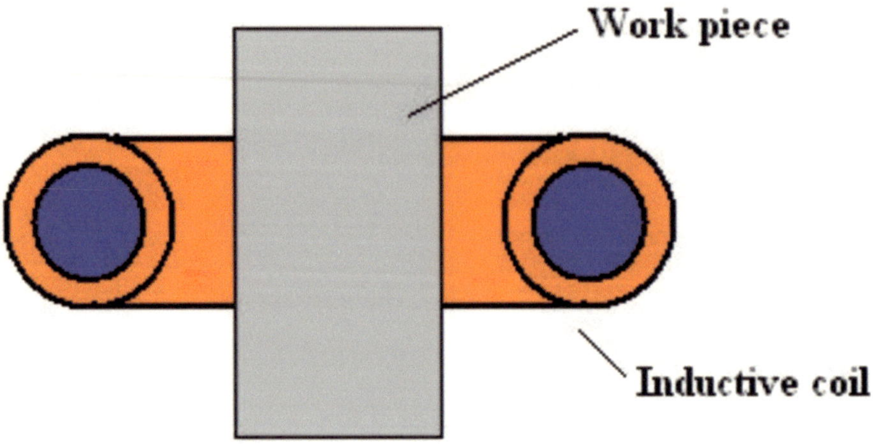

Figure 1. Hypothetical structure of an inductor – work piece of heated

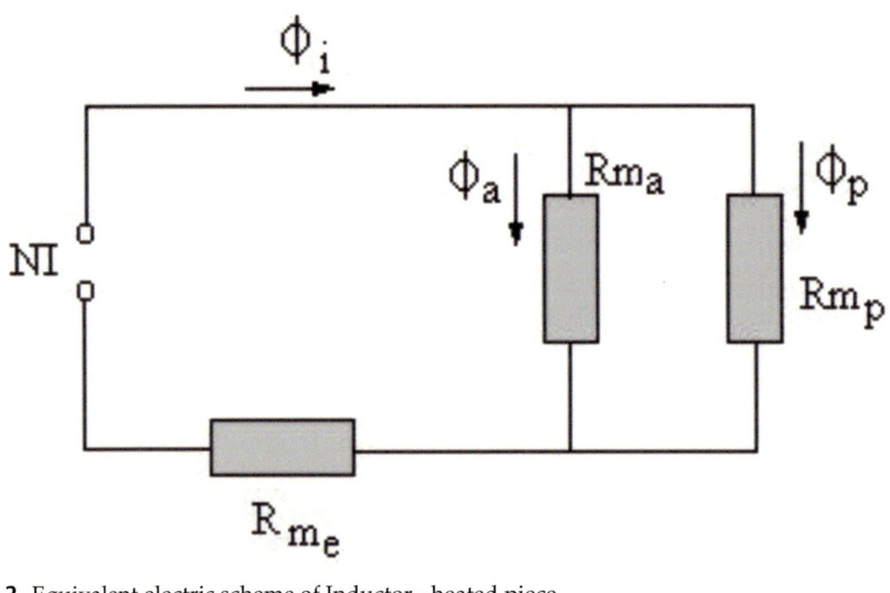

Figure 2. Equivalent electric scheme of Inductor - heated piece

Three groups of magnetic materials can be used to concentrate the magnetic flux: laminates, ferrites and magneto dielectric materials, so-called MDM materials.

Figure 3. present variation curves B = f (H), compared to laminated, ferrite materials and MDM [1,2,6], used in magnetic field concentrator construction.

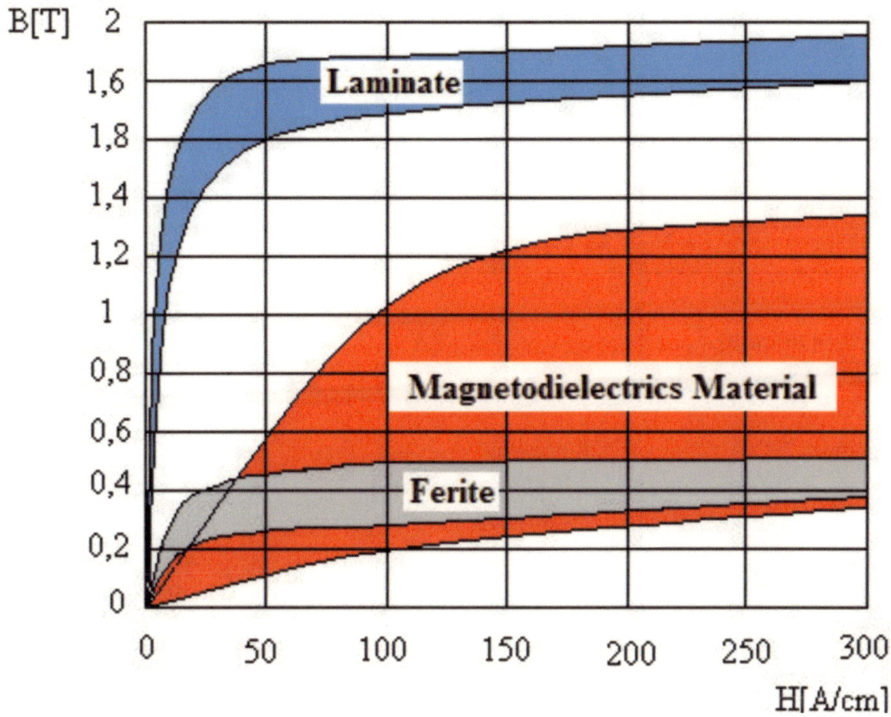

Figure 3. B=f (H) curves [1;2,6]

2. Magnetodielectric materials

These materials are composite materials made of magnetic particles and dielectric materials that serve as links and electrical insulators of magnetic particles. Magnetic properties of magnetodielectric materials (MDM), depend on constituent particle properties, their shape and volume. Mechanical and thermal characteristics depend mainly on the ratio of magnetic material, and dielectric material. The general properties of materials depend by manufacture technology, their achievement, given the fact that their production involves a pressing process, who making certain properties, especially magnetic to manifest differently on different axes.

The world leader in the manufacture of MDM is FLUXTROL Company, which made several types of magnetodielectric materials [6].

In those that follow will present the results of several studies made by the authors, to obtain a magnetodielectric material by using it in inductive heating processes.

To achieve magnetodielectric material was used dielectric material, consisting of two components manufactured epoxipoliamidic resin without solvents, that drying to 80 ° C.

Mixing ratio of components A / B is 2/1 parts by weight. The product is used in electrical engineering, as mass of hardware, building and construction fill in some details of engines and generators, to strengthen the drum and clutch electromagnetic coil for sticking carcasses and broken cylinders from large and small electrical transformators.

Technical characteristics of electro-mass (dielectric material):

a. Features delivery

Features delivery	A Component	B Component
Aspect	Mush mass	Mush mass
Colour	gray	gray
Specific weight [g/cm3]	1.7 - 1.5	1.25-1.30

Table 1.

Curing time at: 23°C is 8 hours
80°C is 1 hours
120°C is 30 minute

Processability time: 30 min.

b. Mechanical properties of hardened product

Mechanical properties	
Tensile resistance [kgf/cm]	190
Compresion resistance[kgf/cm]	800
Bend resistance [kgf/cm]	200

Table 2.

c. Electrical properties

Electrical properties	
dielectric rigidity [kV/mm]	14
Surface resistivity [Ω]	5*1012
Volume resistivity [Ωcm]	3*1014

Table 3.

Characteristics were determined on specimens cold hardened for 7 days

3. Experimental determination of magnetodielectric material resistivity

To achieve the intended purpose of the topic to increasing power conversion efficiency in the heat study was done to achieve magnetodielectric material samples based on ferromagnetic particles, embedded in electro-mass.

In the first phase of research was done a rectangular plaque, the dimension of them it is shown in Figure 4.

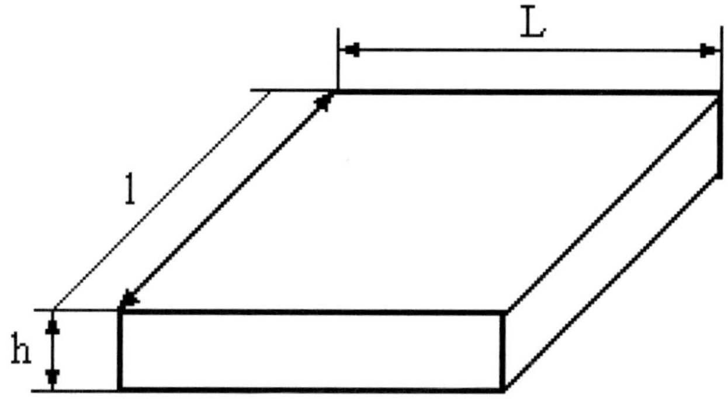

Figure 4. The dimension of the rectangular plaque

The dimension of the rectangular plaque are :

L = 46,6 mm
l = 16,5 mm
h = 8,2 mm

The magnetodielectric material was made by mixing the ferromagnetic metal powder 68%, with 32% resin of the type shown in the previous subsection. Material sample is placed between rectangular form electrodes of a measuring device whose scheme is shown in Figure 5

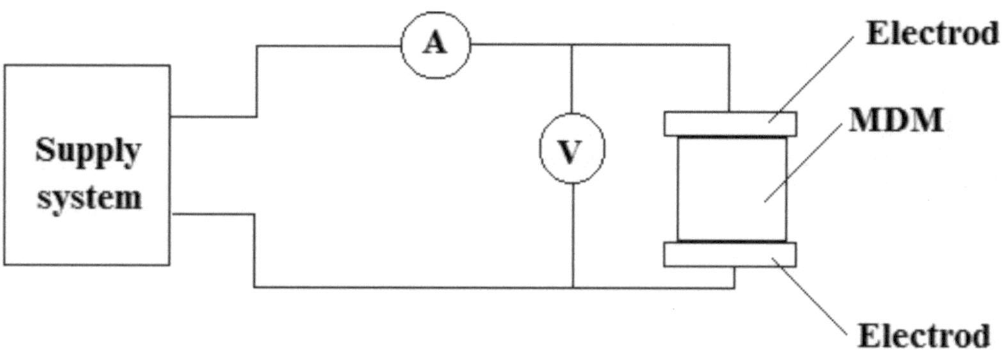

Figure 5. Module scheme used to determine resistivity value

The supply voltage is set at 100V and it was make determinations of the resistance of the sample and on the base of the sample size it was calculated the material resistivity. The determination, at different temperatures of the ambient, by heating the specimen of the MDM in a thermostatic oven, it was repeat.

The results are presented in the following table.

Average values of the coefficient of variation of resistivity with temperature, is $\alpha = 51{,}10$

Curve of variation of resistivity with temperature, obtained for this type of material is shown in Figure 6.

Nr. det	U [V]	I [mA]	R [Ω]	ϱ [Ωm]	T [⁰C]	α
1	100	3,14	31840	738,05	26	
2	100	1.14	87710	2033,11	40	24,54
3	100	1,02	98030	2272	50	46,32
4	100	0,12	833000	19308,9	60	74,98
5	100	0,017	5714280	132457	70	58,59

Table 4.

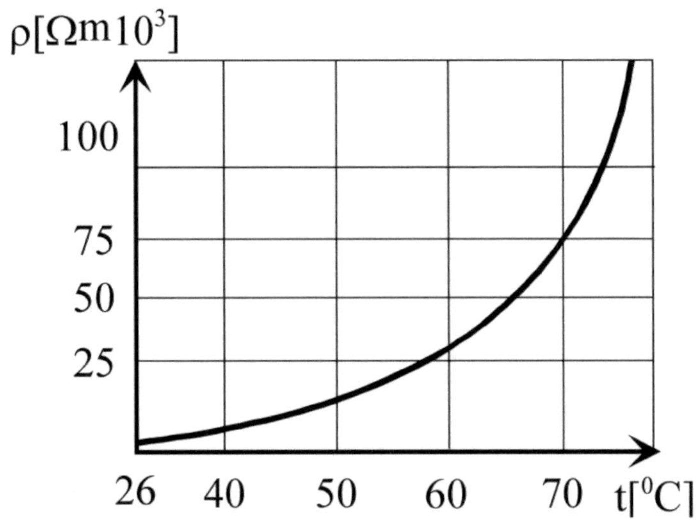

Figure 6. Variation of resistivity with temperature

4. Determination of the curve B = f (H) for magnetodielectric material

To determine this curve was used fluxmeter method. For this, was done a sample of material with thoroidal shape and that were winding two coils. Figure 7 is showing the shape and the dimension of the MDM sample.

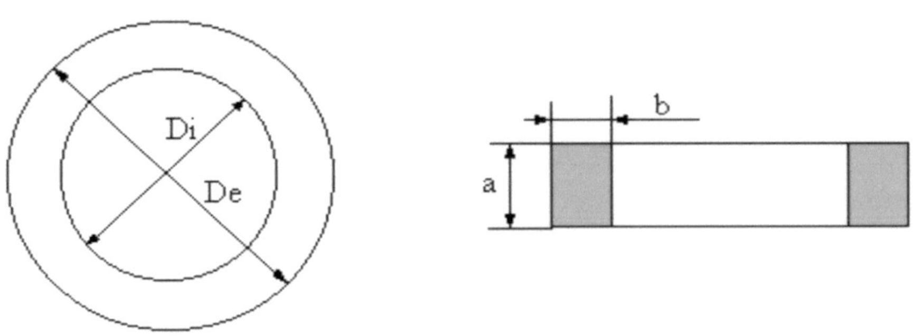

Figure 7. The dimension of the MDM sample

The dimension of the thoroidal sample is:

De = 46 mm
Di = 34 mm
Dm = 40 mm
a = 6mm
b = 6mm

Number of turns in primary :N1= 368
Number of turns in secondary :N2= 1

Scheme used to determine the dependence B = f (H) is shown in Figure 8

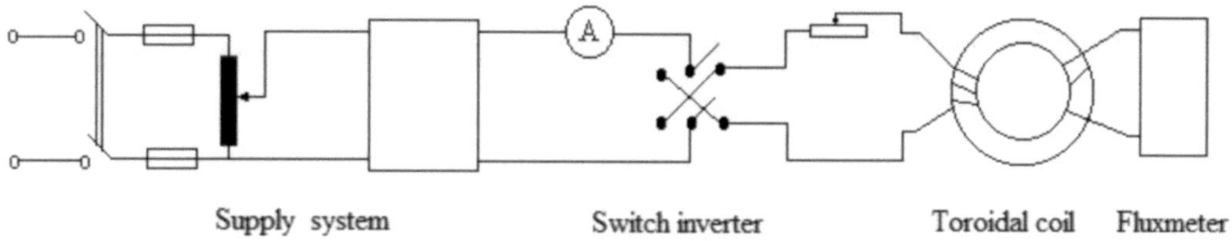

Figure 8. Scheme used to determine the dependence B = f (H)

Magnetic field is calculated with relation:

$$H = \frac{N_1 \cdot I}{\pi \cdot D_m} \left[\frac{A}{m}\right] \tag{2}$$

and magnetic flux density is given by the expression:

$$B = \frac{\Phi}{2N_2 \cdot S} \tag{3}$$

where S is the section of the toroidal core magnetodielectric material.

Measured and calculated values are presented in the table 5.

Figure 9 show the first magnetization. With [*] are the values which have been determined from measurements

5. Inductor modeling with field concentrator

Given the advantages of concentrator field magnetodielectric materials, in this case study was modeled an inductor, witch use a concentrator field made by magnetodielectric material, by the type obtained, by using modeling and numerical simulation with FLUX2D software[4].

The field concentrator made of magnetodielectric material, have the following materials data:

- Relative permeability - constant scalar model, the value 108
- Resistivity - constant scalar model, the value 40 000 Ωm.

I [A]	Φ [mWb]	B[T]	H [A/cm]	I[A]	Φ [mWb]	B[T]	H [A/cm]	I[A]	Φ [mWb]	B[T]	H [A/cm]
1	0,15	0,1	21	1	0,15	0,1	21	1,5	0,01	0,08	31,52
1,5	0,3	0,2	31,52	1,5	0,15	0,1	31,52	2	0,01	0,08	42
2	0,45	0,3	42	2	0,3	0,2	42	2,5	0,15	0,1	52,54
2,5	0,75	0,6	52,54	2,5	0,37	0,3	52,54	3	0,15	0,1	63
3	0,9	0,7	63	3	0,37	0,3	63	3,5	0,15	0,1	73,56
3,5	1,2	1	73,56	3,5	0,6	0,5	73,56	4	0,15	0,1	84
4	1,35	1,1	84	4	0,6	0,5	84	4,5	0,3	0,2	94,58
4,5	1,35	1,1	94,58	4,5	0,6	0,5	94,58	5	0,3	0,2	105
5	1,05	0,9	105	5	0,52	0,4	105				
4,5	0,9	0,7	94,58	4,5	0,45	0,3	94,58				
4	0,75	0,6	84	4	0,6	0,5	84				
3,5	0,75	0,6	73,56	3,5	0,6	0,5	73,56				
3	0,6	0,5	63	3	0,52	0,4	63				
2,5	0,45	0,3	52,54	2,5	0,3	0,2	52,54				
2	0,3	0,2	42	2	0,15	0,1	42				
1,5	0,15	0,1	31,52	1,5	0,15	0,1	31,52				
1	0,01	0,08	21	1	0,01	0,08	21				

Table 5. Measured and calculated values

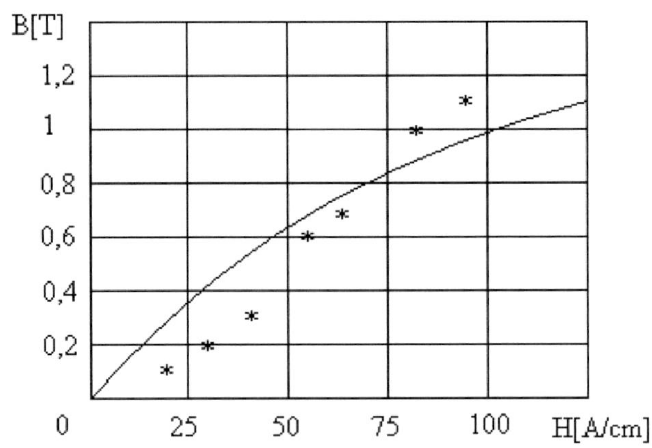

Figure 9. Dependence B = f (H).

Geometry made for this case is shown in Figure 10.

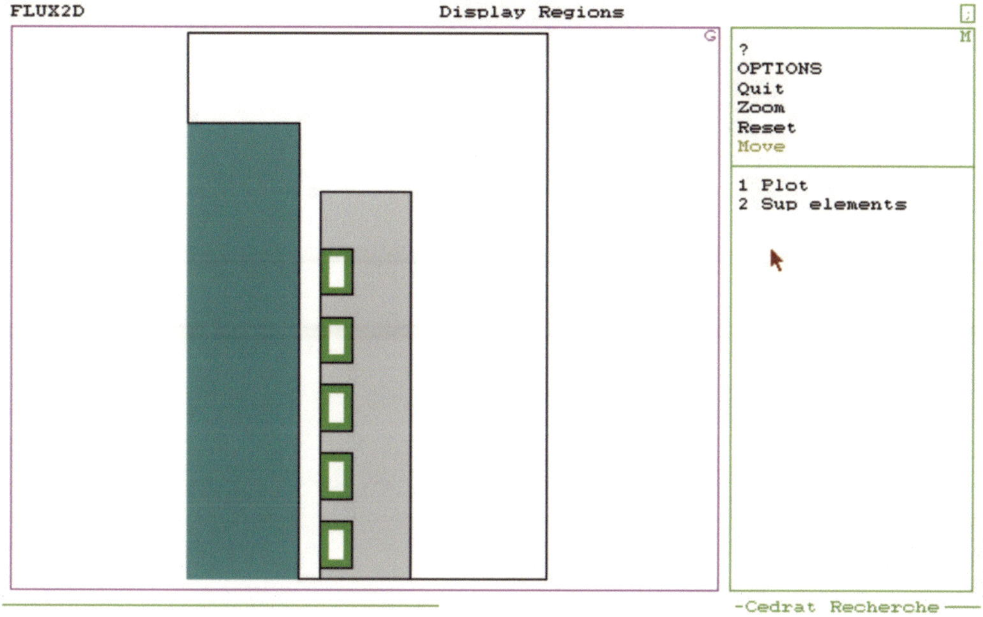

Figure 10. Inductor geometry for monolayer inductor with field concentrator

Was analyzed for this case, the power density distribution in the work piece - Figure 11. , and temperature variation on a point on the surface during the heating process - Figure 12.

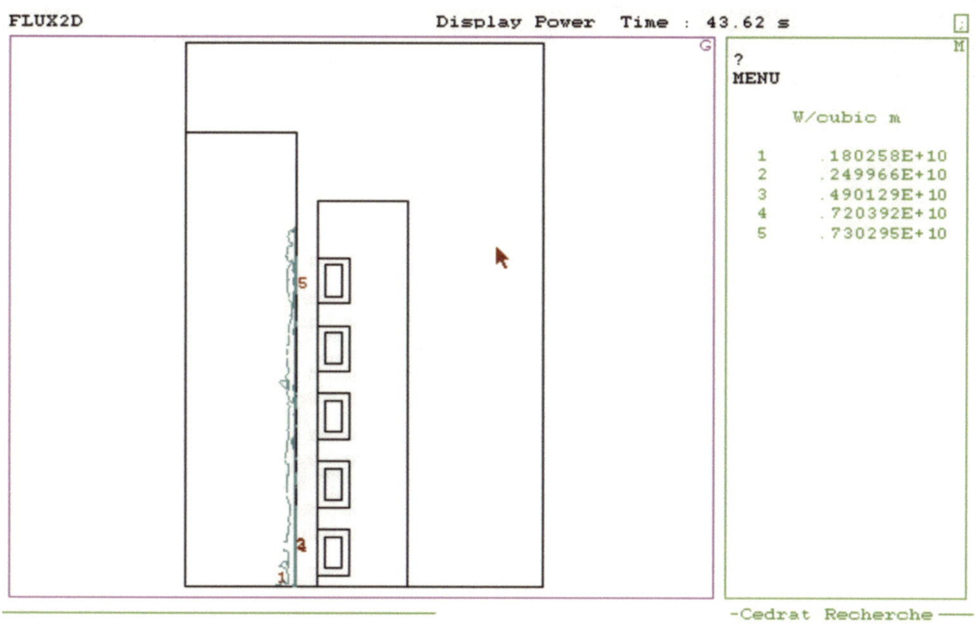

Figure 11. The power density distribution

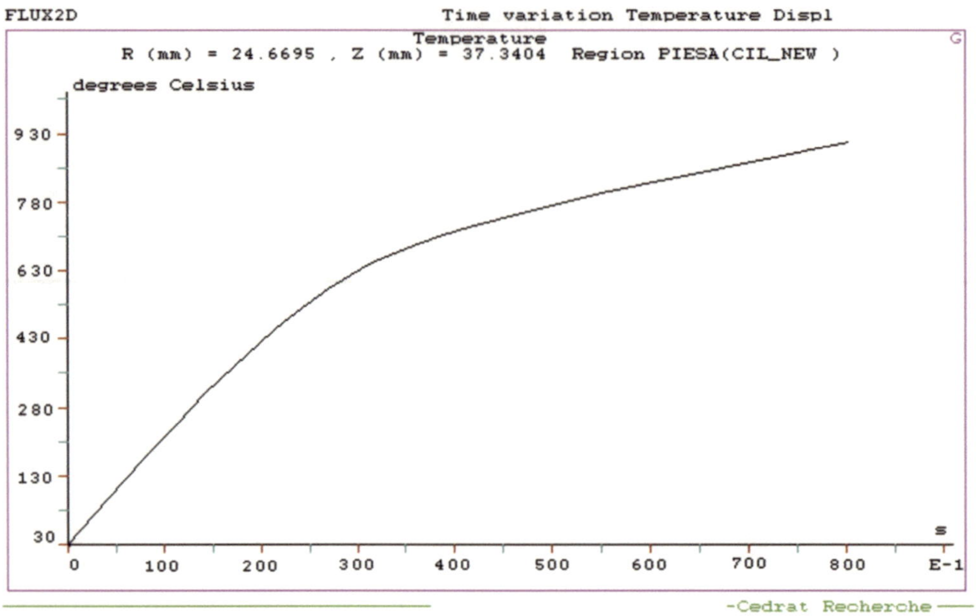

Figure 12. Temperature variation

Based on the results of modeling and simulation was made an inductor with magnetodielectric material concentrator. The image of this inductor during the heating process is shown in the figure below.

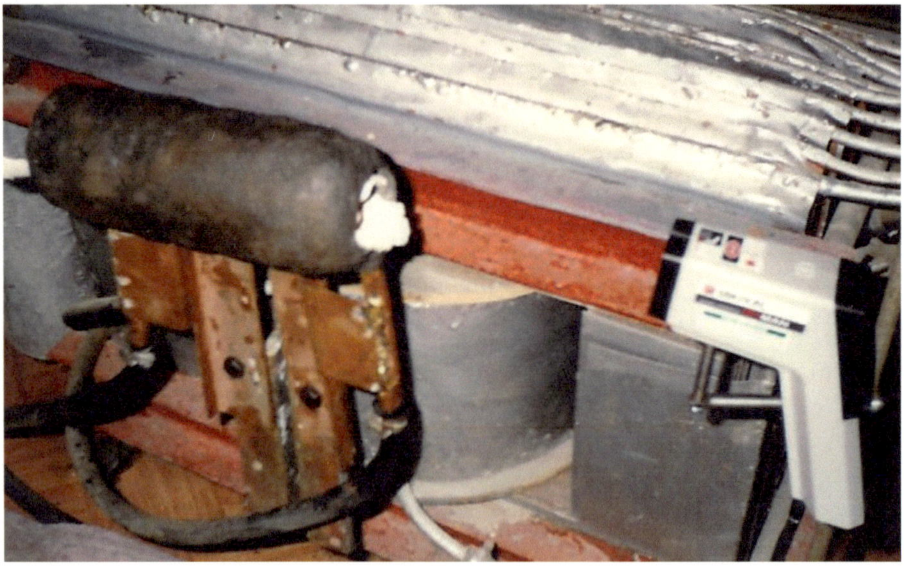

Figure 13. The inductor during the heating process

Current drawn from the supply of inductive heating during the process was 60A, and the measured line voltage was 378V.

The initial temperature of the work piece and the concentrator of magnetodielectric material used were 12°C. Final temperature of the work piece at the end of heating processing was 840°C, temperature reached during the 80s. Final temperature of the outer surface of the concentrator field at the end of the heating process was 14°C.

6. The electrical efficiency in inductive heating processes who magnetodielectric field concentrators use

The primary endpoint of this chapter is evidence increased electrical energy conversion efficiency, in the inductor-piece ensemble, that use a magnetodielectric material field concentrator of the type obtained as explained above.

Efficiency for this case, determined from data modeled is:

$\eta_{\text{i could reg}} = 96\%$
$\eta_{\text{i hot reg}} = 69\%$

For the heating inductive process, without magnetodielectric field concentrator calculated is:

$\eta_{\text{i could reg}} = 81\%$
$\eta_{\text{i hot reg}} = 54\%$,

There is a significant increase of 15% for hot and cold regime of electrical energy conversion efficiency, the inductor-piece ensemble.

These values increase efficiency reveal clearly the advantage of using electromagnetic processing systems for heating that are made up of field concentrators made by magnetodielectric materials

Author details

D.A. Hoble and M.A. Silaghi
University of Oradea, Romania

7. References

[1] Valery I. Rudnev – An objectiv assessmeant of magnetic flux concentrators – heat Treating Progress - 2004
[2] http://www.alpha1induction.com/concentrator_top.html
[3] D.Hoble, C. Stasac - Modern materials utilised in electromagnetic field concentrators - Revista " Acta Electrotehnica", Cluj Napoca, Vol.48, Nr.1-2007 ISSN 1841-3323
[4] Flux2D User,s Guide , CEDRAT , MEYLAN F-38246 , France , 2006
[5] D.A.Hoble – Contributions to increase energy conversion efficiency of inductive heating inductors - PhD Thesis 2001.
[6] htpp://fluxtrol.com

[7] Hoble, D.A., Claudia Staşac - The analysis of the inductive heating process in cylindrical structures.- Revista de tehnologii neconvenţionale nr.2 ICNcT ., ISSN1454-3087, 2009

[8] Di Barba P., Forghani B., Louwther D.A. – Discrete value deseing optimisation of a multiple coil inductor for uniform surface heating. – HES-04 , Padua , Italy

[9] Bitoleanu Alex., Mihai Dan, Popescu Mihaela, Constantin Calin – Convertoare statice şi structuri de comandă performante – Ed. Sitech 2000 Craiova

[10] S. Lupi – Modelling for research and industrial development in induction heating – 4th Int.Conf. on EM Processing of Material EPM 2003 Lyon, France.

[11] Fireţeanu V.,Monica Popa, T. Tudorache, Ecaterina Vladu – Numerical Analysis of Induction trought Heating Processes and Optimal Parameters Evaluation – Proc. 6th International Symposium EMF 2003 Aachen.

Chapter 5

Alternatives for PCB Laminates: Dielectric Properties' Measurements at Microwave Frequencies

Wee Fwen Hoon, Soh Ping Jack,
Mohd Fareq Abd Malek and Nornikman Hasssan

Additional information is available at the end of the chapter

http://dx.doi.org/10.5772/50718

1. Introduction

1.1. Objective

To determine dielectric properties of paddy waste as a potential alternative material for conventional PCB laminate materials. Paddy waste residues, which already possess the characteristic of these conventional laminates, can be strengthened by fabricating it in a tightly-compacted, highly-dense package. These PCB laminates need to be insulated to avoid short circuit, and be physically rigid to mechanically provide stability for the placement of the copper. These properties will then enable a low-cost, sustainable, and renewable solution, with a comparable performance to PCB laminates available in the market.

1.2. Background history

Perlis is the smallest state in Malaysia with agriculture as its main economic activity. Rich rice fields cover most part of the state, enabling easy access to paddy wastes for the fabrication of these particle boards. The raw materials used in this work are rice husk and rice straw, gathered after the harvest season. It is known that fibers with the smallest particle size exhibits the highest tensile strength and hardness [1], hence its increased usage and demand. Within the 2010-2011 period only, about 577 million tonnes of rice (*Oryzae Sativa*) was produced worldwide. Malaysia is one of the more than 80 countries contributing to this sum, with 100,000 tonnes produced annually. Relative to the large quantity of produced agricultural residues (rice husk and rice straw), only a minor portion is reserved as animal feed, while the rest are burnt openly, causing concerns of air pollution [2]. Various other suitable application for such residues are such as mat production, pedestrian bridge, microwave electronic design application, etc [3-4].

Dielectric properties of a material define the physical-chemical properties related to the storage and loss of energy contained in a material or substance. The knowledge of a material's dielectric property is necessary in determining its suitability for a specific application. This property, which includes complex permittivity and dissipation factor, is unique for every material type. These unique sets of electrical characteristics are dependent on electromagnetic properties of the materials. Measurement of dielectric properties involves measurement of the complex relative permittivity (ε_r) and complex relative permeability (μ_r). A complex relative permittivity (ε_r) consists of a real part and an imaginary part. The real part of the complex permittivity, also known as dielectric constant is a measure of the amount of energy from an external electrical field stored in the material. The imaginary part is zero for lossless materials and is also known as loss factor. It is a measure of the amount of energy loss from the material due to an external electric field. The term tangent loss (tan δ) represents the ratio of the imaginary part to the real part of the complex permittivity, and is also known as loss tangent, dissipation factor or loss factor. Accurate measurements of these properties enable scientists and engineers to incorporate the material for the suitable application, for more solid designs or to monitor a manufacturing process for improved quality control [5].

2. Permittivity and tangent loss definitions

The dielectric properties are, by definition, a measure of the polarizability of a material when subjected to an electric field. To evaluate materials, the dielectric properties are represented by the relative complex permittivity, $\varepsilon_r = \varepsilon' - j\varepsilon''$, where ε' is the dielectric constant which describes the ability of the material to store energy, ε'', on the other hand, is the dielectric loss factor, which reflects the ability of a material to dissipate the electric-field energy.

Permittivity (Dielectric Constant)

$$K = \frac{\varepsilon}{\varepsilon_0} = \varepsilon_r = \varepsilon_r' - j\varepsilon_r'' \quad (1)$$

ε_0 = is the free space of permittivity interaction of a material in the presence of an external electric field.

Permittivity (ε), also known as the dielectric constant, describes the interaction of a material with an electric field. Dielectric constant (k) is equivalent to the relative permittivity (ε_r) or the absolute permittivity (ε), relative to the permittivity of free space (ε_0). The real part of permittivity (ε'_r) is a measure of how much energy from an external electric field is stored in a material. The imaginary part of permittivity ($j\varepsilon''_r$) is called the loss factor and is a measure of how dissipative or loss of a material is to an external electric field.

Loss Tangent

When complex permittivity is drawn as a simple vector, the real and imaginary components are 90^0 out of phase. The vector sum forms an angle, δ, with the real axis (ε_r'). The relative "loss" of a material is the ratio of the energy lost to the energy stored.

$$\tan\delta \frac{\varepsilon_r''}{\varepsilon_r'} \tag{2}$$

$$\tan\delta = D = \frac{1}{Q} = \frac{\text{Energy Lost per cycle}}{\text{Energy Stored per cycle}} \tag{3}$$

In some cases the term "quality factor or Q – factor" with respect to an electronic microwave material is used.

Agricultural residues have been subjected to increasing interest, study, and utilization for some decades. The increase in environmental concerns rationalizes, the reduction of polymers' usage, not only because of their non-biodegradability, but also due to energy-intensive production. In other words, polymer production and processing requires large amounts of oil as raw material, which is notoriously not renewable. All these issues induce the need for alternatives.

Paddy residues such as rice husk and rice straw, shown in Figure 1, are materials of interest for a wide range of applications. Non-destructive dielectric properties' measurements are essential for proper understanding of their electrical behavior, ensuring that they could be effectively put into applications. The use of paddy residues is advantageous due to its high silica content and thick walled , providing a fire-resistant feature. For a typical paddy waste fiber, burning causes a layer of char to develop on the outer surface, insulating its inner straw [6].

(a) (b)

Figure 1. Paddy residues (a) rice straw (b) rice husk

At radio and microwave frequencies, dielectric properties of paddy waste materials are dependent on frequency, moisture content, and temperature. In fact, at these frequencies, water is the most influential factor due to its polar nature. The effect of temperature is also water-related, where the change in temperature affects the energy state of water molecules, hence influences their aptitude to follow the alternating electric field [6]. Since water inside the paddy plant is bound to the inner structure, the dielectric properties' frequency dependence is not as spectacular as that of liquid water. Samples of agricultural waste products, rice husk and rice straw, of different resin moisture contents were prepared by pouring Urea Formaldehyde (UF) or Phenol Formaldehyde (PF) paddy samples to bring them to the desired moisture level [8]. Measurements at high frequencies are often taken using two different measurement techniques with different degrees of accuracy, given the granular nature of the proposed materials.

At microwave frequencies, various techniques have been developed to determine these complex properties ranging from time-domain or frequency-domain techniques, utilizing one port or two ports, etc. Every technique has its own limitation, either to a specific range of frequencies, materials or applications. Several popular measurement techniques are the transmission line techniques (waveguide, coaxial and free-space), impedance, dielectric probe and cavity methods [1-8]. Amongst these techniques, free-space measurements and high temperature dielectric probe techniques are chosen based on their suitability for paddy waste dielectric properties measurement. Both setups are shown in Figure 2. The free space measurement technique allows reflection and transmission measurements without direct sample contact. On the contrary, the dielectric probe technique is performed by contacting or immersing the probe into the sample. Both techniques do not require any special fixtures or containers, and they are best applied for thin, flat, parallel-faced materials, or other materials which can be formed into this shape. These measurement techniques are non-destructive and can be gathered in real time, allowing them to be used in process analytic technologies.

For the free space measurement system, a pair of horn antennas providing plane waves at a defined distance, are placed at either ends of a material under test (MUT), as shown in Figure 2. Minimal sample preparation is required [9-11]. The sample thickness is selected to ensure at least 10 dB one way attenuation and the time domain gating feature of the PNA is utilized to ensure an accurate measurement. The permittivity of a material's sample can be calculated automatically using the Agilent 85071 E software, with the signals transmitting and reflecting from the sample. The width and height of the sample (perpendicular to the wave propagation direction) must sufficiently be larger than the horn antennas to avoid inaccuracy caused by signal diffraction at sample edges.

Figure 2(a) shows the High Temperature Probe measurement procedure, which is done by contacting the probe to a flat surface of a solid, or immersing it into a liquid or semisolid. The fields at the probe end "fringe" into the material and change as they come into contact with the MUT. The reflected signal (S_{11}) can be measured, and then related to ε_r. On the other hand, Figure 2(b) shows the free space measurement technique, based on the reflection coefficient (S_{11}), transmission coefficient (S_{21}), or both. The popular "S-parameter" approach (Nicolson-Ross or Weir) uses both S_{11} and S_{21} to calculate both ε_r and μ_r where S_{11} and S_{21} are composed of multiple-reflections from both boundaries. The sample must be thick enough to contain the wavelength, ℓ of interest in order to be measurable, ideally 180°, or ½ ℓ. At mm-wave frequencies, samples thicker than 1 ℓ can create multiple root errors. The sample must be far enough, away from the antenna to be out of the reactive region, ideally at least $2d^2/\ell$, where d is the largest dimension of the antenna.

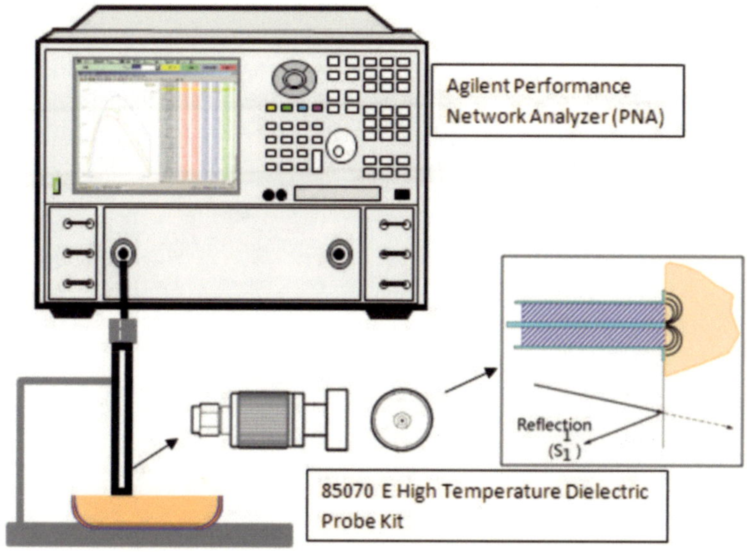

(a)

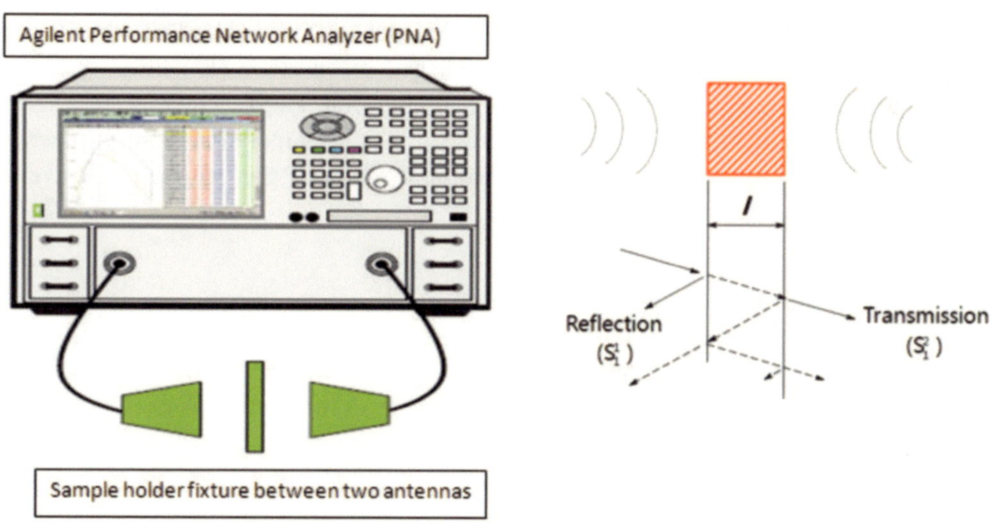

(b)

Figure 2. Measurement techniques for paddy waste (a) High Temperature Dielectric Probe Technique (b) Free Space Measurement Technique.

A microwave laminate is a dielectric material which is usually a poor electrical conductor, commonly used as an insulating layer in building PCBs. Porcelain, mica, glass, plastics and some metal oxides are several examples of these dielectrics. The lower the dielectric loss (proportion of energy lost as heat), the more effective the dielectric material is. If the voltage across a dielectric material becomes too large, and intensifying its electrostatic fields, the material will begin current conduction. Examples of popular PCB laminates are shown in Figure 3, i.e Rogers 5880, Rogers 4350, FR-4, Taconic TLY-5 and etc.

Figure 3. Existing of PCB laminates in the market.

This investigation intends to analyze the dielectric properties of two major paddy waste commodities, i.e., rice husk and rice straw, with each of them representing significant structural and compositional differences. Dielectric properties of paddy wastes are measured using Free Space Measurement Technique and High Temperature Dielectric Probe Kit over a range of microwave frequencies, at room temperature. This is potentially interesting for the microwave industry, as a comparison with the commercially available PCB laminates in the market is also carried out.

Body:

- Problem Statement

This work chooses paddy waste materials due to the sustainable practice of using natural by-products due to the rising environmental concern. About one million tones of these residues are produced over the entire 200,000 hectare per season in Perlis. Thus, besides being available abundantly, its usage avoids open burning activity, which is its common disposal method [6]. This activity obviously deteriorates air quality and raises health risks. With the ban of open burning, beneficial and sustainable alternative methods of such agricultural waste disposal are required. Moreover, the materials for commercial PCB laminates are costly and less environmental-friendly due to the usage of chemicals. The total chemical compounds used in the paddy waste particle boards production, In this case, there is only 9.3 % of the total amount of material used. Thus, characterization of these paddy waste particle boards will be necessary for it to be proposed as an alternative PCB laminate which is low-cost and sustainable.

- Method used

The project is summarized into three major phases, consisting:

Phase 1: raw paddy wastes collection
Phase 2: particle board fabrication,
Phase 3: dielectric properties measurement

Fabrication of paddy waste particle boards

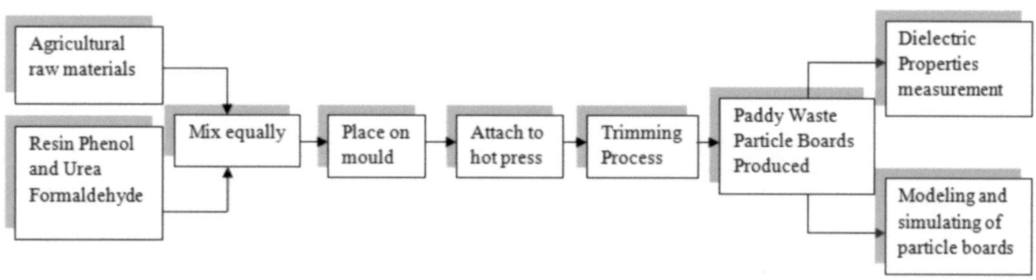

Figure 4. Step of analyzing paddy wastes

The paddy wastes, i.e. rice husk and rice straw, are compressed into a solid board for dielectric measurements, using different percentages of Urea Formaldehyde (UF) and Phenol Formaldehyde (PF) resins. Both are important components in determining its dielectric property. Figure 4 summarizes the rice husk and rice straw particle boards fabrication process. The paddy wastes are first collected from the rice fields before being transformed into particles boards. The boards' dielectric properties are then determined using the Free Space Measurement Technique (FSM) and High Temperature Dielectric Probe Technique.

- Paddy waste raw materials

Figure 5 shows the types of paddy waste raw material which has been used for fabrication.

- Fabrication of Paddy Waste Particle Boards

Firstly, two different bonding agents, Urea Formaldehyde (UF) and Phenol Formaldehyde (PF) are mixed with the respective paddy waste in a basin. UF, also known as urea-methanal - named so for its common synthesis pathway and overall structure - is made from urea and formaldehyde. It possesses a high tensile strength, low water absorption, and high surface hardness. On the other hand, PF is the result of an elimination reaction of phenol with formaldehyde. It is formed by a step-growth polymerization reaction which may be either acid- or base-catalyzed. Figure 6 shows the two types of resin which is important in providing the moisture content in the paddy waste boards, and directly affecting their dielectric properties

(a) (b)

Figure 5. Paddy waste materials (a) rice husk (b) rice straw

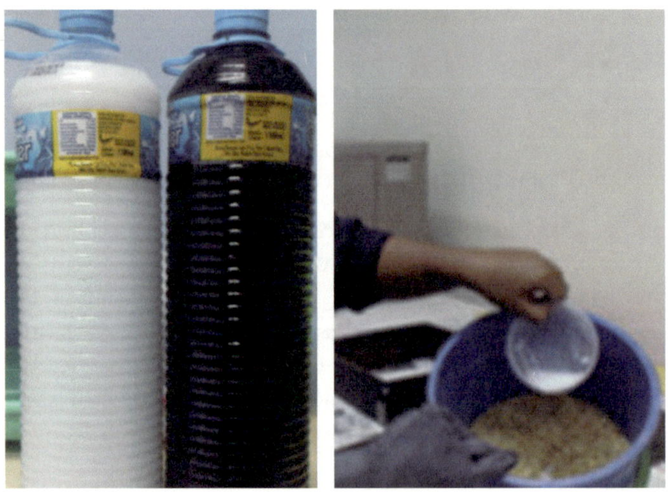

(a) (b)

Figure 6. (a) Urea Formaldehyde and Phenol Formaldehyde resins (b) resin -paddy waste mixture

Paddy Wastes	Resin, Percentage, %	
	Urea Formaldehyde	Phenol Formaldehyde
Rice Husk	10	10
	20	20
	30	30
Rice Straw	10	10
	20	20
	30	30

Table 1. Rice husk and rice straw resin percentage fabrication matrix.

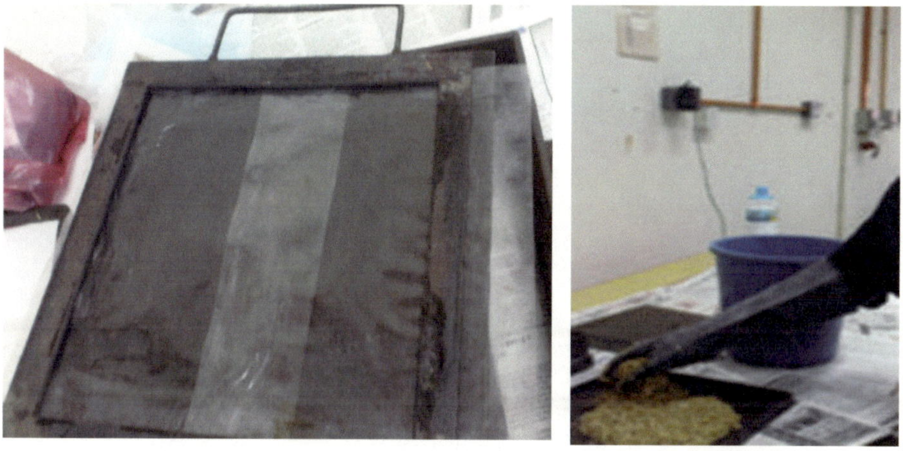

(a) (b)

Figure 7. (a) Rectangular-shaped mould (b) paddy wastes-resin mixture transfer into the mould

First, 500 g of rice husk is weighed and mixed using three different UF resin compositions: 10%, 20% and 30%. This process and procedure is also repeated for rice straw. Another set of similarly composed mixture using UF, rice husk and rice straw is also prepared. Figure 6(b) shows the resins-paddy wastes mixing procedure, while Table 1 presents the fabrication matrix. The mixed material is then placed into a rectangular, 245 x 245 mm mould shown in Figure 7. This mould is chosen to enable the fabrication of a larger particle board for use in the free space measurement technique.

Next, the mixture placed on the shaping mould is then positioned onto the hot press machine, as seen in Figure 8. It is a high pressure, low strain rate material processing machine for compact material forming at high temperature. It consists of an upper and lower mould, the former is higher in temperature for the compression of solid substances, while the latter is designated for the cooling and hardening process. A predefined temperature and compression duration can be set to automate the whole process. This avoids compression overheating which affects the solid substance characteristic.

After cooling, completed particle boards are edge-trimmed for cosmetic reasons. They are shown in Figure 9.

- Hardware Measurement

Hardware measurement involves the setup of free space measurement technique, the high temperature dielectric probe, and their corresponding measurement software (Agilent 85071 E and Agilent 85070 E software). Prior to measurements, calibration of the measurement system (PNA network analyzer, free space measurement system, and high temperature dielectric probe kit) must be taken into consideration, besides the theoretical understanding of both measurement setups.

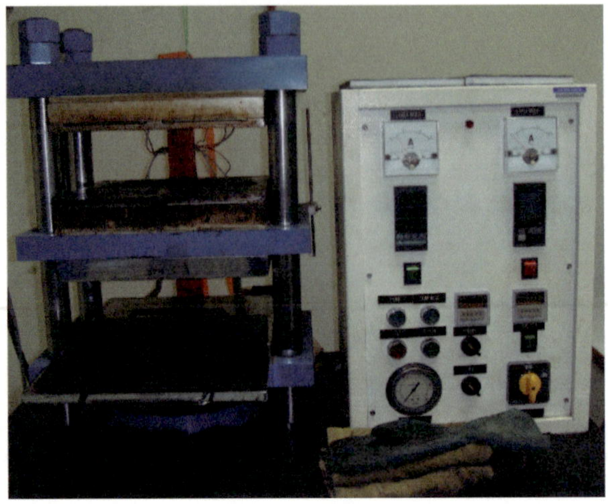

Figure 8. Hot Press Machine

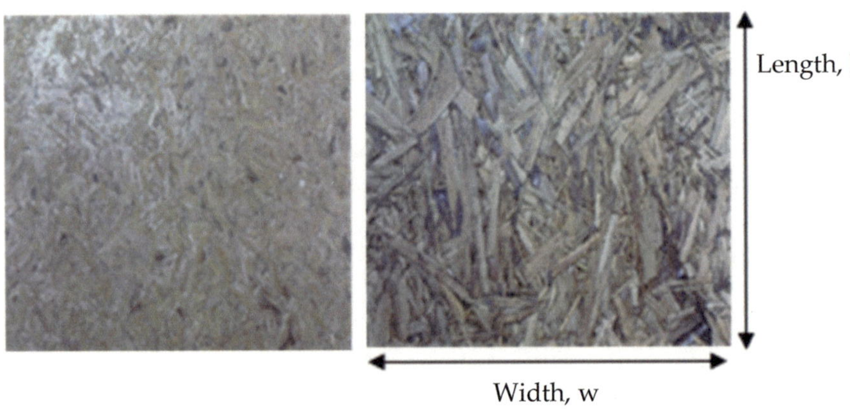

Figure 9. Paddy waste particle boards

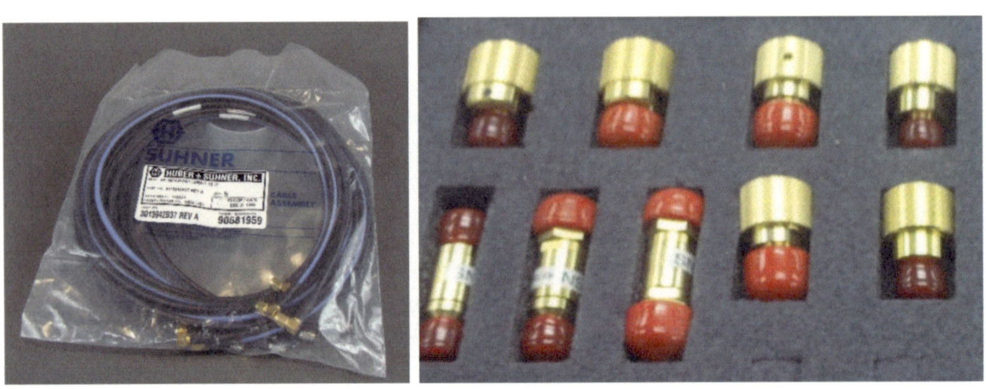

(a) (b)

Figure 10. (a) coaxial cables (b) Agilent 85052 D calibration kit

Before proceeding with the measurement, calibration of coaxial cables using a known dielectric and length reference board is carried out. The calibration is done at both transmitter and receiver to remove undesired errors and ensure measurement accuracy. Full two-port calibrations using SOLT (Short – Open – Load – Through) standard is performed using the Agilent 85052 D 3.5 mm calibration kit, which contains test adapters and a torque wrench [11-16]. Figure 10 shows the coaxial cables and the calibration kit used for this purpose.

- Free Space Measurement Technique

The procedure described in Figure 11 is used for the paddy wastes particle boards measurements. Its setup, shown in Figure 12, consists of the Performance Network Analyzer (PNA), Agilent 85071 E measurement software, horn antennas, coaxial cables, adaptors, and the particle board as the Material under Test (MUT).

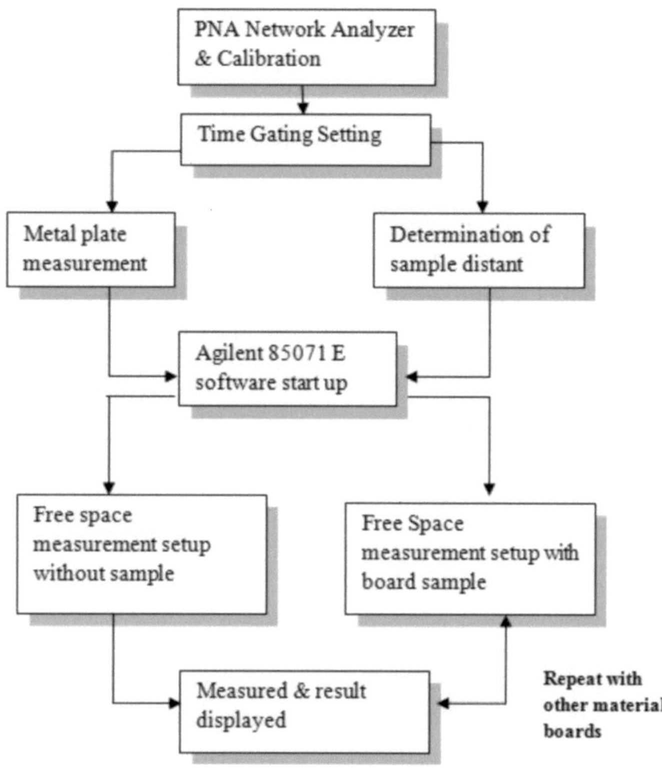

Figure 11. Free Space Measurement technique measurement procedure

For this technique, a MUT calibration setup needs to be performed. A reference board with known dielectric constant is first placed between the two horn antennas. In this case, a copper plate is used as the reference board, with its dielectric constant displayed on the Agilent 85071 E Material Measurement software display. Next, the reference board is removed to ensure the dielectric constant, ε_r is equal or near to 1, similar to air. When both reference board and air are similar to their actual dielectric constants, the calibration setup process is considered complete.

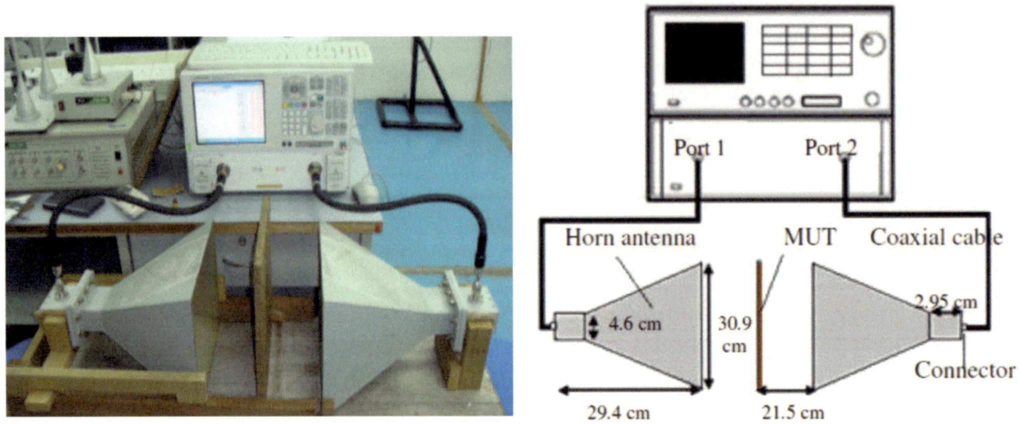

Figure 12. Free Space Measurement Setup

The dimension of the horn antennas used in this setup is 30.9 cm x 23.85 cm x 29.4 cm. The antenna size influences the transmitter-receiver distance, a smaller antenna size results in a shorter distance between the two antennas. The length of the coaxial cable must also be considered, since a longer cable results in a higher attenuation and a weaker signal at the transmitting horn [14-16]. The two antennas are directed into the line of sight (LOS) path and polarized horizontally relative to the MUT to ensure accuracy.

The minimum MUT-antenna distance is another important determining factor to ensure accurate dielectric properties extraction. This distance can be determined by applying the time gating setting at the PNA network analyzer when the sample is placed during calibration, as shown in Figure 13. This feature is useful in lowering the effect of reflections appearing as noise in the time domain response. Average S_{11} measured using a non-metallic and metallic plate must produce at least 40 dB in difference. Three peaks shown in the network analyzer, the first being the response caused by the transmitting horn, the second is for the time domain gating feature while the third is the response caused by the receiving horn. In this example, the difference between the two plates is 52.713 dB, which sufficient to optimize the distance between the two horn antennas. For each MUT measurement, a set of dielectric constant and loss factor, listed in a table, is produced. Besides that, graphical plots can also be viewed using the 85071E measurement software.

- High Temperature Dielectric Probe Measurement method

We have also included the investigation of an Agilent 85070 B High Temperature Dielectric Probe Kit in determining paddy wastes' dielectric properties. The system consists of an Agilent 85070 D High Temperature Dielectric Probe, Agilent Performance Network Analyzer (PNA), and Agilent 85070 B software. This technique is easy to perform, time-effective and simple, without requiring any special fixtures or containers. MUTs, either rice husk board or rice straw board, is pressed using the dielectric probe as shown in Figure 14. This probe propagates signal into the MUT [17], and the resulting measured reflections are then converted into dielectric properties values via Agilent 85070B software. This system is capable of determining dielectric properties up to 20 GHz.

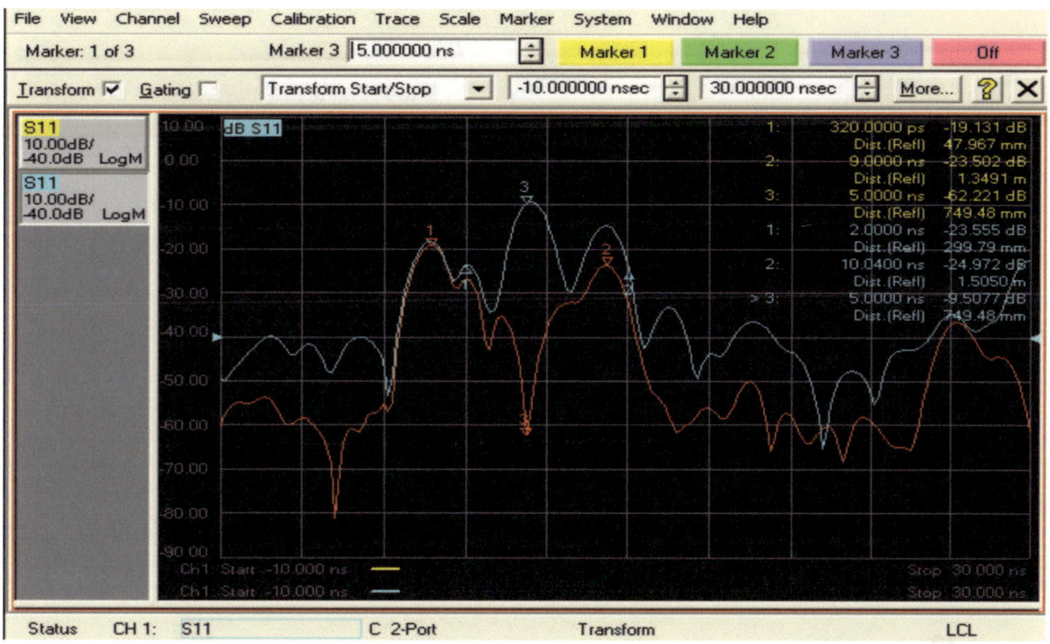

Figure 13. Time domain gating setting

Prior to usage, the High Temperature Dielectric Probe Kit needs to be calibrated using three elements and the software shown in Figure 15(a). The elements are air, a metallic shorting block, and water. This shorting block is shown in Figure 15(b), while for water, users need to ensure that no air bubbles exist. Upon completion of calibration, the MUT can then be placed underneath the probe for measurements.

In this work, the rectangular paddy waste boards and cylindrically-shaped Barium Strontium Titanate (BST) blocks are considered for measurements. Paddy waste boards measurements are conducted at 16 different points, seen in Figure 16 to ensure measurement accuracy. On the other hand, BSTs are measured at five points at its top and bottom, respectively. The middle section of this BST block was not considered due to the high possibility of obtaining inaccurate results. This is mainly due to high possibility of air gap existence between the dielectric probe and the BST material, caused by its curved surface.

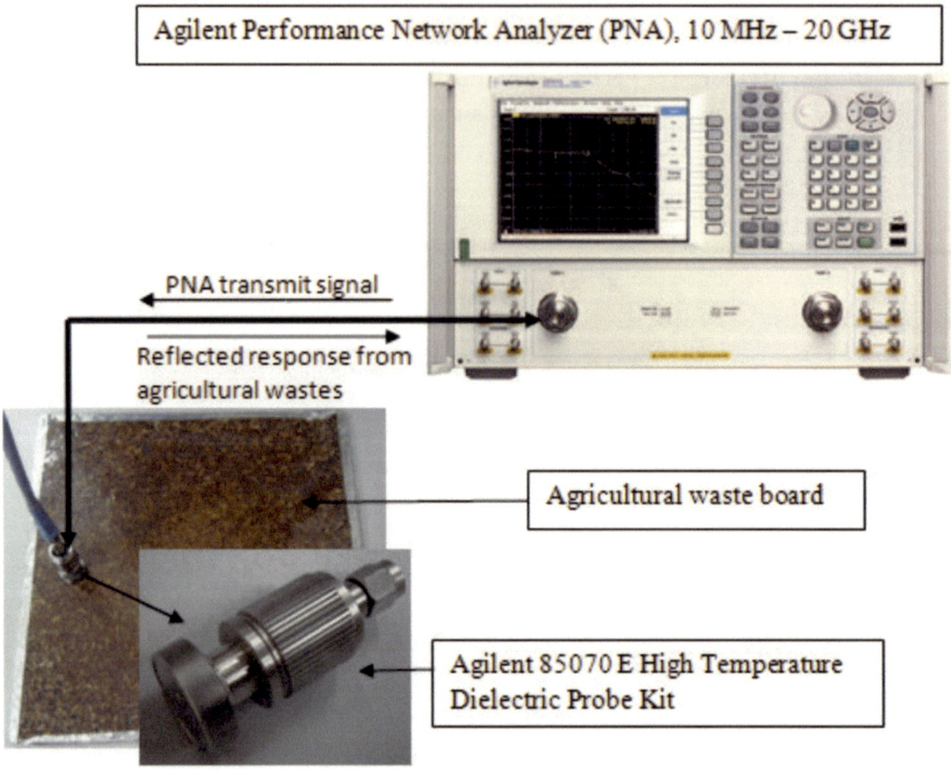

Figure 14. High Temperature Dielectric Probe System

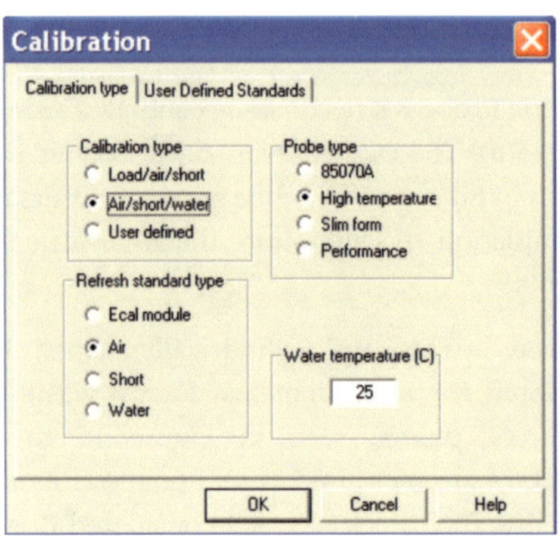

(a)

(b)

Figure 15. Calibration components (a) software window for calibration type selection (b) shorting block

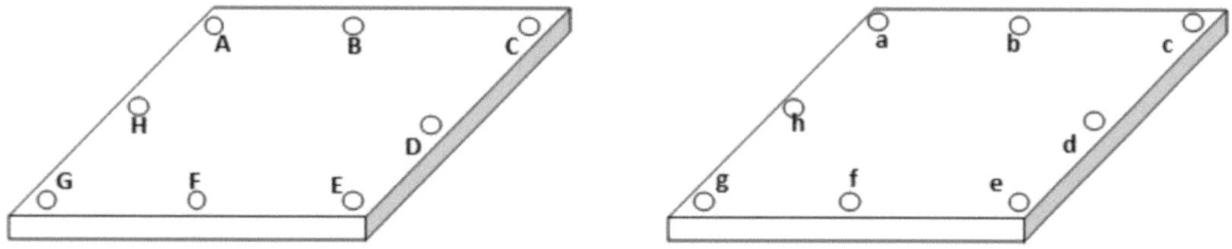

Figure 16. Measurement of 16 different location points on the MUT (a) top view (b) bottom view.

- Results:

Free Space Measurement Technique and High Temperature Dielectric Probe Kit had been utilized to obtain the dielectric properties in different settings, as follows:

1. Dielectric properties of rice husk either PF or UF resins.
2. Dielectric properties for two paddy waste materials (rice husk and rice straw), with different PF resin percentage (10 %, 30 % and 50 %)
3. The accuracy of dielectric properties using different horn-to-MUT distances, investigated using both types of boards.
4. Changes in dielectric properties of rice husk boards with frequency.
5. Comparison Free Space Measurement Technique and High Temperature Dielectric Probe using 50% PF of rice husk and rice straw MUTs.

- Dielectric properties of rice husk using both PF and UF resins

From the Table 2, it can be observed that the dielectric constant, ε′ of the rice husk MUT using PF resin is higher than for UF. For 10% PF and UF content, the dielectric constant of MUT with PF adhesive (3.2355) is larger compared to MUT with UF adhesive (2.8907). This similar observation is also consistently found for 30% and 50% composition for both resins. Besides UF being a low water absorbent, the predominantly higher dielectric constant found in PF-MUTs is due to its higher liquid density, causing a higher moisture level compared to the rice husk boards with UF composite [16].

Resin	Percentage, %	Dielectric Properties	
		ε′	ε″
PF	10	3.2355	0.2742
	30	3.2745	0.2680
	50	3.5813	0.4393
UF	10	2.8907	0.2215
	30	3.4054	0.3286
	50	3.6808	0.2725

Table 2. Dielectric properties of rice husk boards fabricated using different resin percentage.

The loss tangent of the rice husk particle boards are quite similar for both resin PF and UF seen for all 10%, 30%, and 50% UF and PF composition. This is due to densification of rice husk rectangular board. When the electromagnetic radiation is incident on the material board's surface, the amount of reflection experienced by the board is higher than transmission into the material [16]. Hence, this may described as an energy loss process, as more signal is reflected compared to absorbed.

- Dielectric properties for two paddy waste material types (rice husk and rice straw), with different PF resin percentage.

Material	Percentage, %	Dielectric Properties	
		ε'	ε''
Rice Husk	10	3.2355	0.2742
	30	3.2745	0.2680
	50	3.5813	0.4393
Rice Straw	10	1.9061	0.135
	30	2.0127	0.123
	50	2.7358	0.2236

Table 3. Dielectric properties of the two material boards fabricated using different percentage of Phenol Formaldehyde (PF) resin.

Paddy waste is a known non-magnetic material [14]. Measured ε' and ε'' of boards fabricated using a varying UF and PF resins' concentrations (10 %, 30 % and 50 %) are given in Table 3. ε_r increases as the concentration of UF and PF are increased from 10% to 50%. This is due to higher volume fraction of the chemical resin in the composite. ε'' values are expected to be greater than or equal to zero, while the negative ε'' values are caused by measurement uncertainties. Thus, obtained dielectric loss tangents being larger than zero indicate measurement accuracy. To summarize, between the frequency range of 2.2 GHz and 3.3 Ghz, the loss tangent and dielectric constant showed an increasing trend with the rising moisture content provided by PF and UF resins.

- Accuracy of dielectric properties using different horn-to-MUT distances investigated using two types of agricultural waste material boards.

This investigation scope is aimed at identifying the relationship between measured dielectric properties using three different MUT-to-horn distances. Table 4 shows the dielectric constant, ε', and tangent loss, ε'', variation between 2.2 GHz and 3.3 GHz at room temperature of 27°C. The results shows that the dielectric constant, ε', steadily decreases as the distance is increased from 215 mm to 475.5 mm.

This decreasing trend can also be consistently observed for both rice straw and rice husk MUTs, as well as across the different PF resin contents. This is due to the decreased penetration depth into the material boards caused by an increased MUT-to-horn distance, and vice versa for shorter distances. In other words, the longer MUT-to-horn distance caused attenuation and scattering of the emitted signals, leading to less absorption by the MUT [6]. Due to this, ε' is lower when measured using a longer distance, as test signals are unable to reach the MUTs. On the other hand, an opposite trend is seen for ε'' when increasing this measurement distance. In general, it can be said that the paddy waste boards are unable to be measured accurately due to the weak signal when measurement distance is lengthened.

Materials	Percentage, %	Distance (mm)					
		215		377		475.5	
		ε′	ε″	ε′	ε″	ε′	ε″
Rice Husk	10	3.2355	0.2742	3.1386	0.4766	2.8854	0.4964
	30	3.2745	0.2680	3.4727	0.617	3.249	0.6513
	50	3.5813	0.4393	4.4912	0.484	4.0305	0.5389
Rice Straw	10	1.9061	0.135	1.8905	0.2153	1.8138	0.246
	30	2.0127	0.123	1.974	0.2106	1.9068	0.2335
	50	2.7358	0.2236	2.639	0.3722	2.4848	0.3921

Table 4. Dielectric properties of rice husk and rice straw with a varying MUT-to-horn distances.

- Changes in dielectric properties of rice husk material boards with frequency.

Table 5 presents the variation of measured dielectric properties according to frequency. This shows that the frequency significantly affects MUT's loss tangent (ε″), which rises with the increasing frequency. In most cases, this frequency-loss tangent relationship is nearly linear. On the other hand, the ε″ result obtained in the Table 5 are higher at both ends of the test frequency range, and is slightly decreasing in the middle. This tendency is observed for all three different percentages of PF resins for the rice husk particle boards. This discrepancy is caused by the propagation of indirect signal known in this measurement technique, thus degrading the accuracy of the measurement results [18-19].

Freq (GHz)	Percentage of Phenol Formaldehyde								
	10%			30%			50%		
	ε′	ε″	ε′ (%)	ε′	ε″	ε′ (%)	ε′	ε″	ε′ (%)
2.2	3.4897	0.3284	100	3.9276	0.4859	100	5.0302	0.2363	100
2.3	3.4742	0.2888	99.56	3.9074	0.4571	99.49	4.9839	0.1929	99.08
2.4	3.4046	0.2632	97.56	3.8234	0.4165	97.34	4.8626	0.1563	96.65
2.5	3.3571	0.2675	96.16	3.7452	0.4266	95.29	4.7671	0.1809	94.69
2.6	3.311	0.2615	94.79	3.673	0.4222	93.36	4.6695	0.1813	92.64
2.7	3.2514	0.2637	92.99	3.604	0.4253	91.48	4.5813	0.196	90.75
2.8	3.1997	0.263	91.40	3.5342	0.4261	89.54	4.4919	0.2052	88.80
2.9	3.1543	0.2681	89.98	3.4711	0.4341	87.75	4.4135	0.2241	87.05
3.0	3.1154	0.2716	88.75	3.4175	0.4406	86.20	4.3473	0.2358	85.55
3.1	3.0753	0.2737	87.46	3.3596	0.4426	84.50	4.27	0.2509	83.77
3.2	3.299	0.2852	92.46	3.2932	0.4591	82.52	4.1956	0.2813	82.03
3.3	2.9902	0.2999	83.1	3.2353	0.4764	80.76	4.1196	0.3127	80.22

Table 5. Measured rice husk material board's dielectric properties across frequency

Although values of the dielectric properties decreased with the increasing frequency, the linear slope performed differently, depending on the agricultural waste type and moisture content. Besides this trend, its decrease with UF and PF percentage is also evident in Table 5 and Figure 17. However, there exist a set of stray measurement value with 10 % PF content at 3.2 GHz. This error is due to uncertainty of the sample form and size as well as the surface roughness of the material boards [20], which degraded its accuracy. It must be noted that the calculation is only valid for the geometrically ideal sample, which could be avoided by applying a very strict sample preparation process. In short, frequency, loss tangent and dielectric constant will affect the amount of energy that is dissipated in agricultural waste materials. Higher loss tangent results in higher microwave signal absorption by the boards, and conversely, a lower dielectric constant favors higher heat absorption in the fibers [19-21].

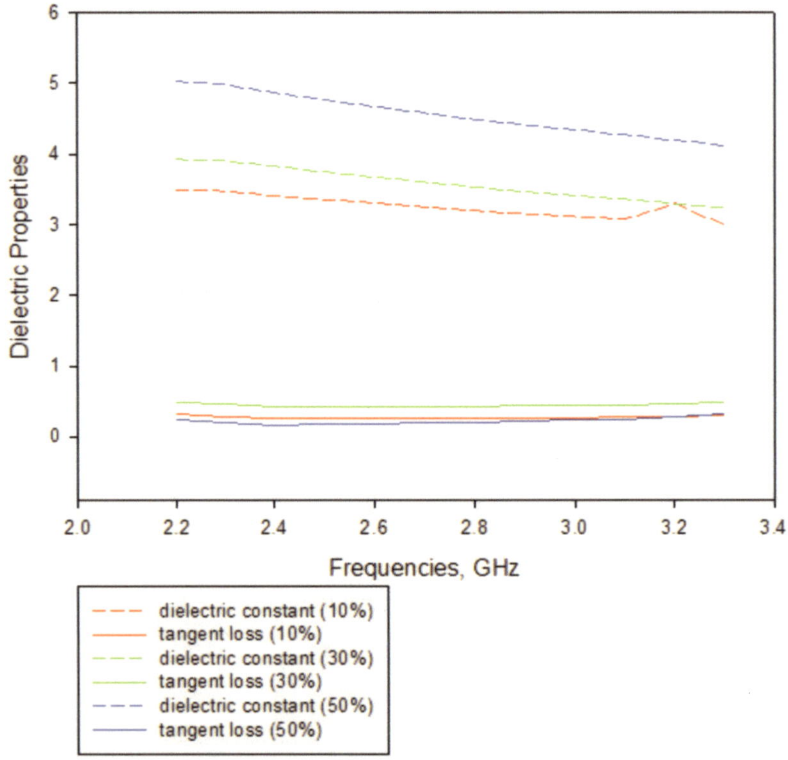

Figure 17. Dielectric constant and tangent loss across frequency

- Comparison Free Space Measurement Technique and High Temperature Dielectric Probe using 50% PF of rice and rice straw MUTs

A comparison of measured rice husk's dielectric properties using Free Space Measurement Technique and High Temperature Dielectric Probe Kit is shown in Figure 18. Tangent loss evaluation of a similar rice husk board (with 50% resin) using these two different techniques is producing an excellent agreement. This is also seen for rice straw board measurement, yielding an almost similar reading. Meanwhile, measurement of dielectric constant using the two measurement techniques shows that the High Temperature Dielectric Probe Kit produced higher dielectric constant values. This is caused by the High Temperature Dielectric Probe's ability to feed test signals directly into the MUT. The MUT-to-horn distance which exists in the Free Space Measurement Technique introduces an additional uncertainty factor, potentially causing loss between signal paths. Besides that, the measurement are also affected by other uncertainties - inconsistent MUTs' geometry, roughness and surface homogeneity. Additional limitations may arise from the systemic uncertainty of the particular instrument, and the imperfections of the test fixture and setup.

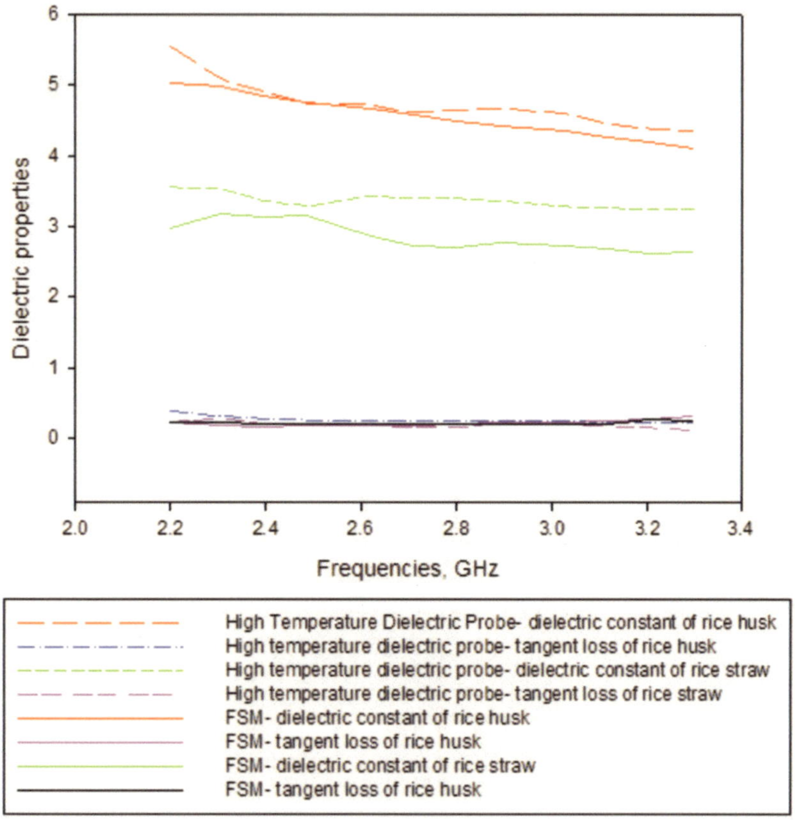

Figure 18. Dielectric property measurements of rice husk and rice straw using two different measurement techniques.

- Status:

In summary, we have described two effective methods to determine the dielectric properties of custom-made paddy waste particle boards Its procedure, advantages, limitations and operation in several different configurations are also carefully investigated and explained. This has laid the foundation for a better understanding of dielectric behavior, which will assist microwave or electronic components' engineers in optimally designing their components, and promoting the use of an alternative, sustainable material.

3. Further research

Other boards fabricated using different agricultural waste materials should also be investigated. This will also depend on the waste material's availability, based on the geographical region and type of agricultural/economic activities. On the other hand, the current scope can be expanded to evaluate boards with different thicknesses, as MUT's thickness could affect measurement accuracy. It is also evident that improvement on the measurement setup is relevant, especially on the free space measurement technique.

4. Conclusion

A systematic procedure to fabricate and evaluate custom particle boards made from two types of agricultural waste product is explained in detail. In total, six variation of waste boards are evaluated, and they are fabricated using different paddy waste types and different bonding resins, i.e. Urea Formaldehyde (UF) and Phenol Formaldehyde (PF). Two standard setup and procedures for measuring their dielectric properties between 2.2 GHz and 3.3 GHz are presented and compared. The results, calibration and factors influencing measurement accuracy are then discussed. Rice husk-based MUTs are measured to be higher in terms of dielectric properties compared to rice straw-based MUTs. This is mainly due to the former's small particle size, enabling a larger surface area which absorbs more test signal. Comparison of two measurement methods presented shows a good agreement, with uncertainties of less than 15 %. The measurements have also proven that MUTs fabricated using PF is higher than UF due to the ability of the former to absorb moisture. The measurement results presented will be potentially useful in encouraging the use of such sustainable, renewable materials at microwave frequencies.

Author details

Wee Fwen Hoon, Soh Ping Jack and Nornikman Hasssan
School of Computer and Communication Engineering,
Universiti Malaysia Perlis (UniMAP), Malaysia

Mohd Fareq Abd Malek
School of Electrical Systems Engineering, Universiti Malaysia Perlis (UniMAP), Malaysia

5. References

[1] Jayanthy, T., & Sankaranarayanan, P.E., (2005) Measurement of Dry Rubber Content in Latex Using Microwave Technique. Measurement Science Review, Volume 5, Section 3.

[2] Gagnon, N., Shaker, J., Roy, L., Petosa, A., & Berini, P., (2004) Low Cost Free Measurement of Dielectric Constant at Ka Band, IEEE Proceedings - Microwaves, Antennas and Propagation, Volume 151, Issue: 3, pg 271-276.

[3] Mohamad Yusof, I., Farid, N.A., & Zainal, Z.A., (2008) Characteristization of Rice Husk for Cyclone Gasifier. Journal of Apllied Sciences 8(4): pg 622 – 628.

[4] Nakhkash, M., Yi Huang, Al-Nuaimy, W., & Fang, M.T.C., (2001) An Improved Calibration Technique for Free Space Measurement of Complex Permitivity, IEEE Transactions on Geoscience and Remote Sensing, Volume 39, Issue 2, pg 453-455.
[5] J. Obrzut, A. Anopchenko and R. Nozaki (2005) Agricultural Fibers For Use in Building Component. Technology Conference Ottawa, Canada, 17-19 May.
[6] Suhardy Daud, Mohd Nazry Salleh, Farizul Hafiz Kasim and Saiful Azhar Saad (2005) Analysis of Chemical Elements in Major Perlis's Agricultural Residue School of Materials Engineering, Kolej Universiti Kejuruteraan Utara Malaysia (KUKUM). Measurement Science Review, Volume 5, Section 3.
[7] Gagnon, N., Shaker, J., Roy, L., Petosa, A., & Berini, P., (2004) Low Cost Free Measurement of Dielectric Constant at Ka Band, IEEE Proceedings - Microwaves, Antennas and Propagation, Volume 151, Issue: 3, pg 271-276.
[8] Ghodgaonkar, D.K., Varadan, V.V., & Varadan, V.K., (2009) A Free Space Method for Measurement of Dielectric Constants and Loss Tangents at Microwave Frequencies, IEEE Transactions on Instrumentation and Measurement, Volume: 38, Issue: 3, pp 789-793.
[9] Maryam, M.I., Ibrahim, N., Shamsudin, R., & Marhaban, M.H., (2009) Sugar Content in Watermelon Juice Based on Dielectric Properties at 10.45GHz, 2009 IEEE Student Conference on Research and Development (SCOReD), pg 529-532.
[10] Ministry of Agriculture, (1995) Malaysian Agricultural Directory & Index 1995/96 (sixth edition), Kuala Lumpur, Malaysia, pg. 107.
[11] Navarkhele, V. V., Nakade, S.T. & Shaikh, A.A., (2006) A Dielectric Approach to Determine Water Content in Soil Using Microwave Transmission Technique. Journal of the Indian Institution of Science, 86, 723–729.
[12] Nelson, S.O., (2003) Measuring Dielectric Properties of Fresh Fruits and Vegetables, 2003 IEEE Antennas and Propagation Society International Symposium, Vol. 4, pg 46-49.
[13] Nelson, S.O., Wen-chuan Guo, Trabelsi, S., & Kays, S.J., (2007) Sensing Quality of Watermelons Trough Dielectric Permittivity, 2007 IEEE Antennas and Propagation Society International Symposium, pg 285-288.
[14] Popovic, D., Okoniewski, M., Hagl, D., Booske, J.H., & Hagness, S.C., (2001) Volume Sensing Properties of Open-ended Coaxial Probes for Dielectric Spectroscopy of Breast Tissue, Antennas and Propagation Society International Symposium, Vol 1, pp 254-257.
[15] Padiberas Nasional Berhad, (2007) Padiberas Nasional Berhad - Annual Report 2007, Petaling Jaya, Selangor.
[16] Pozar, M.D., (2005) Microwave Ebgineering, Third Edition, John Wiley & Sons Inc.
[17] Sheen, J. (2007) Microwave Dielectric Properties Measurements Using the Waveguide Reflection Dielectric Resonator, IEEE Instrumentation and Measurement Technology Conference Proceedings, 2007 (IMTC 2007), pg 1-4.
[18] Shu Chen, Kupershmidt, J., Korolev, K.A., & Afsar, M.N., (2007) A High Resolution Quasi Optical Spectrometer for Complex Permittivity and Loss Tangent Measurements

at Milimeter Wavelength, IEEE Instrumentation and Measurement Technology Conference Proceedings. (IMTC 2007), pg 1-5.

[19] Stutzman, W.L., & Thiele, G.A., (1998) Antenna Theory and Design. John Wiley & Sons.

[20] Trabelsi, S., & Nelson, S.O., (2003) Free Space Measurement of Dielectric Properties of Cereal Grain and Oilseed at Microwave Frequency, Measurement Science and Technology, Institute of Physics Publishing, Vol. 14, pg 589-600.

[21] Venkatesh, M.S., & Raghavan, G.S.V., (2005) An Overview of Dielectric Properties Measuring Techniques. Canadian Biosystem Engineering, Volume 47.

Chapter 6

Microwave Dielectrics Based on Complex Oxide Systems

A.G. Belous

Additional information is available at the end of the chapter

http://dx.doi.org/10.5772/51168

1. Introduction

The development of modern telecommunication systems calls for the creation of novel materials with a high level of electrophysical parameters in the MW range. These materials must have in the MW range a high permittivity ($\varepsilon \geq 20$), a low dielectric loss ($tg\,\delta \leq 10^{-3} - 10^{-4}$) and a high thermostability of electrophysical properties (temperature coefficient of permittivity (TCε) or resonant frequency (τ_f) ~$10^{-6}K^{-1}$). Such materials can be used in the development of resonant elements of radio-frequency filters, solid state oscillators, substrates for hybrid MW circuits, allow the size of communication systems to be greatly reduced and improve their parameters. Moreover, the use of them reduces the manufacturing and operating costs for modern communication systems.

The choice of the permittivity value of MW materials is largely determined by the frequency range of the operation of communication systems, the type of exciting wave and the requirement of the optimal size of dielectric element. The value of ε determines the size of radio components. The influence of microminiaturization is based on the fact that the electromagnetic wavelength in dielectric decreases in inverse proportion to $\sqrt{\varepsilon}$. Therefore, in the decimeter wave band, high-Q thermostable materials with high permittivity value ($\varepsilon \geq$ 80-600) are required, whereas in the centimeter and millimeter wave bands, thermostable materials with $\varepsilon \sim 15$-30 but with extremely high Q ($Q\times f \geq 80000$, where $Q = 1/ tg\,\delta$ and f is frequency in GHz) are needed.

It should be noted that low dielectric loss in the MW range $10^9 - 10^{11}$ Hz is characteristic only of optical and infrared polarization mechanisms. Other polarization mechanisms give rise, as a rule, to considerable dielectric loss [1].

In the case of optical polarization, dielectrics are characterized by a low negative temperature coefficient of permittivity (TCε ~ $10^{-5}K^{-1}$). However, the dielectric contribution of optical polarization is usually small. Therefore, large permittivity values together with high temperature stability of dielectric parameters and low dielectric loss can be observed only in the dielectrics where the main contribution to polarization is made by infrared polarization mechanism [1, 2]. This mechanism is bound up with cation and anion sublattice displacement in electric field, which is only possible in ionic crystals. The contribution of infrared polarization mechanism to permittivity may be $\Delta \varepsilon_{ir} = 1 - 10^4$ in the MW range. The temperature instability of ε increases, as a rule, with increasing ε. The large magnitude of infrared polarization is usually due to the presence of a soft mode in crystal, whose frequency varies by the critical law $\omega_T = A\sqrt{T - Q}$; this leads in accordance with the Liddein–Sax–Teller (LST) relation:

$$\frac{\varepsilon_{MW}}{\varepsilon_{Opt}} = \prod_i \frac{\omega_L^2}{\omega_T^2} \qquad (1)$$

to the Curie – Weiss law for permittivity:

$$\varepsilon_{MW} = \varepsilon_L + \frac{C}{T - Q} \qquad (2)$$

where ω_L and ω_T are the frequencies of longitudinal and transverse optical phonons in the center of Brillouin zone (one of the transverse phonons is soft), C is a constant, Q is Curie – Weiss temperature, ε_L is dielectric contribution, which depends only slightly on temperature [1].

The frequency of transverse and longitudinal optical phonons can be calculated from the equations:

$$\omega_T^2 = \frac{c}{m} - \frac{nq^2}{3\varepsilon_0 m} \cdot \frac{\varepsilon_\infty + 2}{3} \text{ and } \omega_L^2 = \frac{c}{m} + \frac{2nq^2}{3\varepsilon_0 m} \cdot \frac{\varepsilon_\infty + 2}{3} \qquad (3)$$

where c is the elastic coupling parameter of phonons; m is reduced mass; q, n are ion charge and concentration; ε_0 is an electric constant [1].

The parameters c, n, ε_{opt} are temperature-dependent; they decrease with rising temperature due to thermal expansion (lattice anharmonicity). It is evident from system (3) that ω_L is a weak function of temperature because it is determined by the sum of two terms, whereas the dependence ω_T (T) may be strong since ω_T depends on the difference of two terms (see Eqs (3)). The variation of this difference as a function of temperature depends on which effect predominates: the variation of c/m (minuend) or the variation of the subtrahend, which depends on n, ε_{opt}. Depending on ω_T, the dielectric contribution $\Delta \varepsilon_{ir}$ also varies with temperature [1]:

$$\Delta\varepsilon_{IR} = \varepsilon_{MW} - \varepsilon_{Opt} = \frac{nq^2}{m\omega_T^2\varepsilon_0} \cdot \left(\frac{\varepsilon_{Opt} + 2}{3} \right).$$

The above analysis shows that the chemical composition can influence, in principle, the contribution of different polarization mechanisms and hence the value of permittivity and loss, as well as their variation in the MW range as a function of temperature.

The present paper considers the structure peculiarities, electrophysical properties and possible applications of inorganic microwave (MW) dielectrics based on oxide systems.

2. MW dielectrics based on (La, Ca)(Ti, Al)O_3 solid solutions

MW dielectrics are often synthesized on the basis of solid solutions, e.g. Ba(Zn, $Mg)_{1/3}$(Nb, $Ta)_{2/3}O_3$ [3, , 5], (La, Ca)(Ti, Al)O_3 [6, 7], etc. The substance of this approach is that solid solutions are formed by the interaction of phases belonging to the same crystal structure, which have in the MW range a different trend of the plot of permittivity against temperature and a low dielectric loss. Paraelectric is characterized by a low dielectric loss; for example, $CaTiO_3$, which crystallizes in perovskite structure, can be used in the $CaTiO_3$ – $LaAlO_3$ system as a phase with negative temperature coefficient of permittivity (TCε < 0) [6]. At the same time, $LaAlO_3$ can be used as a phase with perovskite structure having TCε > 0 [7]. By varying the ratio $CaTiO_3$ /$LaAlO_3$, one can control the value of TCε. Positive TCε in dielectrics in the MW range usually indicates the presence of a high-temperature phase transition, which is connected with the existence of spontaneously polarized state (ferroelectrics, antiferroelectrics). However, the materials in which spontaneous polarization exists have, as a rule, a considerable dielectric loss in the MW range, which is inadmissible for the creation of high–Q dielectrics. In $LaAlO_3$, there is no spontaneous polarization. It should be noted that there are very few materials having TCε > 0 in the MW range and a low dielectric loss. Therefore, the development of high-Q MW dielectrics with high ε and positive temperature coefficient of permittivity (TCε > 0) is of independent scientific and practical interest.

3. Control of the TCε value by influencing the phonon spectrum

As follows from the analysis of expressions (3), the trend of the plot of ε against temperature in the MW range can be controlled by influencing the phonon spectrum. One of the ways of influencing the phonon spectrum in some types of structures can be iso- and heterovalent substitutions in cation sublattices. As an example, we chose $La_{2/3-x}$ (Na, $K)_{3x}$ TiO_3 materials, which crystallize in defect-perovskite structure in a wide x range (Fig 1).

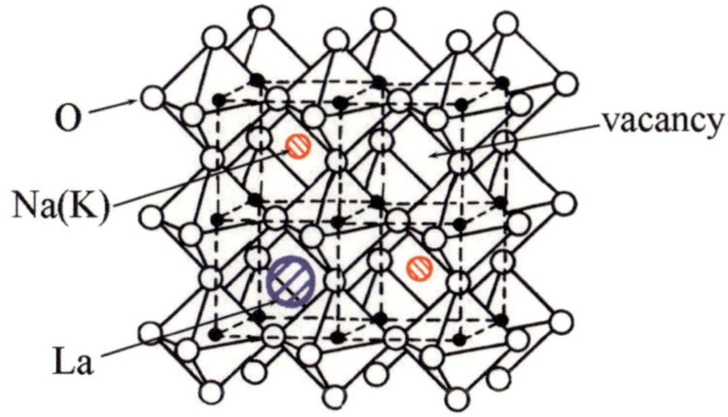

Figure 1. Crystal structure of $La_{2/3-x}$(Na, $K)_{3x}TiO_3$ perovskite

In this system, lanthanum ions in the oxidation state +3 are partially substituted by alkali (sodium or potassium) ions in the oxidation state +1. Substitution is performed so that the electroneutrality condition is satisfied. The value of x was varied from 1/24 to 1/6. In this case, lanthanum ions, alkali metal (sodium or potassium) ions and a structural vacancy (vacant crystal site) could be in one crystal sublattice at the same time. We hoped that in the case of such heterovalent substitution, the phonon spectrum, and hence the trend of the plot of ε against temperature in the MW range, had to change.

$La_{2/3-x}$ (Na, K)$_{3x}$ TiO_3 materials are not characterized by high temperature stability of dielectric parameters. The dielectric properties of these materials had been studied in a wide frequency range [8, 9].

It had been found that by decreasing the number of structural vacancies, using heterovalent substitution in sublattices and locating different ions in vacant crystal sites, one can influence greatly the dielectric loss level (Fig 2) [10].

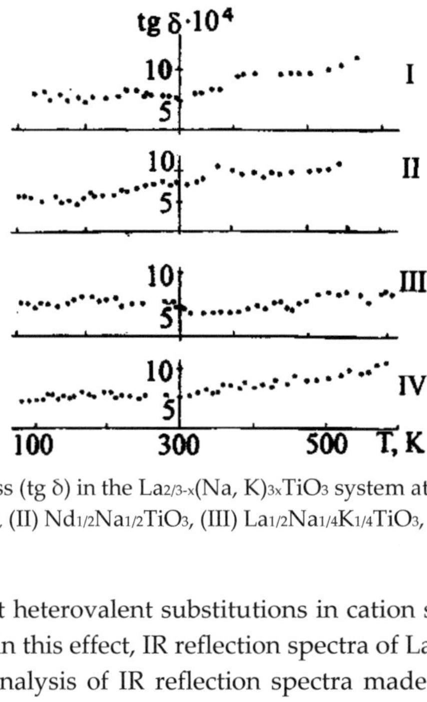

Figure 2. Plots of dielectric loss (tg δ) in the $La_{2/3-x}(Na, K)_{3x}TiO_3$ system at 1.2×10^{10} Hz against temperature: (I) $La_{1/2}Na_{1/2}TiO_3$, (II) $Nd_{1/2}Na_{1/2}TiO_3$, (III) $La_{1/2}Na_{1/4}K_{1/4}TiO_3$, (IV) $La_{7/12}Na_{1/4}•_{1/6}TiO_3$, where • is the structural vacancy

Investigations showed that heterovalent substitutions in cation sublattices affect greatly the value of TCε too. To explain this effect, IR reflection spectra of $La_{2/3-x}$ M_{3x} TiO_3 materials have been analyzed [11]. The analysis of IR reflection spectra made it possible to calculate the parameters of dispersion oscillators (Table 1). It is known that in the materials that crystallize in perovskite structure, a low-frequency lattice vibration exists which is responsible for the high ε value in the MW range [12].

The partial heterovalent substitution of ions in crystal sublattice gives rise to a low-frequency vibration, which affects noticeably the value of permittivity. In this case, the temperature stability of dielectric parameters increases greatly. Thus, the proposed method of influencing the phonon spectrum can be employed in the development of novel MW dielectrics with high temperature stability of dielectric properties.

$La_{1/2}Na_{1/4}K_{1/4}TiO_3$($\varepsilon$ = 106)				$La_{7/12}Na_{1/4}$ • $_{1/6}TiO_3$(ε = 87)			
ω_{TO}	$\omega_{\_LO}$	ΔE	g	ω_{TO}	ω_{LO}	ΔE	g
cm^{-1}				cm^{-1}			
116	163	74.0	0.75	133	178	55.0	0.64
198	258	22.0	0.44	201	224	19.5	0.31
237	334	3.0	0.21	230	265	4.1	0.30
336	375	0.2	0.33	270	343	1.8	0.26
381	489	0.4	0.16	345	460	0.2	0.10
554	747	1.2	0.10	563	694	1.0	0.11
785	816	0.1	0.12	789	860	0.3	0.30

*ε_∞ = 5.1.

Table 1. Parameters of ($La_{2/3-x}$ M_{3x} • $_{1/3-2x}$)TiO_3 dispersion oscillators*

4. MW dielectrics based on $Ba_{6-x}Ln_{8+2x/3}Ti_{18}O_{54}$ (Ln = La-Gd)

The $Ba_{6-x}Ln_{8+2x/3}Ti_{18}O_{54}$ materials (Ln = La - Gd) (BLTss) have promise in the development on their basis of thermostable high-Q MW dielectrics with high permittivity ($\varepsilon \approx$ 80 - 100) [, 14]. They crystallize in KW bronze structure (Fig 3), which includes elements of perovskite structure [15, , 17]. In this structure, the octahedra are linked, as in perovskite, by their apices into parallel rectilinear chains. Unlike perovskite structure, however, the oxygen octahedra are linked so that they form pentangular, quadrangular and triangular channels, in which A ions can be, having in this case the coordination numbers 15, 12 and 9 respectively. This structure allows one to perform iso- and heterovalent substitutions in cation sublattices in a wide range, to control the number of vacant crystal sites in the A sublattice, to influence the partial redistribution of A ions among the pentangular, quadrangular and triangular channels and hence to control the electrophysical properties in the MW range, including the temperature dependence of ε.

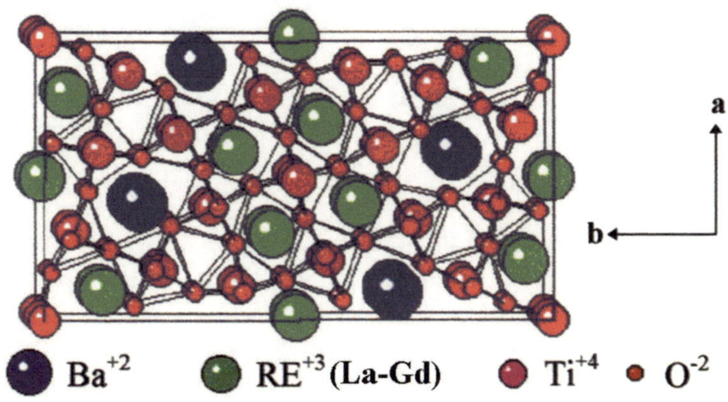

Figure 3. Unit cell of Ba_{6-x} $Ln_{8+2x/3}Ti_{18}O_{54}$ [17]

When investigating $Ba_{6-x}Ln_{8+2x/3}Ti_{18}O_{54}$ materials (Ln = La - Gd), which crystallize in KW bronze structure, a special attention was given to the study of formation reaction and the anomalous behavior of the temperature characteristics of dielectric parameters. It should be noted that the knowledge of the reactions proceeding during the synthesis of compounds can allow dielectric loss to be reduced. The formation of $Ba_{6-x}Ln_{8+2x/3}Ti_{18}O_{54}$ materials (Ln = Nd, Sm) when using the solid-state reaction method was studied on compositions with x = 0.75, 1.5, 2.0. $BaCO_3$, Sm_2O_3 and Tio_2 were used as starting reagents. It was shown that it is a multistage process, which is accompanied by the formation of intermediate phases, e.g. $Ln_2Ti_2O_8$, $BaTi_4O_9$, $BaTiO_3$ [18]. It has been found that independent of x value, the phase $Ba_{3.9}Ln_{9.4}Ti_{18}O_{54}$ is formed at first, which belongs to $Ba_{6-x}Ln_{8+2x/3}Ti_{18}O_{54}$ solid solutions and corresponds to the maximum x value. The formation of the other phases, which belong to the region of $Ba_{6-x}Ln_{8+2x/3}Ti_{18}O_{54}$ – type solid solutions, takes place as a result of interaction between intermediate $Ba_{3.9}Ln_{9.4}Ti_{18}O_{54}$ phases and barium metatitanate ($BaTiO_3$). The phase $Ba_{3.9}Ln_{9.4}Ti_{18}O_{54}$ crystallizes, as $Ba_{6-x}Ln_{8+2x/3}Ti_{18}O_{54}$ materials, in KW bronze structure, which makes the identification of the phase $Ba_{3.9}Ln_{9.4}Ti_{18}O_{54}$ only by the data of X-ray phase analysis impossible. Therefore, EDS and TEM analyses were used additionally [18]. The latter analysis showed that even when all $BaTiO_3$ had reacted, the homogeneity of materials was not reached yet (Fig 4(a)). If the ceramic sintering time was relatively short, a phase, e.g. $Ba_{3.9}Ln_{9.4}Ti_{18}O_{54}$, was present (Fig 4(a)), which had low and high x values within the limits of formation of $Ba_{6-x}Ln_{8+2x/3}Ti_{18}O_{54}$ solid solutions. The structural-defect concentration decreased, and the homogeneity of $Ba_{6-x}Ln_{8+2x/3}Ti_{18}O_{54}$ materials increased only in the case of long ceramic sintering time ($t \geq 3h$) (Fig 4(b)).

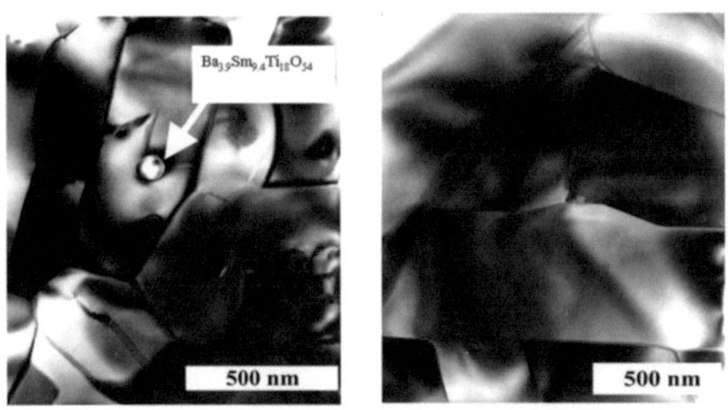

Figure 4. Results of a TEM analysis of $Ba_{6-x}Ln_{8+2x/3}Ti_{18}O_{54}$ ceramic; sintering time: 1 h (a), over 3 h (b)

The electrophysical characteristics of $Ba_{6-x}Ln_{8+2x/3}Ti_{18}O_{54}$ depend largely upon ions in the A sublattice (r_A) [14, 19, 20]. When the rare-earth ion in the A sublattice is changed from La to Gd, the permittivity (ε) value and loss-angle tangent in BLTss decrease. At the same time, the temperature coefficient of permittivity, TCε, increases and changes its sign in the series of rare-earth ions, which are in the A sublattice, in going from Nd to Sm [20, 21]. When investigating $Ba_{6-x}Ln_{8+2x/3}Ti_{18}O_{54}$ materials (where x = 1.5), we had found for the first time an anomaly on the plot of permittivity against temperature [21]. Later, anomalies on the plots of dielectric parameters (ε, tg δ) against temperature were detected in the other $Ba_{6-x}Ln_{8+2x/3}Ti_{18}O_{54}$ materials too [22]. The nature of these anomalies remained uncertain.

Moreover, there was no information about the existence of temperature dependence anomalies of dielectric parameters in other barium-lanthanide analogs, including La-, Nd-, Gd-containing $Ba_{6-x}Ln_{8+2x/3}Ti_{18}O_{54}$. Therefore, we tried to find out the cause of the temperature dependence anomalies of permittivity and dielectric loss since this makes it possible to establish the nature of the thermostability of electrophysical properties in these systems.

It has been found that temperature dependence anomalies of dielectric parameters in La- and Nd-containing BLTss are observed at low temperatures in a wide frequency range, including the submillimeter-wave region (Fig 5) [23], the position of these anomalies on the temperature scale depending not on measurement frequency, but on chemical composition.

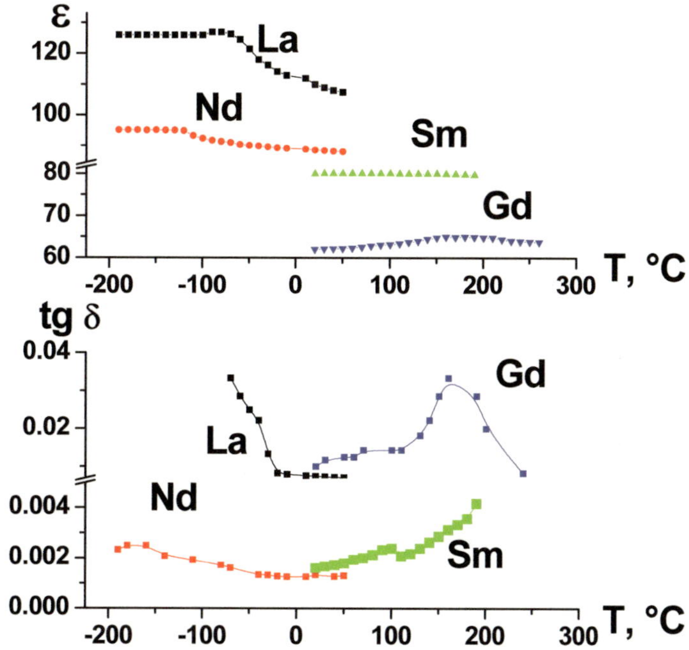

Figure 5. Plots of the dielectric parameters of $Ba_{6-x}Ln_{8+2x/3}Ti_{18}O_{54}$ solid solutions (Ln = La, Nd, Sm, Gd) at 10 GHz against temperature

In Sm- and Gd-containing systems, anomalies of dielectric parameters appear at temperatures above room temperatures (Fig 5). The plots of dielectric parameters against temperature have a similar trend when ferroelectric or antiferroelectric ordering occurs. In the case of BLTss, however, no hysteresis loops were observed, and the temperature dependence of ε did not obey the Curie-Weiss law, which indicated the absence of spontaneously polarized state in these materials. It may be supposed that the appearance of dielectric anomalies is due to the presence of unknown phase transitions. Therefore, $Ba_{6-x}Ln_{8+2x/3}Ti_{18}O_{54}$ systems (x = 1.0), in which dielectric anomalies of ε and tg δ were observed at 100-120 ^{0}C, have been studied by low-temperature differential scanning calorimetry (LT-DSC) and high-temperature X-ray structural analysis of samples. We did not find any phase transitions [22], which was confirmed by the authors of [24], who carried out synchrotron X-ray diffraction studies of $Ba_{4.5}Sm_9Ti_{18}O_{54}$ samples in the temperature range 10-295 K. These data indicate the absence of structural transitions in the temperature ranges where anomalies of dielectric parameters were observed. It can be concluded that the temperature dependence anomalies of dielectric parameters are not coupled with the peculiarities of sample preparation and the presence of structural transitions.

On the basis of an analysis it was assumed that the nature of the anomaly of dielectric parameters is coupled with harmonic and anharmonic BLTss lattice vibration, which is different in the character of influence on the temperature behavior of dielectric parameters [23].

Investigations showed that the plots of ε and tg δ against temperature depend largely upon harmonic and anharmonic lattice vibration modes. Therefore, using different hetero- and isovalent substitutions in cation sublattices, one can influence the lattice phonon spectrum and hence obtain materials with high temperature stability of dielectric parameters, which are used in modern decimeter and centimeter wave band communication systems [25, , , 28].

5. MW dielectrics with "mobile sublattice"

The authors of [29] reported the development of novel MW dielectrics with high temperature stability of dielectric parameters based on solid solutions, where one of the phases (lithium-containing $La_{1/2}Li_{1/2}TiO_3$) had a positive temperature coefficient of permittivity (TCε > 0). However, there was no explanation of the nature of this fact in literature.

The phase $La_{1/2}Li_{1/2}TiO_3$, which crystallizes in perovskite structure, belongs to $Ln_{2.3-x}M_{3x}Tio_3$ solid solutions (Ln = rare-earth elements, M = alkali metal ion) (Fig 6). In this system, the M ions partially substitute for rare-earth ions. In this case, the electroneutrality condition is satisfied [30, 31]. If M = Na, K, materials crystallize in perovskite structure and are characterized by low dielectric loss (tg δ < 10^{-3}) and high permittivity in the MW range (about 100) and TCε < 0 [32].

At the same time, when M = Li, a system of solid solutions is formed, in which rare-earth ions, lithium ions, structural vacancies, which are characterized by TCε > 0, are in one sublattice at the same time. Lithium ions can move along structural channels, ensuring a high lithium-ion conductivity [33, 34], positive temperature coefficient of permittivity (TCε > 0) and causing considerable dielectric loss in the MW range. The latter is inadmissible in the creation of high-Q dielectrics. Investigations showed, however, that the dielectric loss in the MW range can be greatly reduced by decreasing lithium ion conductivity. The latter is achieved by substituting rare-earth ions with smaller radius for lanthanum ions. This leads to a decrease in the size of structural channels, in which lithium ions are; this reduces dielectric loss in the MW range through a decrease in lithium ion mobility in the structure. In this case, TCε > 0 is retained. As a result, solid solutions have been obtained, in which rare-earth ions are simultaneously substituted by alkali metal ions with large (Na, K) and small (Li) radius and which have a high temperature stability of dielectric parameters in the MW range (Table 2) [35].

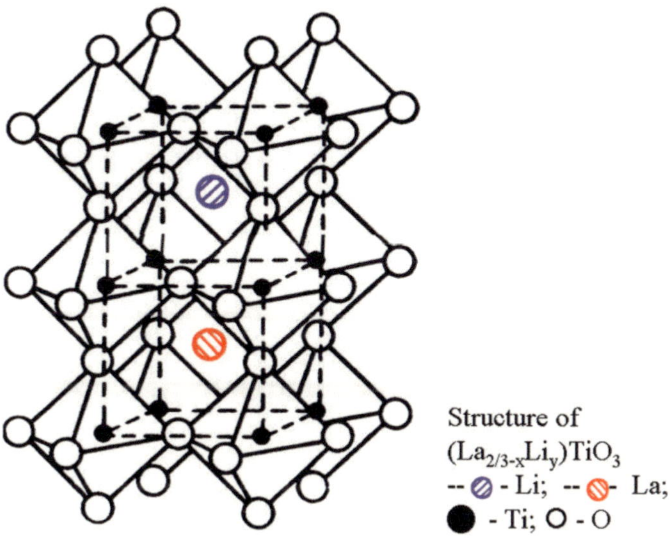

Figure 6. Perovskite structure of $La_{2/3-x}Li_{3x}TiO_3$

x	ε	TCε, ppm/°C (20–100°C)	Q (10 GHz)
0	80	–520	1300
0.30	95	–240	100
0.50	90	–140	150
0.55	85	–50	200
0.58	80	–5	200
0.60	75	+60	200
0.1	52	+580	100

Table 2. Dielectric parameters of the $Sm_{1/2}Li_{1/2}TiO_3$ – (1-x) $Sm_{1/2}Na_{1/2}TiO_3$ system at 10 GHz

6. MW dielectrics based on antiferroelectrics-paraelectrics

The permittivity value of the above-mentioned thermostable MW dielectrics is not over 80 – 100. To achieve higher permittivity values, use must be made of other polarization mechanisms connected with spontaneously polarized state. One of the possible ways of developing thermostable MW resonant elements is the creation of two-layer systems. Each of the layers must have a high Q ($Q = 1/\text{tg}\,\delta$), a high permittivity and TCε of different sign. A large number of high-Q dielectrics which have TCε < 0 in the MW range is known (mainly paraelectrics), whereas there are few of them with TCε > 0 ($LiNbO_3$, $LiTaO_3$, $LiAlO_3$ single crystals).

Besides, they are characterized by relatively low ε values (ε < 50) in the MW range, and the use of them in this range makes it possible to obtain two-layer resonant dielectric elements with an effective permittivity (ε_{eff}) of not over 50. Therefore, we examined the possibility to use as high-Q materials with TCε > 0 antiferroelectrics based on tellurium-containing

Pb_2BTeO_6-type perovskites (B = bivalent metal ions) [36, 37]. It has been found that the plot of ε against temperature for lead-cobalt tellurate (Pb_2CoTeO_6) in the MW range passes through a maximum (phase transition from the antiferroelectric to the paraelectric state) at 380 K (Fig 7). At room temperature, the materials of this group have high permittivity values (ε > 110). In the temperature range 220-350 K, the dependence ε (T) is close to linear one and has TCε ~ 700×10^{-6} K^{-1}. Two-layer resonant elements: Pb_2CoTeO_6 – TiO_2, Pb_2CoTeO_6 – $CaTiO_3$, Pb_2CoTeO_6 – $SrTiO_3$ have been prepared on the basis of paraelectrics (TiO_2, $CaTiO_3$, $SrTiO_3$) and the antiferroelectric Pb_2CoTeO_6. Known two-layer resonant dielectric $LiNBO_3$ – TiO_2 elements have been investigated for comparison. It can be seen from Table 3 and Fig 8 that Pb_2CoTeO_6 - $SrTiO_3$ based two-layer resonant elements have in the MW range a high effective permittivity (ε_{eff} ≈ 135), a high Q (Q_{10GHz} ~ 900) and a high temperature stability of dielectric parameters (TCε tends to zero) [37]. Of course, two-layer systems have a number of demerits since they require mechanical bonding of different layers.

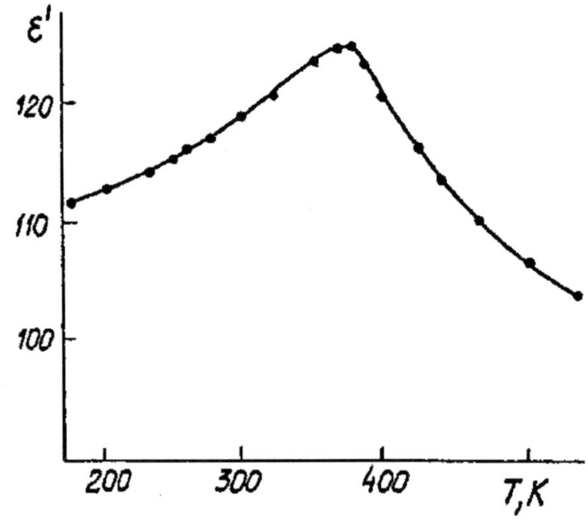

Figure 7. Plot of permittivity against temperature for Pb_2CoTe_6 at 10GHz

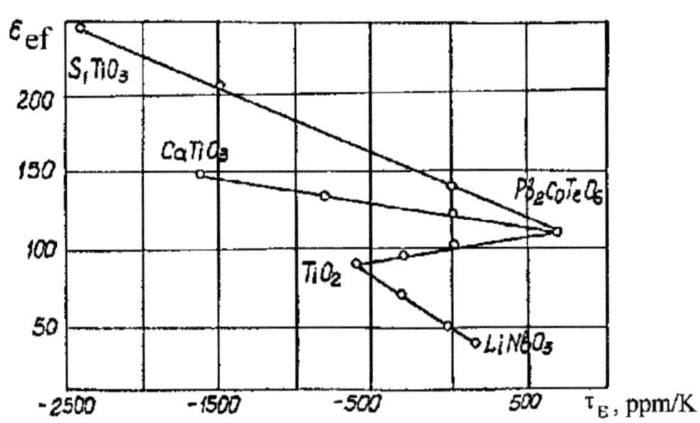

Figure 8. Dependence of the effective permittivity (ε_{ff}) of two layer resonant elements on the temperature coefficient of frequency (τ_ε) at 10 GHz

Materials	$Q_{ef.}$	$Q_{ef.}$
Pb_2CoTe_6–TiO_2	1000	110
Pb_2CoTe_6–$CaTiO_3$	900	125
Pb_2CoTe_6–$SrTiO_3$	900	135
$LiNbO_3$–TiO_2	3000	49

Table 3. Properties of two-layer resonant elements at 10GHz

Thermostable MW dielectrics with high permittivity (ε ~ 430) and Q (Q_{1GHz} ~ 700) have been obtained by the authors of [38] on the basis of the Ag (Nb, Ta) O_3 system, which is characterized by the spontaneously polarized state. However, in the materials in which the spontaneously polarized state is present, increase in permittivity is always accompanied by an increase in dielectric loss, which impairs the technical characteristics of MW elements made on their basis.

7. MW dielectrics based on spontaneously polarized phases

The materials in which there is a phase transition from the spontaneously polarized to the unpolarized state at high temperatures are characterized by positive TCε. In the phase transition region, the tg δ values in the MW range are, as a rule, large, which is due to the presence of mobile domain walls (ferroelectrics). A salient feature of antiferroelectrics is the immobility of domain walls. This results in the fact that antiferroelectric (e.g. Pb_2CoTeO_6) is characterized by a relatively low tg δ value at room temperatures in the MW range [37]. In some cases, ferroelectrics, too, have low tg δ values, e.g. single-domain $LiNbO_3$ single crystal, in which phase transition is observed at high temperature (> 1200 ^{0}C).

It was investing to find out whether it is possible to create thermostable MW dielectrics on the basis of solid solutions firmed by ferroelectrics and/or antiferroelectrics, which are characterized by high phase transition temperature, and materials having a defect crystal structure (with vacancies). To this end, we investigated $Ln_{2/3-x}Na_{3x}Nb_2O_6$ materials (Ln = La, Nd), which were formed by interaction between the $La_{2/3} \bullet_{4/3}Nb_2O_6$ phase with defect-perovskite structure (Fig 9) and the $NaNbO_3$ phase with perovskite structure (Fig 10), in which transition from the spontaneously polarized to the unpolarized state is observed at high temperature (> 520 ^{0}C).

An analysis of X-ray data for polycrystalline $La_{2/3-x}Na_{3x} \bullet_{4/3-2x}Nb_2O_6$ samples (Ln = La, Nd) showed that depending on x, solid solutions having three different space groups are formed. The space group changes in the order Pmmm → Pmmn Pbcn with increasing x. In the interval $0 \leq x \leq 0.24$, the solid solutions have, independent of Ln, a defect-perovskite structure ($La_{2/3} \bullet_{4/3}Nb_2O_6$), where is a vacancy in the cation sublattice with the space group Pmmm [39].

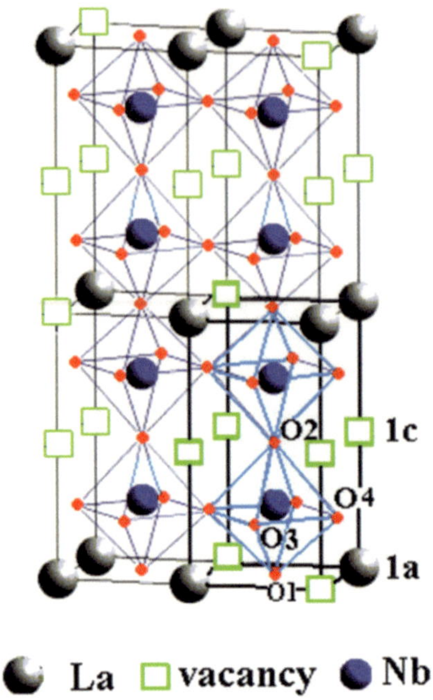

Figure 9. Defect-perovskite structure $Ln_{2/3} \bullet {}_{4/3}Nb_2O_6$ (space group Pmmm). Atomic positions: La (1a) 000; Nb (2t) ½ ½ z; 0 (1) (1*f*) ½ ½ 0; 0 (2)

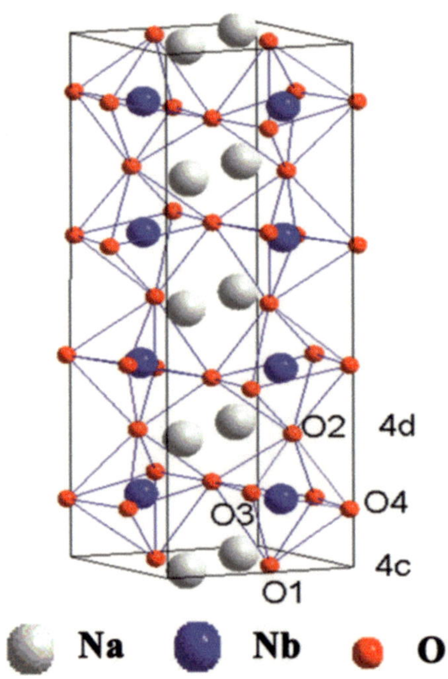

Figure 10. Perovskite structure of $NaNbO_3$ at room temperature (space group Pbcm). Atomic positions: Na (1) (4c) x ¼ 0; Na (2) (4d) x y ¼; Nb (8c) x y z; O (1) (4c) x ¼ 0; O (2) (4d) x y ¼ ; O (3) (8c) x y z; 0 (4) (8c) x y z

When x is increased, the space group changes, independent of the kind of Ln, from Pmmm to Pmmn. In neodymium-containing solid solutions, the change of space group doespractically lead to change in normalized unit cell volume V/Z; in lanthanum-containing solid solutions, monotonic variation of V/Z as a function of x is accompanied by a slight inflection in the sodium concentration range corresponding to the change from Pmmm to Pmmn. The observed differences in the trend of plots of V/Z (x) are probably due to the fact that the ionic radius of Na^+ is smaller than that of the substituted La^{3+} ion, whereas the difference in the ionic radii of Na^+ and Nd^{3+} is small. The closeness of the ionic radius values of Na^+ and Nd^{3+} leads to the existence of a wider concentration range ($0.24 \leq x \leq 0.54$), which corresponds to the space group Pmmm , for neodymium-containing solid solutions as compared with lanthanum-containing ones ($0.24 \leq x \leq 0.45$).

In the intervals $0.54 \leq x \leq 0.66$, (for neodymium) and $0.45 \leq x \leq 0.66$ (for lanthanum), the crystal structure of $La_{2/3\text{-}x}Na_{3x} \bullet {}_{4/3\text{-}2x}Nb_2O_6$ solid solutions has the space group Pbcn, which is typical of $NaNbO_3$ at room temperature [40]. In fact, when the sodium content of the system is increased, a decrease in the symmetry of solid solutions from Pmmm to Pmmn and to Pbcn is observed.

$La_{2/3}Nb_2O_6$ and $Nd_{2/3}Nb_2O_6$ materials (x = 0) are characterized by a high permittivity value (130 and 160 respectively) and a relatively low dielectric loss (in both cases, tg δ is of the order of $2 - 5 \times 10^{-3}$ at a frequency of 10 GHz). There is no permittivity dispersion. The plot of ε (T) for $La_{2/3}Nb_2O_6$ and $Nd_{2/3}Nb_2O$ materials exhibits deflections in the MW range. In the low-temperature range, the ε value varies only slightly with rising temperature. As the x values in $La_{2/3\text{-}x}Na_{3x} \bullet {}_{4/3\text{-}2x}Nb_2O_6$ materials (Ln = La, Nd) increases, TCε changes its sign from negative to positive. Plots of dielectric parameters against concentration in the MW range are shown in Fig 11. In the interval $0 \leq x \leq 0.24$ (space group) Pmmm, increasing the sodium concentration leads to a slight increase in permittivity independent of the kind of rare-earth element (La or Nd), which is accounted for by increase in cation vacancy concentration.

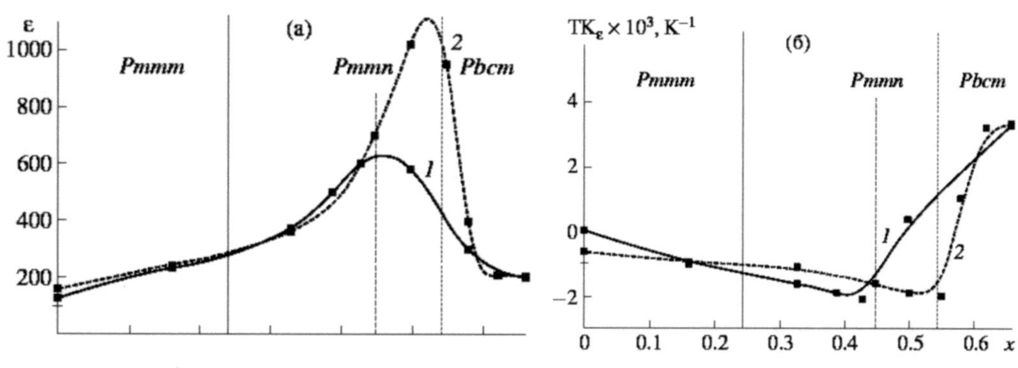

Figure 11. Plots of permittivity (a) and the temperature coefficient of permittivity (b) against the sodium content of the solid solutions $La_{2/3\text{-}x}Na_{3x} \bullet {}_{4/3\text{-}2x}Nb_2O_6$ (1) and $Nd_{2/3\text{-}x}Na_{3x} \bullet {}_{4/3\text{-}2x}Nb_2O_6$ (2)

Within the limits of the space group Pmmn, the permittivity value decreases greatly with increasing x, passing through a maximum, when the space group changes from Pmmn to Pbcn . Investigations showed that it is possible to create thermostable dielectrics based on the system in $Ln_{2/3}Na_{3X}$ $\bullet_{4/3-2X}Nb_2O_6$ (Ln = La, Nd), which have a high permittivity ($\varepsilon \sim 300$-600) and a relatively low dielectric loss (tg $\delta \sim 2 - 7 \times 10^{-3}$) in the MW range [41, , , 44].

The systems considered above have a relatively high thermostability of electrophysical properties ($TC\varepsilon \sim 10^{-5} - 10^{-6}K^{-1}$), $Q \times f \leq 12000$ and relatively high permittivity values ($\varepsilon \geq 80$ 600) in the MW range. This makes it possible to develop on their basis elements for decimeter wave band communication systems, where the problems of microminiaturization, for the solution of which high ε values are required, are especially important.

In the centimeter wave band and especially in the millimeter wave band, however, materials with relatively low permittivity (10 -30) are required, which must possess very high Q valies ($Q \times f \geq 80000$ - 100000). Let us consider some systems, which have promise in gaining these purposes.

8. $M_{1+x}Nb_2O_6$ – Based MW dielectrics (M = Mg, Co, Zn) with columbite structure

Among the MW dielectrics known to date, M^{2+} Nb_2O_6 niobates (M = Mg, Co, Zn) are of considerable interest. The crystal structure of $A^{2+}Nb_2O_6$ columbite is infinite zigzag chains of oxygen linked by shared edges (Fig 12) [45]. For this structure, redistribution of crystal sites, which are in the oxygen octahedral, among A^{2+} ion size and the cation ratio $Nb^{5+}:A^{2+}$ in the unit cell will affect the crystallographic distortions of the columbite structure and hence the phase composition and electrophysical properties of synthesized materials.

The $ZnNb_2O_6$ and $MgNb_2O_6$ materials have a high Q (3000 and 9400 respectively) and permittivity (23 and 20 respectively) [46, 47, 48]. In contrast to magnesium- and zinc-containing niobates, the literature data on the dielectric properties of cobalt niobate ($CoNb_2O_6$) are very contradictory. For instance, Ref [46] reported low Q values for $CoNb_2O_6$ ($Q \times f = 40000$). Earlier it was shown [49, 50, 51, 52, 53] that the making of single-phase M^{2+} Nb_2O_6 materials depends largely upon their synthesis conditions. For instance, the difficulty of making single-phase magnesium niobate with columbite structure is accounted for, in particular, by the simultaneous formation of two phases: $MgNb_2O_6$ columbite and $Mg_4Nb_2O_9$ corundum [51]. It should be noted that in the Mg-Nb-O system, a number of compounds: $MgNb_2O_6$, $Mg_4Nb_2O_9$, $Mg_5Nb_4O_{15}$, $Mg_{1/3}Nb_{11}$ $_{(1/3)}$ are formed [52, 53]; however, only the phases $MgNb_2O_6$ (columbite structure) and $Mg_4Nb_2O_9$ (corundum structure) are stable at room temperature [54]. Therefore, even after long heat treatment at highj temperature (T > 1100 ^{0}C) [50], there were intermediate phases in the end product (generally a corundum phase). In such cases, the phase composition and electrophysical properties can be greatly affected even by a small deviation from stoichiometry. In view of this, we have studied the effect of small deviations from stoichiometry in A^{2+} Nb_2O_6 materials (A = Co, Mg, Zn) with columbite structure on phase composition, microstructure and MW properties [55, 56, 57].

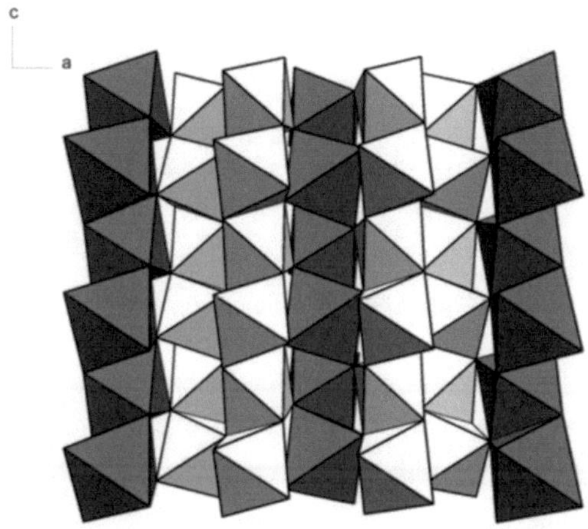

Figure 12. Columbite structure of $A^{2+}Nb_2O_6$

We have shown that when the solid-state reaction method is used, the formation of cobalt- and magnesium-containing niobates with columbite structure is a multistage process. In this case, two concurrent processes of formation of niobates with columbite structure ($A^{2+}Nb_2O_6$) and corundum structure ($A_4^{2+}Nb_2O_9$) ($A^{2+} = Co^{2+}, Mg^{2+}$) take place:

$$2A_3^{2+}O_4 + 6Nb_2O_5 \rightarrow 6A^{2+}Nb_2O_6 + O_2;$$

$$4A_3^{2+}O_4 + 3Nb_2O_5 \rightarrow 3A_4^{2+}Nb_2O_9 + 2O_2.$$

At higher temperatures (> 1000 °C), the formation of columbite structure took place by interaction between the $A_4^{2+}Nb_2O_9$ phase and unreacted Nb_2O_5:

$$A_4^{2+}Nb_2O_9 + 3Nb_2O_5 \rightarrow 4A^{2+}Nb_2O_6 \ (A^{2+} - Co^{2+}, Mg^{2+}).$$

At the same time, the synthesis of Zn Nb_2O_6 with columbite structure takes place in the temperature range 500-800 °C without formation of intermediate products.

In the case of deviation from stoichiometry in the $A_{1+x}^{2+}Nb_2O_6$ system ($A^{2+} = Mg^{2+}, Co^{2+}, Zn^{2+}$), when $x < 0$, samples contained two phases: the main phase $A^{2+}Nb_2O_6$ with columbite structure and the Nb_2O_5 phase, whose concentration increased with x (Fig 13). At $x > 0$, a narrow concentration range, in which samples are single-phase ones, exists in all three systems. On further deviation from stoichiometry in the direction of increasing excess of cobalt, magnesium or zinc, extra phases are formed.

The results of investigating electrophysical properties in the MW range turned out unlooked-for. At $x < 0$, when there were traces of the minor phase Nb_2O_5, the samples had a low Q. At the same time, extremely high Q values ($Q \times f$) were observed at $x > 0$ (Fig 14). For example, in $Mg_{1+x}Nb_2O_6$, $Q \times f$ reached a value of 128000 at $x \geq 0.03 - 0.05$ in multiphase samples, in which the phase $Mg_4Nb_2O_9$ (corundum structure) was present together with the main phase $MgNb_2O_6$ (columbite structure).

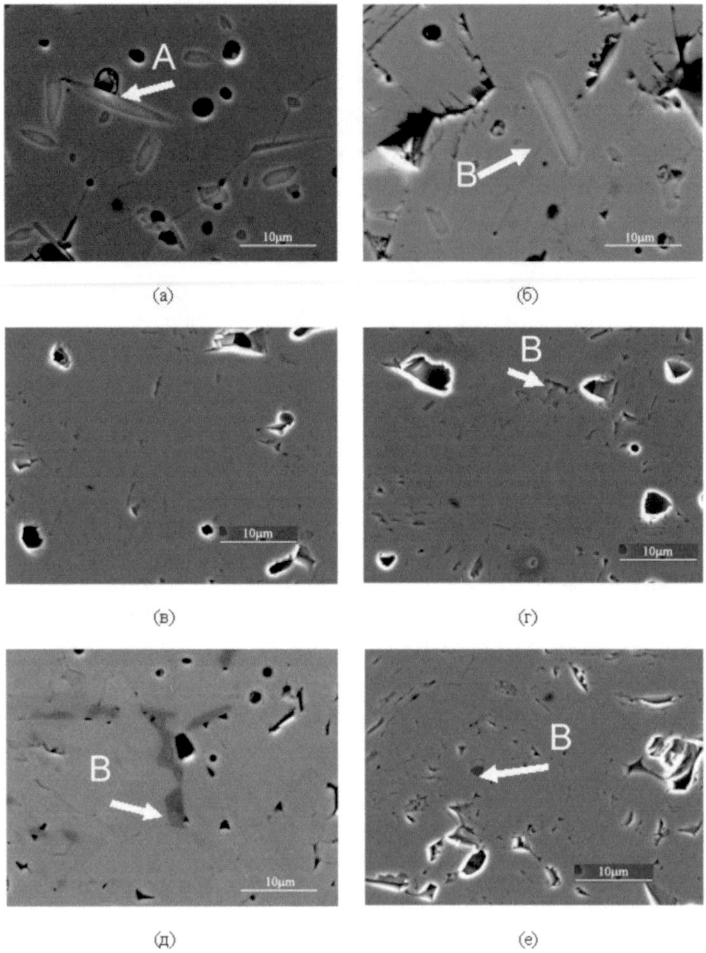

Figure 13. Micrographs of microsections of polycrystalline $Co_{1-x}Nb_2O_{6-x}$ samples with x = 0.05 (a), x = 0.03 (b), x = 0 (c), x = - 0.03 (d), x = - 0.05 (e, f), sintered at 1500 ^{0}C for 1h (a - f), 6 h (e): A = Nb_2O_5, B = $Co_4Nb_2O_9$

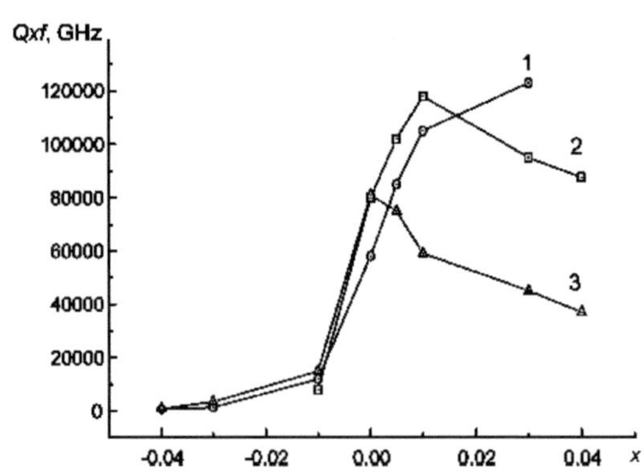

Figure 14. Plots of the product Q × *f* of $A_{1-x}Nb_2O_6$ samples (where A = Mg (1), Zn (2), Co(3)) against concentration. The samples were sintered in air for 8 h at 1400 ^{0}C (1 and 3) and 1300 ^{0}C (2).

This is accounted for by the high Q values (Q × f ~ 230000) of the extra phase $Mg_2Nb_4O_9$. However, further deviation from stoichiometry leads to a considerable decrease in permittivity, which is due to the fact that ε of the corundum phase was 11.

This instance is interesting in that each of the phases $MgNb_2O_6$ and $Mg_2Nb_4O_9$ characterized by certain merits and demerits. Only multiphase materials based on them have both a relatively high permittivity (ε ~ 20) and a highQ(Q × f ~ 128000) in the MW range.

9. ZrO_2-TiO_2-SnO_2 – Based Mw dielectrics

The first information about $ZrTiO_4$ as a promising high-Q dielectric was presented in Ref [58]. Later, in the 1950s, investigations of solid solutions in the ZrO_2-TiO_2-SnO_2 system were carried out [59]. It was shown that the composition $Zr_{0.8}Sn_{0.2}TiO_4$ has the highest Q values [60, 61].

$ZrTiO_4$ crystallizes in orthorhombic structure (space group Pbcn) [62, 63, 64] with the space lattice parameters: a = 4.806Å, b = 5.447Å, c = 5.032Å. The unit cell contains two formula units, theoretical density 5.15 g/cm³. It should be noted that in $ZrTiO_4$ there is an order-disorder phase transition in the temperature range 1100-1200°C [65, , , , 69]. When the temperature is decreased, this transition is from an α-PbO_2 – type high-temperature phase of which disordered arrangement of Zr and Ti ions is typical, to a low-temperature phase with ordered arrangement of Zr and Ti ions [70, 71]. Addition of Sn to $ZrTiO_4$ results in the stabilization of the disordered distribution of cations. The variation of the lattice parameters in the $Zr_{1-x}Sn_xTiO_4$ system with increasing x is shown in Fig15. As is seen from Fig 15, there are no noticeable changes in the behavior of the parameters a and b in the phase transition region (1100-1200°C) with increasing Sn content. At the same time, as Sn ions are added, a noticeable change in the dependence of the parameter c in the phase transition region is observed [69].

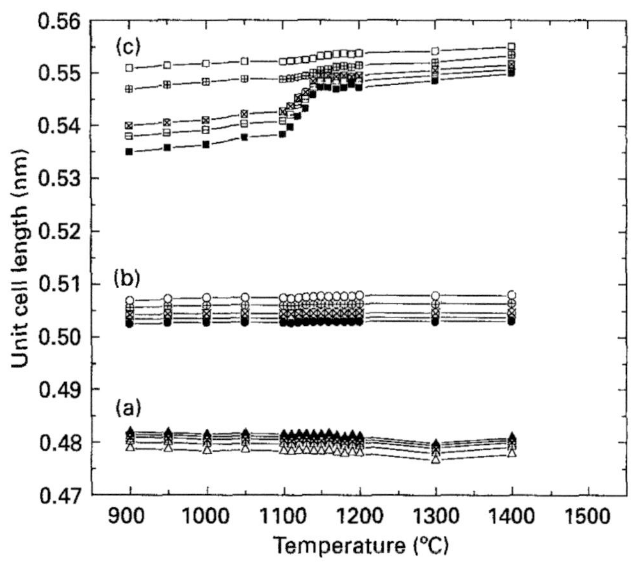

Figure 15. Variation of the lattice parameters of $Zr_{1-x}Sn_xTiO_4$ materials as a function of temperature: (■) - $ZrTiO_4$; (⊟) $Zr_{0,95}Sn_{0,05}TiO_4$; (⊠) $Zr_{0,9}Sn_{0,1}TiO_4$; (⊞) $Zr_{0,8}Sn_{0,2}TiO_4$; (□) $Zr_{0,7}Sn_{0,3}TiO_4$ [69].

The phase transition is greatly affect by the cooling rate of $Zr_{1-x}Sn_xTiO_4$ samples (Fig. 16.). At high cooling rates, the high-temperature disordered phase is frozen in the sample. As the cooling rate is decreased, a noticeable change in the behavior of the parameter *c* is observed, which is connected with increase in the degree of cation ordering [69].

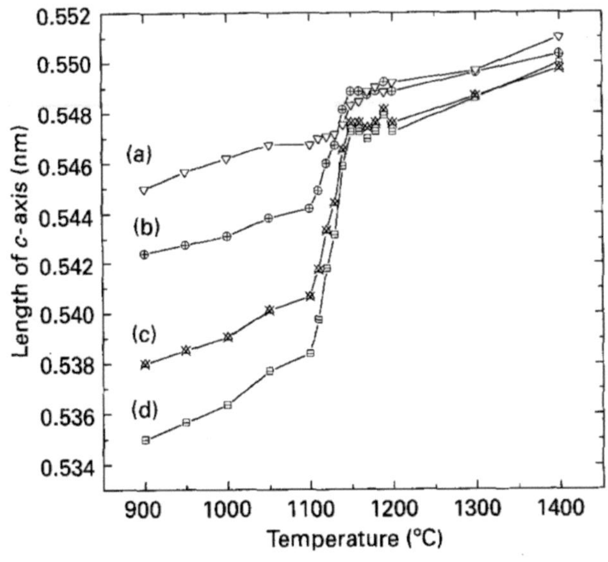

Figure 16. Variation of the parameter *c* of $Zr_{1-x}Sn_xTiO_4$ ceramic as a function of temperature at different cooling rates: (a) 100°C/h, (b) 15°C/h, (c) 5°C/h, (d) 1°C/h [69].

In the system $Zr_xSn_yTi_zO_4$, where x+y+z = 2, single-phase materials are formed in a limited region [62]. The partial substitution of Sn ions for Zr ions stabilized the high-temperature phase with disordered distribution of cations [72] and extends the temperature range of phase transition in $ZrTiO_4$ [73, 74].

Many authors studied the dielectric properties of $ZrTiO_4$ in the MW range and showed it to the have the following parameters: ε = 42, $Q \times f$ = 28 000 GHz, τ_f = 58 ppm/°C [62, 63, 75, 76]. The partial substitution of Sn ions for Zr ions affect greatly the dielectric properties of $Zr_{1-x}Sn_xTiO_4$ materials.

The substitution of Sn ions for Zr ions results in the formation of $Zr_{0.8}Sn_{0.2}TiO_4$, which has good dielectric properties in the MW region: ε = 38, $Q \times f$ = 49 000 GHz, τ_f = 0 ppm/°C [77]. This made is possible to use widely this composition in engineering. The use of anatase as a starting reagent instead of rutile made it possible to increase the value of $Q \times f$ [78]. It should be noted that cation ordering in the $ZrTiO_4$ structure leads to an increase $Q \times f$. It was shown that addition of tin leads to a decrease in cation ordering [62, 72, 79, , 81], but in spite of this, the value of $Q \times f$ increases greatly. Cation ordering is also observed in tin-containing samples, but the ordering domain size decrease with increasing tin content [82]. These data indicate that the increase in $Q \times f$ on the partial substitution of tin ions for zirconium ions cannot be attributed to cation ordering. The authors of [80] suggested that segregation of tin ions takes place at domain boundaries, which decrease their contribution to dielectric loss in ceramic.

$Zr_{0.8}Sn_{0.2}TiO_4$ – based dielectric materials are widely used in the manufacture of various MW equipment elements [63, 83, , 85]. How ever, the manufacture of high-density $Zr_{0.8}Sn_{0.2}TiO_4$ ceramic is a big technological problem even at high temperatures (above 1600^{0}C). Therefore, many authors studied the effect of small additions on the sintering temperature and density of $Zr_{0.8}Sn_{0.2}TiO_4$ [85,88]. The dopants ZnO, CuO, Y_2O_3 are generally used. Addition of small amounts, e.g. of ZnO, results in the formation of a liquid phase at grain boundaries, which increases greatly the ceramic density thanks to fast mass transport through the liquid phase and considerable decrease in sintering temperature.

10. $Ba(M^{2+}{}_{1/3}\ M^{5+}{}_{2/3})O_3$ – Based MW dielectrics (M^{2+} = Mg, Zn, Co, Ni; M^{5+} = Ta, Nb) with extremely high Q

$Ba(B^{2+}{}_{1/3}\ B^{5+}{}_{2/3})\ O_3$ compounds where B^{2+} = Mg, Zn, Co, Ni; B^{5+} = Ta, Nb (perovskite crystal structure) had been synthesized for the first time by the authors of [89,, , 92]. In these compounds, a 2:1- type ion ordering in the B sublattice is observed, in which two layers filled with B^{5+} ions alternate with a layer filled with B^{2+} ions. The authors [93, 94] showed that tantalum-containing materials possess a high Q value in the MW range. It should be noted that the synthesis of these materials involves many problems. Ceramics sinter at a high temperature, which may result in considerable evaporation of constituents (cobalt, zinc) and hence in the impairment of electrical properties. In the case of synthesis by the solid-state reaction method, extra phases $Ba_5Ta_4O_{15}$, $Ba_4Ta_2O_9$ are often present in ceramics [95], which affect adversely the Q value. To prevent this, the authors of [96] carried out synthesis from solutions, where solutions containing Mg^{2+}, Ta^{5+} were used as starting substances, to which a solution of ammonia with oxyquinoline was added.

It had been found that in this case, the single-phase product $Ba(Mg_{1/3}Ta_{2/3})O_3$ is formed above 1300 ^{0}C without intermediate phases. Single-phase $Ba(Mg_{1/3}Ta_{2/3})O_3$ had been obtained by the solid-state reaction method too, using highly active reagents as starting substances [95-97].

When synthesizing tantalum-containing materials, the preparation of high-density ceramics was a difficult problem. Therefore, Nomura with coauthors [98] proposed to prepare dense ceramics ($\varrho \approx 7520\ kg/m^3$) by using additionally manganese impurities. Matsumoto and Hinga [99] used fast heating (330 ^{0}C/min) for the same purposes, which made it possible to achieve 96% of the theoretical density. To increase the rate of sintering and to order ions in the B sublattice, the authors of [100, 101] proposed preliminarily synthesized MTa_2O_6(M = Mg, Zn) as starting reagents. Renoult with coworkers [102] had synthesized fine-grained $Ba(MgTa)O_3$ by the sol-gel method. In that case, dense ceramics could be obtained without additives.

The Q value in $Ba(B^{2+}{}_{1/3}\ B^{5+}{}_{2/3})O_3$ perovskites is greatly affected by the type and degree of ion ordering in the B sublattice [103]. It had been found that by the partial substitution of Zr^{4+},

Ti^{4+}, W^{6+} ions for ions in the B sublattice, one can increase the degree of 1:2 - type ion ordering and hence increase the Q value [104, 105].

Ion ordering in the B sublattice is also affected by slight substitutions of ions in the A sublattice. For instance, Ref [104] showed that partial substitution of La^{2+} ions for Ba^{2+} ions in magnesium-barium tantalate results in the change of ion ordering in the B sublattice from the 2:1 type (space group Pm3I) to the 1:1 type (space group Fm3m). Similar changes of the type of ion ordering in the B sublattice were found in the case of partial substitution of lanthanum ions for barium ions in zinc-barium niobate with perovskite structure [105]. This result shows that the type of ion ordering in the B sublattice is very sensitive to chemical composition and preparation technique. The 1:1-type ion ordering in the B sublattice in $Ba(B^{2+}_{1/3} B^{5+}_{2/3})O_3$ compounds was explained in terms of a "space-charge" model [106, 107], according to which only B^{2+} and B^{5+} ions can occupy the sites in the ordered $Ba(\beta'_{1/2} \beta''_{1/2})O_3$ structure (1:1 type). Since in this case the electroneutrality condition is not satisfied in the crystalline phase, it may be assumed that the domains of the ordered crystalline phase (ordering type 1:1), which has an uncompensated charge, are in a disordered matrix rich in B^{5+} ions, as a result of which the electroneutrality condition is satisfied throughout the sample volume.

However, when investigating the system of solid solutions (1-x)Ba $(Zn_{1/3}Nb_{2/3})O_3$-$xLa((Zn_{1/3}Nb_{2/3})O_3$ ($0 \leq x \leq 0.6$), it was shown that 1:2 –type ordering persists in the interval $0 \leq x \leq 0.5$ [105], whereas in the interval $0.05 \leq x \leq 0.6$, 1:1 ion ordering in the B sublattice is observed. In this case, there is no aggregation. The investigation of the microstructure did not reveal the existence of disordered perovskite phase region, which was assumed in the "space-charge" model. Therefore, to describe the 1:1 ion ordering in the B sublattice, a "random-site" model was proposed [105, 108, , 110]. According to this model, 1:1 ordering in the above systems is described as follows. There are two alternating crystal planes, in which the B sublattice ions reside. One of them is occupied with B^{5+} cations and the other with B^{2+} cations and the remaining B^{5+} cations, which are disordered in this crystal plane.

Thus, in $Ba(M^{2+}_{1/3} M^{5+}_{2/3})O_3$ compounds with perovskite structure, the B sublattice ions may be fully disordered as well as have 1:2 or 1:1 ordering. A calculation of the lattice energy of ordered and disordered structures showed [111] that ordered structure is characterized by lower Madelung energy, indicating this structure to be stable. Investigations of the electrophysical properties of $Ba(B^{2+}_{1/3} B^{5+}_{2/3})O_3$ compounds showed that ion ordering in the B sublattice affects greatly the Q value [112].

It should be noted that $Ba(B^{2+}_{1/3} B^{5+}_{2/3})O_3$ compounds with perovskite structure and 1:1 cation ordering in the B sublattice have, as a rule, a relatively low Q [113]. The largest Q value is observed in the $Ba(Mg_{1/3}Ta_{2/3})O_3$ and $Ba(Zn_{1/3}Ta_{2/3})O_3$ compounds. In these compounds, the B cations are stoichiometrically ordered in the hexagonal unit cell Pm3I (2:1ordering), in which the layers of Ta^{5+} and $Zn(Mg)^{2+}$ cations are sequentially arranged along the (111) crystal plane (Fig 17). The layers of cations are separated by oxygen layers, which are displaced in the direction of small pentavalent tantalum cations [112]. The Q value is very sensitive to ordering in the B sublattice [93, 114]. It can be greatly increased by using additional annealing. For instance, the Q × f value increased from 60000 to 168000 after additional annealing at 1350 ^{0}C for 120 h [93].

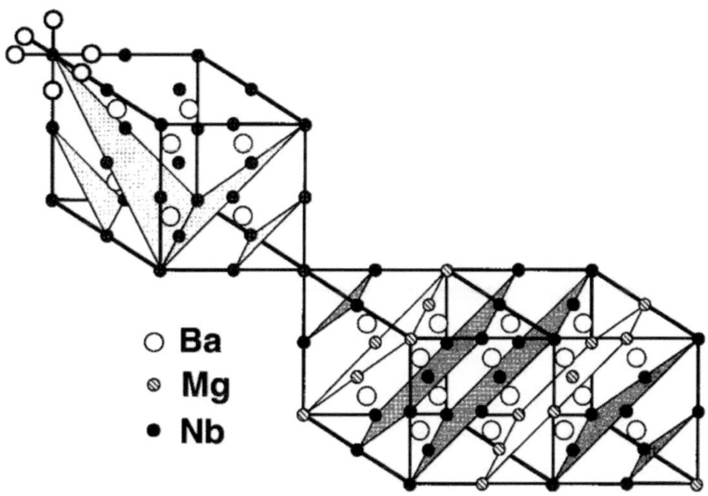

Figure 17. Schematic representation of 1:2 cation ordering in $Ba(Zn_{1/3}Ta_{1/3})O_3$. On the top left are shown two possible (111) directions for Zn and Ta orientation in the perovskite structure; on the right below is shown one of the possible variants of 1: 2 ordering. The oxygen ions were omitted for clarity [108].

Tamura et al [115] showed that Q in $Ba(Zn_{1/3}Ta_{2/3})O_3$ can be improved by adding $BaZrO_3$(BZ) of low concentration (< 4 mol %). In this case, the ceramic sintering time is greatly reduced, which is required for the attainment of high Q values (Q × f = 105000 had been attained by adding 4 mol % BZ).

This was accounted for by the formation of defects in the B sublattice, the presence of which increased the rate of cation ordering in this sublattice. As the BZ concentration was increased, the type of ordering changed (when substitution reached 4 mol %), the system came to have 1:1 ordering and a double cell. Similar regularities were observed when small amounts of $BaWO_4$ [116] and $BaSnO_3$ [117] were added to $Ba(Mg_{1/3}Ta_{2/3})O_3$. The individual compounds $Ba(Mg_{1/3}Ta_{2/3})O_3$ and $Ba(Zn_{1/3}Ta_{2/3})O_3$ allow one to achieve high Q values, but their electrophysical properties have a low thermostability. Therefore, to increase the thermostability of electrical properties, materials are synthesized on the basis of solid solutions, where the end members have temperature dependences of permittivity of different sign. The materials based on solid solutions possess a higher thermostability, but the Q value is lower as compared with individual compounds, which may be attributed, in particular, to decrease in cation ordering in the B sublattice. On the basis of a solid solution, e.g. $Ba(Zn_{1/3}Ta_{2/3})O_3$, materials have been obtained which have a high level of electrophysical properties: ε = 30-40, $TC\varepsilon = (0\text{-}28) \times 10^{-6}\ {}^0C^{-1}$, $Q_{8GHz} = 15000$.

The materials based on tantalum-containing perovskites possess today the highest Q values in the MW range among dielectrics with increased ε value. However, the difficulty of their synthesis: high sintering temperatures (T_{sint} > 1600 ^{0}C), the necessity of long additional annealing ($T_{ann} \approx 1500$ ^{0}C, 20-120 h), low reproducibility of properties, as well as the high price of tantalum (in 2000, the price of reagents containing tantalum increased by 500%) calls for search for new promising systems. Therefore, in recent years, attention has been given just to niobium-containing $Ba(Nb_{2/3}B^{2+}_{1/3})O_3$ compounds (where B^{2+} = Mg, Zn, Co), which crystallize in perovskite structure [118, 119, 120].

These materials sinter at lower temperature, and cheaper reagents are needed for them. However, their main disadvantage is a lower *Q* value as compared with tantalum-containing perovskites [121, 122]. We have shown for the first time that $Ba(Mg_{1/3}Nb_{2/3})O_3$ – based materials can have in some cases an extremely high *Q* ($Q \times f \approx 150000$) [123]. New papers appeared [5, 124, 125], which show the creation of high-Q MW dielectrics based on niobium-containing perovskites to be worth-while.

$Ba(Co_{1/3}Nb_{2/3})O_3$ (BCN) offers a particularly attractive combination of properties. Polycrystalline $Ba(Co_{1/3}Nb_{2/3})O_3$ has high dielectric permittivity (ε =32) and *Qf* = 40000–60000 GHz [126, 127]. At the same time, centimeter and millimeter wave applications require higher *Q* values. The properties of BCN are very weak functions of temperature. In particular, the temperature coefficient of its resonant frequency (τf) lies in the range –10 to –7 ppm/K, suggesting that it can be used in the production of high-Q/low-τf microwave materials [128, , 130]. The electrical properties of BCN, especially its *Q* (or its *Qf* product, where *f* is frequency), strongly depend on preparation conditions, in particular on the sintering temperature, heat-treatment time, and heating/cooling rate.

It is reasonable to assume that the observed variations in the properties of BCN are related to the ceramic microstructure evolution during the fabrication process, cation ordering, and lattice distortions. Such distortions can be produced in BCN via slight changes in its cation composition.

Therefore, we have studied the effect of partial nonstoichiometry in cation sublattices on the phase composition, microstructure and electrophysical properties of BCN.

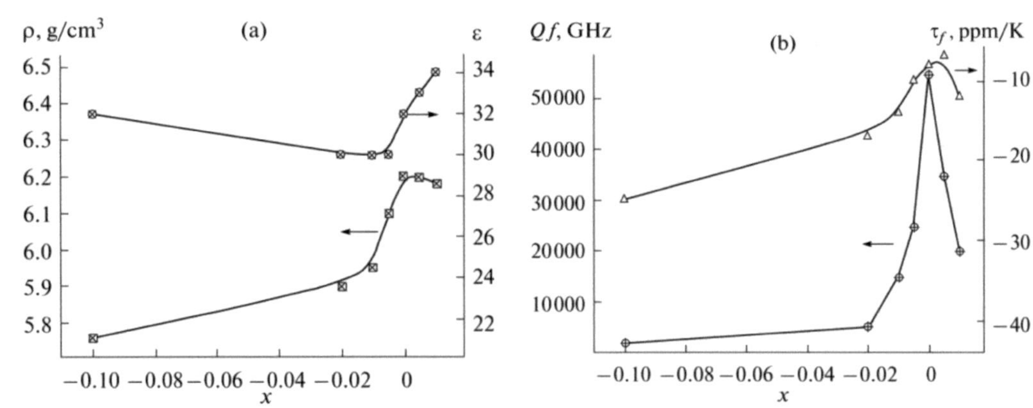

Figure 18. Composition dependences of the (a) apparent density (ϱ), permittivity (ε), (b) Qf product, and temperature coefficient of resonant frequency (τ_f) for $Ba_{3+3x}CoNb_2O_{9+3x}$ materials; measurements at10 GHz.

When studying nonstoichiometry in barium sublattice ($Ba_{1+x}Co_{1/3}Nb_{2/3}O_{3+x}$ at $-0.03 < x < 0.03$), an extra phase $Ba_9CoNb_{14}O_{45}$ appears, which makes a noticeable decreases greatly (Fig. 18(a)). Therefore, the highest Q values are observed for stoichiometric composition ($BaCo_{1/3}Nb_{2/3}O_3$) (Fig. 18(b)).

Interesting properties are observed in the case of nonstoichiometry in the cobalt sublattice ($BaCo_{1/3+y}Nb_{2/3}O_{3+y}$, where $-0.05 < y < 0.01$).

Increasing the cobalt content of this system with reference to stoichiometry leads to a monotonic increase in relative density and corresponding slight increase in permittivity ε in the range 32 to 34 (Fig. 19a). At higher cobalt deficiencies ($y < -0.02$), the samples contained an additional phase $Ba_8CoNb_6O_{24}$. In the range $-0.03 \le y \le 0.01$, the quality factor varies nonlinearly, with a maximum at $-0.03 \le y \le -0.02$. The Qf product exceeds that of stoichiometric BCN by 30–50%, reaching 80000–85000 GHz (Fig. 19b).

Electron diffraction data for the $BaCo_{1/3+y}Nb_{2/3}O_{3+y}$ samples indicates that cobalt deficiencies in the range $-0.03 \le y \le -0.02$ are favourable for 1 : 2 B-site cation ordering in the cobalt-containing perovskite (Fig. 20). As mentioned above, cation ordering is accompanied by an increase in quality factor, as observed in this system (Fig. 18b). The reduction in quality factor at large deviations from stoichiometry ($y < -0.1$) is due to the presence of a significant amount of $Ba_8CoNb_6O_{24}$.

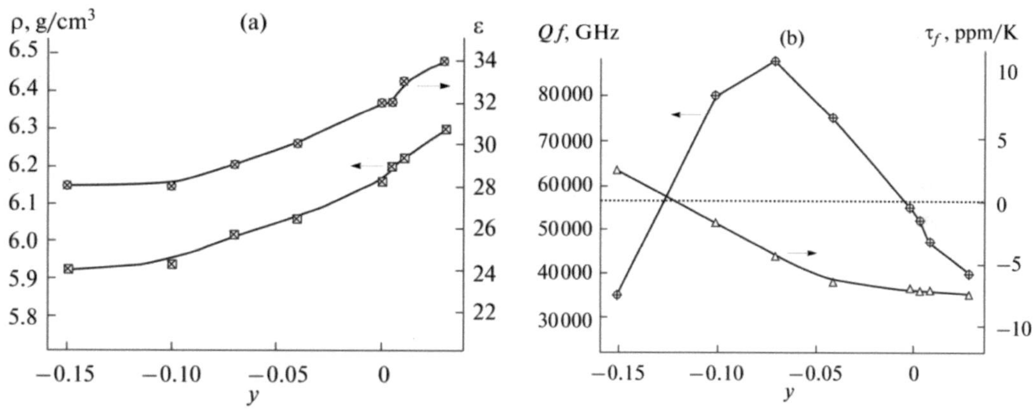

Figure 19. Composition dependences of the (a) apparent density (ϱ), permittivity (ε), (b) Qf product, and temperature coefficient of resonant frequency (τ_f) for $Ba_3Co_{1+y}Nb_2O_{9+y}$ materials; measurements at 10 GHz.

The increase in $Ba_8CoNb_6O_{24}$ content with increasing cobalt deficiency ($y < 0$) leads to a result of practical interest: the temperature coefficient of resonant frequency switches sign (Fig. 18 b). The reason for this is that $BaCo_{1/3}Nb_{2/3}O_3$ and $Ba_8CoNb_6O_{24}$ differ in the sign of τ_f: –7 and +16 ppm/K, respectively [129, 131]. Because of this, the $BaCo_{1/3+y}Nb_{2/3}O_{3+y}$ materials exhibit a temperature compensation effect, whose magnitude can be tuned by varying the cobalt content. The present results, therefore, suggest that cobalt-deficient $Ba(Co_{1/3}Nb_{2/3})O_3$ is an attractive host for engineering advanced temperature-stable microwave dielectric materials with Qf on the order of 80000–90000 GHz and τ_f = –2 to +3 ppm/K.

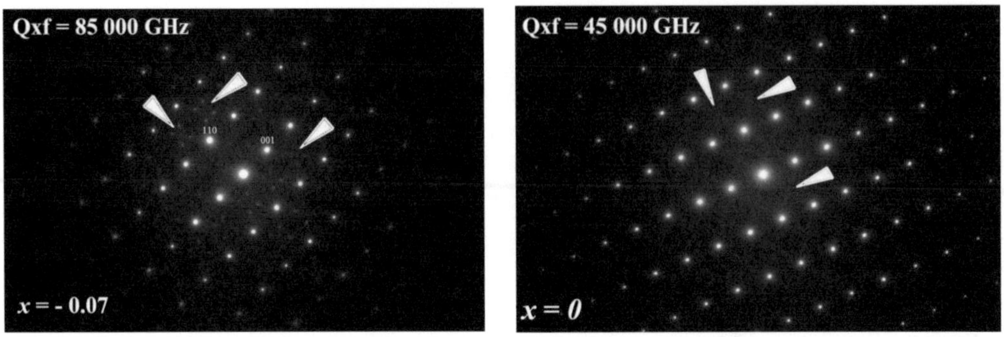

Figure 20. [110] electron diffraction patterns of the $Ba_3Co_{1+y}Nb_2O_{9+y}$ samples with y = (a) –0.07 and (b) 0. The arrows mark superlattice reflections.

11. Multiphase MW dielectrics

High Q value is usually observed in single-phase systems. In the case of complex cation sublattices, it is necessary that the ions should be ordered by a definite type [105, 106]. In multiphase systems, which are chemically inhomogeneous, considerable dielectric loss (relatively low Q) is generally observed. When investigating barium polytitanates, however, we showed that multiphase systems having a high Q and thermostability of electrophysical properties can be formed. When zinc oxide is added to barium polytitanates, an extra $BaZn_2Ti_4O_{11}$ phase is formed [132] which does not interact chemically with the main phase. A multiphase system is formed, in which the main and extra phases have the dependence ε(T) of different sign, which ensures the realization of the volume temperature compensation effect and hence a high thermostability of electrophysical properties (TCε = ±2 × $10^{-6}K^{-1}$) in the MW range. The multiphase dielectrics obtained have a high Q (Q_{10GHz} ~ 6500-7000).

One more example of multiphase MW dielectrics is TiO_2 materials, viz the compounds $MgTiO_3$ and Mg_2TiO_4, which have a high Q (Q_{10GHz} ~ 5000-10000) and permittivity (14 and 16 respectively) [133]. A demerit of these materials is the temperature instability of electrophysical parameters (TCε = (40-50) × $10^{-6}K^{-1}$). To increase the Q value, cobalt ions were partially substituted for magnesium ions, and to increase the temperature stability of electrophysical properties, small amounts of the paraelectric phase $CaTiO_3$, which has a high negative value of TCε, were added. Investigations showed that in this case multiphase systems of chemically noninteracting phases are formed (Fig 21) [134, , 136].

This made it possible to obtain MW dielectrics with a permittivity of 18-20, high Q values (Q × f ≥ 5000-10000) and thermostable electrical properties. It may be supposed that high Q values are due to the fact that the size of chemical inhomogeneity is much smaller than the electromagnetic wavelength in dielectric and does not cause, therefore, noticeable electromagnetic scattering.

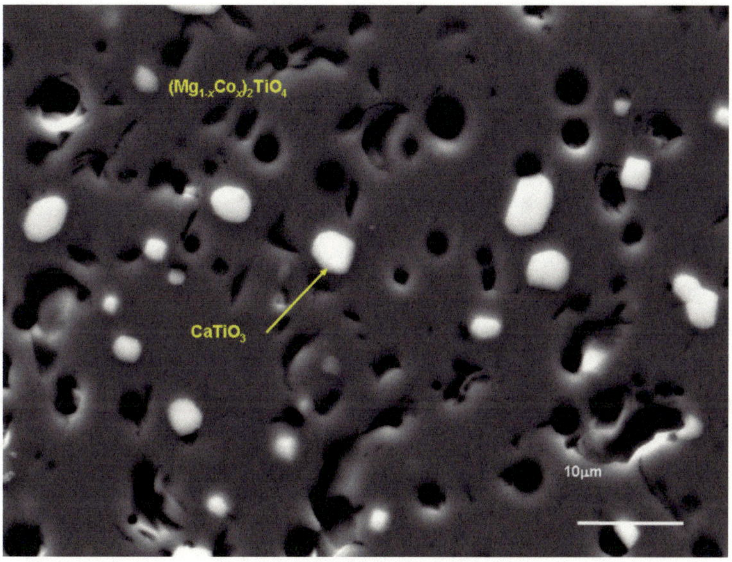

Figure 21. Micrograph of the microsection of 0.93 [0.98Mg_2TiO_4 – 0.02 Co_2 TiO_4] -0.07$CaTiO_3$ ceramic

12. Analysis of the physical properties of MW dielectrics

The main attention in the analysis of the physical properties of MW dielectrics is given to the temperature coefficient of permittivity (TCε) and to dielectric loss (tg δ). The expression for TCε can be derived directly from the Clausius-Mossotti equation [137]:

$$TC\varepsilon = \frac{(\varepsilon-1)(\varepsilon+2)}{\varepsilon}\left[\frac{v}{\alpha}\left(\frac{\partial\alpha}{\partial v}\right)_T a_l + \frac{1}{3\alpha}\left(\frac{\partial\alpha}{\partial T}\right)_v - a_l\right] \qquad (4)$$

where a_l is linear thermal expansion coefficient.

It should be noted that polarization a in Eq (4) is equal to the sum of polarizations of all atoms of cell, whose volume is v, only if all atoms of the structure have a cubic environment. This is the case, e.g. for alkali halide crystals. In more complex structures, effective polarization a_{eff} is used [138]. For example, for perovskite structure, effective polarization a_{eff} is obtained by introducing an ionic component of polarization, Δ a^i , in addition to the electronic and ionic polarization of all atoms in the unit cell. In this case, the last term a_l in expression (4) must be written as a_l Δ a^i [137]. An analysis of the TCε value of different materials as a function of chemical composition showed [139] that the large positive value of TCε in alkali halide crystals probably arises from high ionic polarizability a^i and great thermal expansion a_l. There are compounds, e.g. $LaAlO_3$, $SrZrO_3$, the behavior of whose ε(T) differs from that of paraelectrics. This is attributed to the presence of nonferroelectric phase transitions [137], which are coupled with the rotation of oxygen octahedra. Similar nonferroelectric phase transitions were found, e.g. in the system $(Ba_xSr_{1-x})(Zn_{1/3} - Nb_{2/3})O_3$ [140]. It was shown that structural transitions are accompanied by the reversal of the TCε sign, though the ε quantity does not undergo noticeable changes, which are typical of spontaneously polarized state. Similar dependences were also found in the system $(Ba_xSr_{1-x})(Mg_{1/3}-Nb_{2/3})O_3$ [141]. When x is changed, a correlation between the value of tolerance factor (t) and the inclination of oxygen octahedra is observed, which results in the reversal of the TCε sign.

When MW dielectric loss is studied, three sources of loss are considered: 1) loss in perfect crystal, which is coupled with anharmonicity, which is due to interaction between crystal phonons, resulting in optical-phonon attenuation; such loss is usually called intrinsic loss; 2) loss in real homogeneous material, which is caused by deviation from lattice or defect periodicity (point defects, dopant atoms, vacancies or twin defects, which give rise to quasi-bound states); such defects give rise to phonon scattering; 3) loss in real inhomogeneous crystalline materials, which is caused by the existence of dislocations, grain boundaries, including minor phases; this loss is usually called extrinsic loss.

ε Dispersion for infrared polarization is usually described by the Drude-Lorentz equation:

$$\varepsilon(\omega) = \varepsilon_\infty + \frac{\varepsilon_0 - \varepsilon_\infty}{1 - \left(\frac{\omega}{\omega_T}\right)^2 + i\Gamma\frac{\omega}{\omega_T}} \tag{5}$$

where ε_∞ is permittivity at optical frequencies, ω_T is transverse optical mode frequency, $(\varepsilon_0 - \varepsilon_\infty)$ is dielectric oscillator strength, Γ is relative attenuation.

Using Eq (5), we can estimate dielectric loss (tg δ) pertaining to phonon attenuation in the MW range (intrinsic loss) ($\omega \gg \omega_T$):

$$tg\delta \approx \Gamma \frac{\omega}{\omega_T^4} \frac{\varepsilon_0 - \varepsilon_\infty}{\varepsilon_0} \tag{6}$$

Even in perfect crystal, loss may arise from the anharmonicity of vibration. In this case, three-phonon and four-phonon interactions may predominate, in which $\Gamma \sim T$ and $\Gamma \sim T^2$ respectively [142]. Equation (6) shows that if intrinsic loss predominates, the product of Q (Q = 1/tg δ) and frequency f(ω = 2πf) is a constant, which is employed for the analysis of MW dielectrics.

To determine intrinsic loss, IR spectroscopy is usually used [143, 144] since the value of intrinsic loss at IR frequencies is much larger than that of extrinsic loss. Having determined intrinsic loss at IR frequencies, one can approximate the quantity $Q \times f$ = const to MW frequencies and calculate thereby extrinsic loss, which is due to ceramic imperfection; this loss can be reduced by improving the technology.

One of the important questions concerning the determination of dielectric loss in MW ceramics is the possibility of determining intrinsic loss solely from IR spectroscopic data. To this end, the authors of [145, 146] investigated several $Ba(B'_{1/2}B''_{1/2})O_3$ compounds in a wide frequency (10^2 – 10^{14} Hz) and temperature (20-600 K) range. On the basis of the data obtained, they came to the following conclusions:

- MW loss calculated from IR spectra and Eq (6) can be interpreted as a lower limit of intrinsic loss.
- MW loss extrapolated from IR data is systematically lower than that determined in accordance with the microscopic theory [147].
- The value of intrinsic loss correlates with tolerance factor (t); increase in intrinsic loss with decreasing t indicates that the main contribution to it is made by low-frequency vibration mode.
- The value of intrinsic loss varies with the ε value as $tg\,\delta \sim \varepsilon^a$, where a = 4.

It is not less important to elucidate the effect of cation ordering on MW loss. It was shown earlier [58] that cation ordering in $Ba(Zn_{1/3}Ta_{2/3})O_3$ and $Ba(Mg_{1/3}Ta_{2/3})O_3$ allows MW loss to be reduced. When investigating $(Zr_{1-x}Sn_x)$-TiO_4 materials, however, it was found that the substitution of Sn^{4+} ions for Zr^{4+} ions results in the suppression of cation ordering [148], and that the Q value increases in this case [140]. It is likely that in the system $(Zr_{1-x}Sn_x)TiO_4$, the increase in Q on the substitution of Sn^{4+} ions with smaller radius for Zr^{4+} ions with larger radius is due to a decrease in intrinsic loss.

13. Applications of MW dielectrics

MW dielectrics are used in modern communication systems, for the manufacture of dielectric resonators (DR) of various types, substrates for MW hybrid integrated circuits. On the basis of DRs, radio-frequency filters are developed; they are also used in the manufacture of solid state oscillators. In the frequency range 150 MHz – 3 GHz, coaxial resonators are often used, whose surface is metallized (Fig 22). The height of quarter-wave coaxial resonator is determined from the formula:

$$l = \frac{\lambda_0}{4}\frac{1}{\sqrt{\varepsilon}} \tag{7}$$

where λ_0 is free-space electromagnetic wavelength, ε is permittivity in the operating frequency range.

At higher frequencies, open resonators are generally used (Fig 23), whose diameter is determined from the formula:

$$D = \frac{\lambda_0}{\sqrt{\varepsilon}} \tag{8}$$

The characteristic modes for coaxial and open dielectric resonators are usually the modes TE_{01s} and H_{01s}.

The Q value of coaxial resonator is determined both by dielectric loss and by loss in the resonator metal coating, which may be high; therefore, the Q value of coaxial resonators is, as a rule, under 1000, which is their demerit. At the same time, Q of open resonators is determined only by dielectric loss (intrinsic and extrinsic). Therefore, Q of open resonators is over 1000. Since open resonators have a high Q value, they can be used in the decimeter wave band (about 1 GHz), though in this case their size becomes large.

Figure 22. Coaxial dielectric resonators

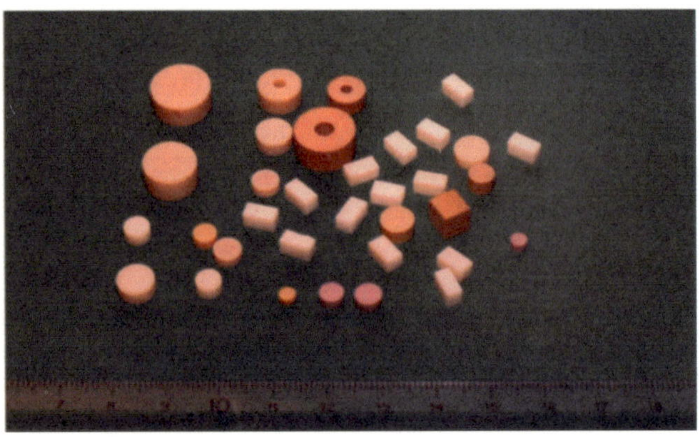

Figure 23. Open dielectric resonators

The size of open dielectric resonators operating on the characteristic modes TE_{01s} and H_{01s} becomes very small at frequencies above 30 GHz, which makes their use in this band impossible. Therefore, at frequencies above 30 GHz, it is expedient to use extraordinary vibration modes such as whispering gallery modes.

The modern communication systems in which dielectric resonators are used operate in a temperature range of – 40-80 ^{0}C. Therefore, high thermostability of dielectric resonator resonance frequency is required. It is necessary that the temperature coefficient of frequency should tend to zero; it is defined as:

$$TK\varepsilon = \frac{1}{f_p}\frac{\Delta f_p}{\Delta T} \tag{9}$$

where f_r is the resonance frequency of dielectric resonator, Δf_r is change in the resonance frequency of dielectric resonator in the temperature range ΔT.

Dielectric resonators are used for the frequency stabilization of oscillators, which are used, in turn, in radars, various communication systems. At frequencies below 200 MHz, quartz resonators are often used; in the frequency range 1000-3000 MHz, coaxial resonators and above 3000 MHz open resonators are used.

On the basis of dielectric resonators, miniature bandpass filters, frequency separators are developed (Fig 24) [149].

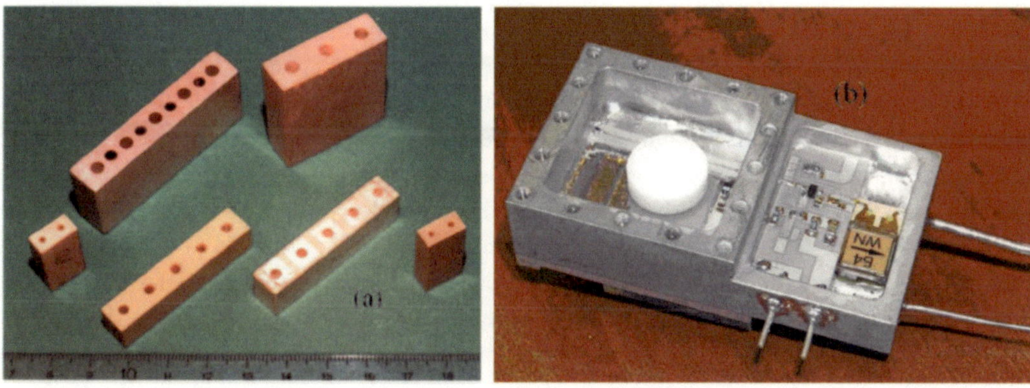

Figure 24. Monolithic ceramic blocks for the utilization in radiofilters operating in the decimetre wavelength band (a), and low-noise microwave oscillator for the frequency of around 9 GHz (b)..

Dielectric resonators are also used in the creation of antennas of the new generation. The advantages of such antennas are: small size, simplicity, relatively broad emission band, simple scheme of coupling with all commonly used transmission lines; possibility to obtain different radiative characteristics using different resonator modes.

As was mentioned above, the use of dielectric resonators operating on the TE_{01s} and H_{01s} modes is limited in the millimeter wave band since the size of resonators becomes too small. Therefore, it is expedient to use in the millimeter wave band dielectric resonators operating on whispering gallery modes [150]. Besides, it is relatively easy to suppress spurious modes in such resonators. It should be noted that the Q value in the resonators using whispering gallery modes is limited only by intrinsic loss in the material in contrast to coaxial and open resonators.

14. Conclusion

High-Q MW dielectrics with high thermostability of electrophysical properties can be developed on the basis of single-phase and multiphase systems. Single-phase MW dielectrics are produced on the basis of solid solutions [6, 15, 98] using heterovalent substitutions in one of the crystal sublattices and influencing thereby the phonon spectrum [10] and by making one of the sublattices "mobile" [151]. At the same time, high-Q thermostable MW dielectrics based on multiphase systems are developed using the volume temperature compensation effect [55-57, 124, 132, 135].

During the last decade, MW dielectrics with increased permittivity ($\varepsilon \geq 10$) contribute, to a larger measure than other factors, to considerable miniaturization and reduction in the price of modern communication systems. It should be noted that there is still a great potential for further microminiaturization and reduction of prices of modern communication systems thanks to the use of components made on the basis of MW dielectrics.

Depending on the frequency range of modern communication systems, MW dielectrics with different properties are needed. In the decimeter wave band, high permittivity values ($\varepsilon \geq 100$) are required along with the high thermostability of electrophysical properties and high Q, which enables effective solution of microminiaturization problems. At the present time, solid solutions based on barium-lanthanide titanates ($Ba_{6-x}Ln_{8+2x/3}Ti_{18}O_{54}$ (Ln = La-Gd)), which have a potassium-tungsten bronze structure and $\varepsilon \approx 80$-100, meet best these requirements. However, the nature of the thermostability of the electrophysical properties of these solid solutions has not been elucidated definitively; there are only qualitative explanations, which greatly restrains the search for new promising MW dielectrics with high permittivity ($\varepsilon \geq 100$). The presence of spontaneous polarization in dielectrics causes, along with increase in ε, a considerable increase in dielectric loss (the Q value decreases), which impairs greatly the technical characteristics of communication system elements based on them. Therefore, the acquirement of fundamental knowledge, which is required for obtaining thermostable high-Q materials with $\varepsilon \geq 150$-200, is the most important problem in developing modern decimeter wave band communication systems.

In the centimeter and millimeter wave bands, where the electromagnetic wavelength is much smaller as compared with the decimeter wave band, thermostable MW dielectrics with extremely high Q values are required. In this case, permittivity values may be relatively low ($\varepsilon = 15$-30). To date, tantalum-containing perovskites possess the highest Q values. However, the difficulty of their preparation and the high price call for search for new promising compounds, and it is going on in several directions. In particular, research is now under way to develop niobium-containing perovskites and to create multiphase systems, in which volume temperature compensation effect is realized. It is these directions that will probably be major directions in the next few years in developing high-Q centimeter and millimeter wave band MW dielectrics, though the search for new promising compounds will always be vital. Quite a number of problems pertaining to solid-state physics and chemistry will have to be solved. For instance, it is necessary to investigate the nature of extrinsic loss, which is coupled with various structural defects, as well as with the presence of grain boundaries, ordering of crystal sublattices and domain nanostructure. Research aimed at developing thermostable dielectrics, which will be used as millimeter wave band dielectric resonators using whispering gallery modes, will be of special scientific and practical interest. This requires considerable increase of the chemical and structural homogeneity of dielectrics.

An important problem is the creation of retunable resonant elements. To this end, multilayer bulk and film materials, which will contain a thermostable dielectric phase and a nonlinear magnetic or electrical phase at the same time, will probably have to be developed.

Thus, the synthesis of novel high-Q MW dielectrics and the investigation of their structure and properties are an important scientific-technical trend in solid-state chemistry.

Author details

A.G. Belous
V.I. Vernadskii Institute of General and Inorganic Chemistry of the Ukrainian NAS, Kyiv, Ukraine

15. References

[1] Poplavko Yu.M., (1980) Physics of Dielectrics (in Russian), Vyshch. Shk., Kyiv 325.
[2] Poplavko Yu.M., Belous A.G., (1984) Physical background of the temperature stability of microwave dielectrics. Dielektriki I Poluprovodniki. 25: 3-15.
[3] Chool-Woo Ahu, Hyun-Jung Jang, Sahn Nahm et al. (2003) Effects of microstructure on the microwave dielectric properties of $Ba(Co_{1/3}Nb_{2/3})O_3$ and $(1-x)Ba(Co_{1/3}Nb_{2/3})O_3$–$xBa(Zn_{1/3}Nb_{2/3})O_3$ ceramics J. Eur. Ceram. Soc. 23: 2473-2474.
[4] J-Nan Lin, Chih-Ta Chia, Hsiang-Lin Liu et al. (2002) Dielectric Properties of $xBa(Mg_{1/3}Ta_{2/3})O_3$-$(1-x)Ba(Mg_{1/3}Nb_{2/3})O_3$ Complex Perovskite Ceramics. Jpn. J. Appl. Phys. 41: 6952-6956.
[5] Cheol-Woo Ahn, Sahn Hahm, Seok-Jin Yoou et al. (2003) Microstructure and Microwave Dielectric Properties of $(1-x)Ba(Co_{1/3}Nb_{2/3})O_3$–$xBa(Zn_{1/3}Nb_{2/3})O_3$ Ceramics Jpn. J. Appl. Phys. 42: 6964-6968.
[6] Nenesheva E. A., Mudroliuba L. P., and Kartenko N. F. (2003) Microwave dielectric properties of ceramics based on $CaTiO_3$–$LnMO_3$ system (Ln=La, Nd; M=Al, Ga). J. Eur. Ceram. Soc. 23: 2443–2448.
[7] Moon J. H., Jung H. M., Park H. S., Shin J. Y., and Kim H. S. (1999) Sintering behaviour and microwave dielectric properties of $(Ca,La)(Ti,Al)O_3$ ceramics. Jpn. J. Appl. Phys. 38: 6821–6827.
[8] Belous A.G., Butko V.I., Novitskaya G.N. et al. (1985) Dielectric spectra of the perovskites $La_{2/3-x}M_{3x}TiO_3$ Fizika Tv. Tela. 27: 2013-2016.
[9] Belous A.G., Butko V.I., Polyanetskaya S.V. (1984) Electrical parameters of the solid solutions of rare-earth titanates Ukr. Khim. Zhurn. 50: 1139-1142.
[10] Belous A.G (1998) Physicochemical aspect of the development of new functional materials based on heterosubstituted titanates of rare-earth elements with the perovskite structure. Teoret. I Eksperim. Khimiya. 34: 301-318.
[11] Butko V.I., Belous A.G., Yevtushenko N.P. (1986) Vibration spectra of the perovskites $La_{2/3-x}M_{3x}TiO_3$ Fizika Tv. Tela. 28: 1181-1183.
[12] Knyazev A.S., Poplasvko Ye.M., Zakharov V.P., Alekseev V.V. (1973) Soft mode in the vibration spectrum of $CaTiO_3$ Fiz. Tverd.Tela. 15: 3006-3010.
[13] Ohsato H., Nishigaki S., Okuda T. Superlattice and Dielectric Properties of BaO-R_2O_3-TiO_2 (R=La, Nd and Sm) Microwave Dielectric Compounds //Jpn. J. Appl. Phys. -1992. -31. –P. 3136-3140.

[14] Negas N., Davies P.K. (1995) Influence of chemistry and processing on the electrical properties of Ba6-3xLn8+2xTi18O54 solid solutions Materials and processing for wireless communications. Ceram. Trans. 53: 170-196.

[15] Matveeva R.G., Varfolomeev M.B., Ilyushenko L.S. (1984) Refinement of the composition, and the crystal structure of $Ba_{3.75}Pr_{9.5}Ti_{18}O_{54}$. Zhurn. Neorgan. Khimii. 29: 31-34.

[16] Rawn C.J., Birnie D.P., Bruck M.A. et al. (1998) Structural investigation of $Ba_{6-3x}Ln_{8+2x}Ti_{18}O_{54}$ (x = 0.27, Ln = Sm) by single crystal x-ray diffraction in space group *Pnma*. J. Mater. Res. 13: 187-196.

[17] Ubic R., Reaney I.M., Lee William E. (1999) Space Group Determination of $Ba_{6-3x}Nd_{8+2x}Ti_{18}O_{54}$ J. Amer. Ceram. Soc. 82: 1336-1338.

[18] Belous A., Ovchar O., Valant M., Suvorov D. (2001) Solid-state reaction mechanism for the formation of $Ba_{6-x}Ln_{8+2x/3}Ti_{18}O_{54}$ (Ln = Nd, Sm) solid solutions J. Mater. Res. 16: 2350-2356.

[19] Ohsato H., T. Ohhashi, S. Nishigaki, T. Okuda, K. Sumiya, and S. Suzuki. (1993) Formation of solid solutions of new tungsten bronze type microwave dielectric compounds $Ba_{6-3x}R_{8.2x}Ti_{18}O_{54}$ (R=Nd,Sm 0_x_1). Jpn. J. Appl. Phys. 32: 4323–4326.

[20] Valant M., Suvorov D., Rawn C.J. (1999) Intrinsic Reasons for Variations in Dielectric Properties of $Ba_{6-3x}R_{8+2x}Ti_{18}O_{54}$ (R= La–Gd) solid solutions. Jpn. J. Appl. Phys. 38: 2820-2826.

[21] Butko V.I., Belous A.G., Nenasheva Ye. A. et al. (1984) Microwave dielectric properties of barium lanthanide tetratitanates Fizika Tv. Tela. 26: 2951-2955.

[22] Belous A., Ovchar O., Valant M., Suvorov D. (2000) Anomalies in the temperature dependence of the microwave dielectric properties of $Ba_{6-x}Sm_{8+2x/3}Ti_{18}O_{54}$ Appl. Phys. Lett. 77: 1707-1709.

[23] Belous A., Ovchar O., Valant M., Suvorov D. (2002) Abnormal behavior of the dielectric parameters of $Ba_{6-x}Ln_{8+2x/3}Ti_{18}O_{54}$ (Ln=La–Gd) solid solutions J. Appl. Phys. 92: 3917-3922.

[24] Tang C.C., Roberts M.A., Azough F. et al. (2002) Synchrotron X-ray Diffraction Study of $Ba_{4.5}Nd_9Ti_{18}O_{54}$ Microwave Dielectric Ceramics at 10–295 K. J. Mater. Res. 17: 675-682.

[25] Declaration patent no 58005 A, H 01B 3/12. Published 15.07.2003.

[26] Declaration patent no 58007 A, H 01B 3/12. Published 15.07.2003.

[27] Declaration patent no 58009 A, H 01B 3/12. Published 15.07.2003.

[28] Declaration patent no 58008 A, H 01B 3/12. Published 15.07.2003.

[29] Takahashi H., Baba Y., Ezaki K. et al. (1992) Dielectric Characteristics of $(A_{1/2}^{1+}A_{1/2}^{3+})TiO_3$ ceramics at microwave frequencies. Jpn. J. Appl. Phys. 30: 2339-2342.

[30] Belous A.G., Novitskaya G.N., Polyanetskaya S.V. (1987) Study of the oxides $Ln_{2/3-x}M_{3x}TiO_3$ (Ln - Gd-Lu, M-Li,Na,K). Izv. AN SSSR. Ser. Neorgan. Materialy. 23: 1330-1332.

[31] Belous A.G., Novitskaya G.N., Polyanetskaya S.V., Gornikov Yu.I. (1987) Investigation ox complex oxide of the composition $Ln_{2/3-x}Li_{3x}TiO_3$. Izv. AN SSSR. Ser. Neorgan. Materialy. 23: 470-472.

[32] Belous A.G., Novitskaya G.N., Polyanetskaya S.V., Gornikov Yu.I. (1987) Crystal-chemical and electro-physical properties of complex oxides $Ln_{2/3-x}M_{3x}TiO_3$. Zurn. Neorgan. Khimii. 32: 283-286.
[33] Belous A.G., Butko V.I., Novitskaya G.N. et al. (1986) Electrical conductivity of the perovskites $La_{2/3-x}M_{3x}TiO_3$. Ukr. Fiz. Zurn. 31: 576-581.
[34] Belous A.G., Gavrilova L.G., Polyanetskaya S.V. et al. (1984) Stabilization of the perovskite structure of lanthanum titanate Ukr. Khim. Zurn. 50: 460-461.
[35] Belous A., Ovchar O. (2003) Temperature compensated microwave dielectrics based on lithium containing titanates. J. Eur. Ceram. Soc. 23: 2525-2528.
[36] Belous A.G., Poplavko Yu.M. (1976) Dielectric properties of the tellurium containing perovskites in the microwave region Fizika Tv. Tela. 18: 2248-2451.
[37] Belous A.G., Politova Ye.D., Venevtsev Yu.N. et al. (1981) Lead-cobalt telluride – a material for microwave dielectric resonators Elektronnaya Tekhnika. Ser. Elektronika, SVCh. 331: 45-46.
[38] Valant M., Suvorov D., Hoffman C., Sommariva H. (2001) Ag(Nb,Ta)O3-based ceramics with suppressed temperature dependence of permittivity. J. Eur. Ceram. Soc. 21: 2647-2654.
[39] Trunov V.K., Frolov A.M., Averina I.M. (1981) Refinement of the structure $La_{0.33}NbO_3$ Kristallografiya. 26: 189-191.
[40] Sakowski-Cowley A.C., Lurasszewich K., Megaw H.D. (1969) The structure of sodium niobate at room temperature, and the problem of reliability in pseudosymmetric structures Acta Crystallogr. Sect. B. 25: 851-856.
[41] Mishchuk D.O., Vyunov O.I., Ovchar O.V., Belous A.G. (2004) Structural and dielectric properties of solid solutions of sodium niobate in lanthanum and neodymium niobates Inorganic Materials. 40: 1324-1330.
[42] Belous A.G. (2006) Microwave dielectrics with enhanced permittivity J. Eur. Ceram. Soc. 26: 1821-1826.
[43] Declaration patent no 54166 A, H 01P 7/10. Published 17.02.2003.
[44] Declaration patent no 54167 A, H 01P 7/10. Published 17.02.2003.
[45] dos Santos C.A., Zawislak L.I., Antonietti V. et al. (1999) Iron oxidation and order-disorder in the $(Fe^{2+},Mn)(Ta,Nb)_2O_6$ to $(Fe^{2+},Mn)Fe^{3+}(Ta,Nb)_2O_8$ transition. J. Phys. Condens. Matter. 11: 7021-7033.
[46] Lee H.J., Hong K.S., Kim I.T. (1997) Dielectric properties of MNb_2O_6 compounds (where M = Ca, Mn, Co, Ni, or Zn) Mater. Res. Bull. 32: 847-855.
[47] Zhang Y.C., Yue Z.X., GuiZ., Li L.T. (2003) Microwave dielectric properties of $(Zn_{1-x}Mg_x)Nb_2O_6$ ceramics. Mater. Lett. 57: 4531-4534.
[48] Pullar R.C., Breeze J.D., Alford N. (2005) Characterization and microwave dielectric properties of $M^{2+}Nb_2O_6$ ceramics // Journal of the American Ceramic Society. 88: 2466-2471.
[49] Joy P.A., and Sreedhar K. (1997) Formation of Lead Magnesium Niobate Perovskite from Niobate Precursors Having Varying Magnesium Content J. Am. Ceram. Soc. 80: 770–772.

[50] Ananta S. (2003) Effect of calcination condition on phase formation characteristic of magnesium niobate powders synthesized by the solid-state reaction. CMU. Journal. 2: 79–88.

[51] Ananta A., Brydson R. N., Thomas W. (1999) Synthesis, formation and Characterisation of $MgNb_2O_6$ Powder in a Columbite-like Phase J. Eur. Ceram. Soc. 19: 355–362.

[52] Norin P., Arbin C.G., Nalander B. (1972) Note on the Phase Composition of the MgO-Nb_2O_5 System. Acta Chim. Scand. 26: 3389–3390.

[53] Paqola S., Carbonio R.E., Alonso J.A. Femandez-Diaz M.T. (1997) Crystal structure refinement of $MgNb_2O_6$ columbite from neutron powder diffraction data and study of the ternary system MgO–Nb_2O_5–NbO, with evidence of formation of new reduced pseudobrookite $Mg_{5-x}Nb_{4+x}O_{15-\delta}$ ($1.14 \le x \le 1.60$) phases. Journal of Solid State Chemistry. 134: 76–84.

[54] You Y.C., Park H.L., Song Y.G., Moon H.S., Kim G.C. (1994) Stable phases in the MgO-Nb_2O_5 system at 1250°C. J. Mater. Sci. Lett. 13: 1487–1489.

[55] A.G. Belous, O.V, Ovchar, A.V, Kramarenko et al, (2007) Synthesis and microwave dielectric properties of $Zn_{1+x}Nb_2O_{6+x}$ Inorganic Materials. 43: 326-330.

[56] Belous A.G., Ovchar O.V., Mishchuk D.O. et al. (2007) Synthesis and properties of columbite-structure $Mg_{1-x}Nb_2O_{6-x}$ Inorganic Materials. 43: 477-483.

[57] Belous A.G., Ovchar O.V., Kramarenko A.V. et al. (2006) Effect of nonstoichiometry on the structure and microwave dielectric properties of cobalt metaniobate. Inorganic Materials. 42: 1369-1373.

[58] Rath W. (1941) Keramische sindermassen fur die elektronik fortschritter auf dem greblet der keramischen isollerstoff fur die electroteknik. Keram. Radsch. 49: 137–139.

[59] Brit Patent No. 692468 (1952).

[60] Wakino K., Nishikawa T., Tamura S., and Ishikawa Y. (1975) Microwave band pass filters containing dielectric resonators with improved temperature stability and spurious response. Proc.IEEE MTT Symposium (New York) 63–65.

[61] Wakino K., Nishikawa T., Tamura S., and Ishikawa Y. (1978) Miniaturised band pass filters using half wave dielectric resonators with improved spurious response. Proc IEEE MTT Symposium 230–232.

[62] Wolfram W. and Gobel H.E. (1981) Existence range, structural and dielectric properties of $Zr_xTi_ySn_zO_4$ ceramics (x.y.z=4). Mater. Res. Bull. 16: 1455–1463.

[63] Tamura H. (1994) Microwave loss quality of $(Zr_{0.8}Sn_{0.2})TiO_4$. Am. Ceram. Soc. Bull. 73: 92–95.

[64] Blasse G. (1966) Compounds with α-PbO_2 structure. J. Anorg. Allg. Chem. 345: 222–224.

[65] Newnham R.E. (1967) Crystal structure of $ZrTiO_4$. J. Am. Ceram. Soc. 50: 216.

[66] Mc Hale A.E. and Roth R.S. (1983) Investigation of the phase transition in $ZrTiO_4$ and $ZrTiO_4$–SnO_2 solid solutions. J. Am. Ceram. Soc. 66: 18–20.

[67] Mc Hale A.E. and Roth R.S. (1986) Low-temperature phase relationships in the system ZrO_2–TiO_2. J. Am. Ceram. Soc. 69: 827-832.

[68] Ikawa H., Iwai A., Hiruta K., Shimojima H., Urabe K., and Udagawa S. (1988) Phase transformation and thermal expansion of zirconium and hafnium titanate and their solid solutions. J. Am.Ceram. Soc. 71: 120–127.

[69] Park Y. (1995) Influence of order disorder transition on microwave characteristics of tin-modified zirconium titanate. J. Mater. Sci. Lett. 14: 873–875.

[70] Ikawa H., Shimojima H., Ukrabe K., Yamada T., and Udagawa S. (1988) Polymorphism in $ZrTiO_4$. Science of Ceramics. D. Taylor (Ed.). Institute of Ceramics, Shelton, Uk. 509–514.

[71] Christerfferson R. and Davies P. K. (1992) Structure of commensurate and incommensurate ordered phase in the system $ZrTiO_4$-$Zr_5Ti_7O_{24}$. J. Am. Ceram. Soc. 75: 563–569.

[72] Han K. R., Jang J.-W., Cho S.-Y., Jeong D.-Y., and Hong K.-S. (1998) Preparation and dielectric properties of low temperature-sinterable $(Zr_{0.8}Sn_{0.2})TiO_4$ powder. J. Am. Ceram. Soc. 81: 1209–1214.

[73] Park Y. and Kim Y. (1996) Order-disorder transition of tin-modified zirconium titanate. Mater.Res. Bull. 31: 7–15.

[74] Park Y., Kim Y., and Kim H. G. (1996) Structural-phase transition and electrical conductivity in tin-modified zirconium titanate. Solid State Ionics. 90: 245–249.

[75] Wakino K., Nishikawa T., Tamura S., and Ishikawa Y. (1978) Miniaturised band pass filters using half wave dielectric resonators with improved spurious response. Proc IEEE MTT Symposium. 230–132.

[76] Azough F., Freer R., Wang C.L., and Lorimer G.W. (1996) The relationship between the microstructure and microwave dielectric properties of zirconium titanate ceramics. J. Mater. Sci. 31: 2539–2549.

[77] Wakino K., Minai K., and Tamura H. (1984) Microwave characteristics of $(Zr,Sn)TiO_4$ and BaO–PbO–Nd_2O_3–TiO_2 dielectric resonator. J. Am. Ceram. Soc. 67: 278–281.

[78] Heiao Y.C., Wu L., and Wei C.C. (1988) Microwave dielectric properties of $(ZrSn)TiO_4$ ceramic. Mater. Res. Bull. 23: 1687–1692.

[79] Khairulla F. and Phule P. (1992) Chemical synthesis and structural evolution of zirconium titanate. Mater. Sci. Eng. B. 12: 327-336.

[80] Christofferson R., Davies P.K., Wei X., and Negas T. (1994) Effect of Sn substitution on cation ordering in $(Zr_{1-x}Sn_x)TiO_4$ microwave dielectric ceramics. J. Am. Ceram. Soc. 77: 1441–1450.

[81] Iddles D. M., Bell A. J., and Moulson A. J. (1992) Relationship between dopants, microstructure and the microwave dielectric properties of ZrO_2–TiO_2–SnO_2 ceramics. J. Mater. Sci. 27: 6303–6310.

[82] Davies P.K. (1994) Influence of internal interfaces on the dielectric properties of ceramic dielectric resonators. Res. Soc. Symp. Proc. 357: 351–361.

[83] Wersing W. (1991) High frequency ceramic dielectrics and their applications for microwave components. In: Electroceramics, B. C. H. Steele (Ed.), Elsevier Applied Sciences, London and New York. 67–119.

[84] Wakino K. (1989) Recent developments of dielectric resonator materials and filters in Japan. Ferroelectrics. 91: 69–86.

[85] Azough F. and Freer R. (1989) The microstructure and low frequency dielectric properties of some zirconium titanate stannate (ZTS) ceramics. Proc. Br. Ceram. Soc. 42: 225–233.

[86] Huang C.-L., Weng M.-H., and Chen H.-L. (2001) Effects of additives on microstructures and microwave dielectric properties of $(Zr,Sn)TiO_4$ ceramics. Mater. Chem. Phys. 71: 17–22.

[87] Takada T., Wang S. F., Yoshikawa S., Jang S.-J., and Newnham R. E. (1994) Effects of glass on $(Zr,Sn)TiO_4$ for microwave applications. J. Am. Ceram. Soc. 77: 2485–2488.

[88] Ioachin A., Banau M. G., Toacsan M. I., Nedelcu L., Ghetu D., Alexander H. V., Toica G., Annino G., Cassettari M., and Martnelli M. (2005) Nickel doped $(Zr_{0.8}Sn_{0.2})TiO_4$ for microwave and millimeter wave applications. Mater. Sci. Eng. B. 118: 205–209.

[89] Galasso F., Pule J. (1963) Ordering in compounds of the $A(B'_{0.33}Ta_{0.67})O_3$ type. Inorg Chem. 2: P. 482–484.

[90] Galasso F., Katz L. (1959) Substitution in the octahedrally coordinated cation positions in compounds of the perovskite type J. Amer. Ceram. Soc. 81: 820-823

[91] Galasso F., Pule J. (1962) Preparation and study of ordering in $A(B'_{0.33}Nb_{0.67})O_3$ perovskite-type compounds J. Phys. Chem. 67: 1561–1562.

[92] Roy R. (1954) Multiple Ion Substitution in the Perovskite Lattice J. Amer. Ceram. Soc. 37: 581-588.

[93] Kawashima S., Nichida M., Ueda I., Oici H. (1983) $Ba(Zn_{1/3}Ta_{2/3})O_3$ Ceramics with Low Dielectric Loss at Microwave Frequencies J. Amer. Ceram. Soc. 66: 421–423.

[94] Nomura S., Kaneta K. (1984) $Ba(Mn_{1/3}Ta_{2/3})O_3$ Ceramic with Ultra-Low Loss at Microwave Frequency Jpn. J. Appl. Phys. 23: 507–508.

[95] Chen X.M., Suzuki Y., Sato N. (1994) Sinterability improvement of $Ba(Mg_{1/3}Ta_{2/3})O_3$ dielectric ceramics. Journal of Materials Science. Materials in Electronics. 5: 244–247.

[96] Kakegawa K., Wakabayashi T., Sasaki Y. (1986) Preparation of $Ba(Mg_{1/3}Ta_{2/3})O_3$ Using Oxine J. Amer. Ceram. Soc. 69: 82–89.

[97] Chen X.M., Wu Y. J. (1995) A low-temperature approach to synthesize pure complex perovskite $Ba(Mg_{13}Ta_{23})O_3$ powders Materials Letters. 26: 237–239.

[98] Nomura S., Toyoma K. and Kaneta K. (1982) $Ba(Mg_{1/3}Ta_{2/3})O_3$ ceramics with temperature-stable high dielectric constant and low microwave loss. Jpn. J. Appl. Phys. 21: 624–626.

[99] Matsumoto K., Hiuga T., Takada K. et al. (1986) BA(M1/3TA2/3)O3 Ceramics with ultra-low loss at microwave-frequencies. IEEE Transactions on ultrasonics ferroelectrics and frequency control. 33: 802-802.

[100] Novitskaya G.N., Yanchevskii O.Z., Polyanetskaya S.V., Belous A.G. (1991)Formation of the phases (phase formation) in the systems $BaCO_3$-$(Nb,Ta)_2O_5$-ZnO. Ukr. Khim. Zhurn. 57: 801-802.

[101] USSR inventor's certificate 1837599, IPC C 04 B 35, H 01 B 3/12. Published 13.10. 1992.

[102] Renoult O., Bollot J.-P., Chaput F., papierik R., Hubert-Pfalzgraf L.G., Leycune M. (1992) Sol–Gel Processing and Microwave Characteristics of $Ba(Mg_{1/3}Ta_{2/3})O_3$ Dielectrics J. Amer. Ceram. Soc. 75: 3337-3340.

[103] Kolodiazhnyi T., Petric A., Johari G. Belous A. (2002) Effect of preparation conditions on cation ordering and dielectric properties of $Ba(Mg_{1/3}Ta_{2/3})O_3$ ceramics J. Eur. Ceram. Soc. 22: 2013–2021.

[104] Youn H.J., Hong K.S., Kim H. (1997) Coexistence of 1 2 and 1:1 long-range ordering types in La-modified $Ba(Mg_{0.33}Ta_{0.67})O_3$ ceramics. J. Mater. Res. 12: 589–592.

[105] Akbas M. A., Davies P. K. (1998) Cation Ordering Transformations in the $Ba(Zn_{1/3}Nb_{2/3})O_3$-$La(Zn_{2/3}Nb_{1/3})O_3$ System J. Amer. Ceram. Soc. 81: 1061–1064.

[106] Chen J., Chan H.M., Harmer M.P. (1989) Ordering structure and dielectric properties of undoped and La/Na-Doped $Pb(Mg_{1/3}Nb_{2/3})O_3$ J. Amer. Ceram. Soc. 72: 593–598.

[107] Hilton A.D., Barber D.J., Randall A., Shrout T.R. (1990) On short range ordering in the perovskite lead magnesium niobate. J. Mater. Sci. 25: 3461–3466.

[108] Akbas M. A., Davies P. K. (1998) Ordering-induced microstructures and microwave dielectric properties of the $Ba(Mg_{1/3}Nb_{2/3})O_3$–$BaZrO_3$ system. J. Amer. Ceram. Soc. 81: 670–676.

[109] Chai L., Akbas M. A., Davies P. K., Parise J.B. (1997) Cation ordering transformations in $Ba(Mg_{13}Ta_{23})O_3$-$BaZrO_3$ perovskite solid solutions Mater. Res. Bull. 32: 1261–1269.

[110] Chai L., Davies P. K. (1997) Formation and Structural Characterization of 1:1 Ordered Perovskites in the $Ba(Zn_{1/3}Ta_{2/3})O_3$–$BaZrO_3$ System J. Amer. Ceram. Soc. 80: 3193–3198.

[111] Djuniadi A., Sagala N. (1992) Lattice Energy Calculations for Ordered and Disordered $Ba(Zn_{1/3}Ta_{2/3})O_3$. Journal of the Physical Society of Japan. 61: 1791–1797.

[112] Mehmet A., Akbas M.A., Davies P.K. (1998) Ordering-Induced Microstructures and Microwave Dielectric Properties of the $Ba(Mg_{1/3}Nb_{2/3})O_3$–$BaZrO_3$ System J. Amer. Ceram. Soc. 81: 670–676.

[113] Mitsuhiro Takata, Keisuke Kageyama (1989) Microwave Characteristics of $A(B^{3+}_{1/2}B^{5+}_{1/2})O_3$ Ceramics (A = Ba, Ca, Sr; B^{3+}= La, Nd, Sm, Yb; B^{5+}= Nb, Ta) J. Amer. Ceram. Soc. 72: 1955–1959.

[114] Matsumoto K., Hiuga T., Takada K., and Ichimura H.. (1986) $Ba(Mg_{1/3}Ta_{2/3})O_3$ ceramics with ultralow loss at microwave frequencies. Proc. 6th IEEE Intl. Symp. On Applications of Ferroelectrics IEEE (NY). 118.

[115] Tamura H., Konoike T., Sakabe Y., and. Wakino K. (1984) Improved High-Q Dielectric Resonator with Complex Perovskite Structure J. Am. Ceram. Soc. 67: C.59–61.

[116] Yon Ki H., Dong P.K., Kim E.S. (1994) Annealing effect on microwave dielectric properties of $Ba(Mg_{1/3}Ta_{2/3})O_3$ with $BaWO_4$ Ferroelectric. 154: 337–342.

[117] Matsumoto H., Tamura H., Wakino K. (1991) $Ba(Mg,Ta)O_3$–$BaSnO_3$ High-Q dielectric resonator. Jpn. J. Appl. Phys. 30: 2347-2349.

[118] Hughes H., Iddles D., Reaney I.M. (2001) Niobate-based microwave dielectrics suitable for third generation mobile phone base stations Appl. Phys. Letters. 79: 2952–2954.

[119] Cheng-Ling Huang, Ruei-jsung Lin (2002) Liquid Phase Sintering and Microwave Dielectric Properties of $Ba(Mg_{1/3}Ta_{2/3})O_3$ Ceramics. Jpn. J. Appl. Phys. 41: 712–716.

[120] Azough F., Leach C., Freeer R. (2006) Effect of nonstoichiometry on the structure and microwave dielectric properties of $Ba(Co_{1/3}Nb_{2/3})O_3$ ceramics. J. Eur. Ceram. Soc. 26: 2877–2884.

[121] Liu H.X., Tian Z.Q., Wang H., Yu H.T., Ouyang S.X. (2004) New microwave dielectric ceramics with near-zero τ_f in the $Ba(Mg_{1/3}Nb_{2/3})O_3$-$Ba(Ni_{1/3}Nb_{2/3})O_3$ system. J. Mater. Sci. 39: 4319–4320.

[122] Seo-Yong Cho, Hyuk-Joon Youn, Kug-Sun Hong (1997) A new microwave dielectric ceramics based on the solid solution system between $Ba(Ni_{1/3}Nb_{2/3})O_3$ and $Ba(Zn_{1/3}Nb_{2/3})O_3$ J. Mater. Res. 12: 1558–1562.

[123] Kolodiazhnyi T., Petric A., Belous A., V'yunov O., Yanchevskij O. (2002) Synthesis and dielectric properties of barium tantalates and niobates with complex perovskite structure. J. Mater. Res. 17: 3182–3189.

[124] Scott R.I., Thomas M., Hampson C. (2003) Development of low cost, high performance $Ba(Zn_{1/3}Nb_{2/3}O_3)$ based materials for microwave resonator applications. J. Eur. Ceram. Soc. 23: 2467–2471.

[125] Davies P.K., Borisevich A., Thirunal M. (2003) Communicating with wireless perovskites: cation order and zinc volatilization. J. Eur.Ceram.Soc. 23: 2461-2466.

[126] Davis P.K., Tong, J., and Negas, T., (1997) Effect of Ordering-Induced Domain Boundaries on Low-Loss $Ba(Zn_{1/3}Ta_{2/3})O_3$-$BaZrO_3$ Perovskite Microwave Dielectrics. J. Am. Ceram. Soc. 80: 1724–1740.

[127] Molodetsky, I. and Davies, P.K., (2001) Effect of $Ba(Y_{1/2}Nb_{1/2})O_3$ and $BaZrO_3$ on the Cation Order and Properties of $Ba(Co_{1/3}Nb_{2/3})O_3$ Microwave Ceramics. J. Eur. Ceram. Soc. 21: 2587–2591.

[128] Endo, K., Fujimoto, K., and Murakawa, K. (1987) Dielectric Properties of Ceramics in $Ba(Co_{1/3}Nb_{2/3})O_3$–$Ba(Zn_{1/3}Nb_{2/3})O_3$ Solid Solutions. J. Am. Ceram. Soc. 70: 215–218.

[129] Ahn C.W., Jang H.J., Nahm S., et al. (2003) Effect of Microstructure on the Microwave Dielectric Properties of $Ba(Co_{1/3}Nb_{2/3})O_3$ and $(1-x)Ba(Co_{1/3}Nb_{2/3})O_3$–$xBa(Zn_{1/3}Nb_{2/3})O_3$ Solid Solutions. J. Eur. Ceram. Soc. 23: 2473–2478.

[130] Davies P.K., Borisevich A., and Thirumal M. (2003) Communicating with Wireless Perovskites: Cation Order and Zinc Volatilization. J. Eur. Ceram. Soc. 23: 2461–2466.

[131] Mallinson, P.M., Allix, M.M., Claridge, J.B., et al. (2005) $Ba_8CoNb_6O_{24}$: A d^0 Dielectric oxide host containing ordered d^7 cation layers 1.88 nm apart. Angew. Chem. Int. Ed. 44: 7733–7736.

[132] Belous A., Ovchar O., Macek-Krzmanc M., Valant M. (2006) The homogeneity range and the microwave dielectric properties of the $BaZn_2Ti_4O_{11}$ ceramics. J. Eur. Ceram. Soc. 26: 3733–3739.

[133] C.-L. Huang, Chung-Long Pan 2002 Low-temperature sintering and microwave dielectric properties of $(1 - x)MgTiO_3$-$xCaTiO_3$ ceramics using bismuth addition. Jpn. J. Appl. Phys. 41: 707–711.

[134] Belous A., Ovchar O., Durilin D., Macek-Krzmanc M., Valant M., Suvorov D. (2006) High-Q Microwave Dielectric Materials Based on the Spinel Mg_2TiO_4 J. Am. Ceram. Soc. 89: 3441–3445.

[135] O.V. Ovchar, O.I. Vyunov, D.A. Durilin et al. (2004) Synthesis and microwave dielectric properties of $MgO–TiO_2–SiO_2$ ceramics. Inorganic Materials. 40: 1116-1121.

[136] Belous A.G., O.V. Ovchar, D.A. Durilin et al. (2006) Microwave composite dielectrics based on the system $MgO–CaO–SiO_2–TiO_2$. Abstr. Book "Microwave Materials and Their Applications" (12-15 June). –Oulu (Finland) 97.

[137] Wersing W. (1991) Electronic Ceramics. Ed. By Steele BCH. –London and New York: Elsevier Appl. Science. 67-119.

[138] Heywang W. (1951) Zur Wirksamen Feldstarke im kubischen Gitter. Zeitschrift fur Naturforschung section a-a journal of physical sciences. 6: 219, 220.

[139] Wersing W. (1996) Microwave ceramics for resonators and filters. Current Ohinion in Solid State and Materials Science. 1: 715-731.

[140] Colla I.L., Reaney I.M., Setter N. (1993) Effect of structural changes in complex perovskites on the temperature coefficient of the relative permittivity. J. Appl. Phys. 74: 3414-3425

[141] Sugiyama M., Nagai T. (1993)Anomaly of dielectric-constant of $(Ba_{1-x}Sr_x)(Mg_{1/3}Ta_{2/3})O_3$ solid-solution and its relation to structural-change. Japanese Journal Of Applied Physics Part 1-Regular Papers Short Notes & Review Papers. 32: 4360-4363.

[142] Gurevich V.L., Tagantsev A.K. (1986) Intrinsic dielectric losses in crystals - low-temperatures. Sov. Phys. JETP. 64: 142-151.

[143] Kudesiak K., Mc Itale A.E., Condrate R.A., Sr. Snyder R.L. (1993) Microwave characteristics and far-infrared reflection spectra of zirconium tin titanate dielectrics. J. Mater. Sci. 28: 5569-5575.

[144] Fukuda K., Kitoh R. (1994) Far-Infrared Reflection Spectra of Dielectric Ceramics for Microwave Applications J. Amer. Ceram. Soc. 77: 149-154.

[145] Zurmuhlen R., Colla E., Dube D.C. et al. (1994) Structure of $Ba(Y^{+3}_{1/2}Ta^{+5}_{1/2})O_3$ and its dielectric properties in the range 102–1014 Hz, 20–600 K J. Appl. Phys. 76: 5864-5863.

[146] Zurmuhlen R., Petzelt J., Komba S. et al. (1995) Dielectric-spectroscopy of $Ba(B_{1/2}'B_{1/2}'')O_3$ complex perovskite ceramics - Correlations between ionic parameters and microwave dielectric-properties .I. Infrared reflectivity study (10^{12}-10^{14} HZ). J. Appl. Phys. 77: 5341-5350.

[147] Gurevich V.L., Tagantsev A.K. (1991) Intrinsic dielectric loss in crystals. Adv. Phys. 40: 719-767.

[148] Christoffersen R., Davies P.K., Wie X. (1994) Effect of Sn Substitution on Cation Ordering in $(Zr_{1-x}Sn_x)TiO_4$ Microwave Dielectric Ceramics J. Amer. Ceram. Soc. 77: 1441-1450.

[149] Tsykalov V.G., Belous A.G., Ovchar O.V., Stupin Y.D. (1997) Monolithic filters and frequency-separation devices based on the ceramic resonators. 27th Europ. Microwave Conf. Proceed. 544-600.

[150] Han Q., Kogami Y., Tomabechi Y. (1994) Resonance characteristics of circularly propagating mode in a coaxial dielectric resonator. IEICE Trans Electron. 77: 1747-1751.

[151] Belous A.G., Ovchar O.V. (1995) The origin of the temperature stabilization of dielectric permittivity in the system $x(Sm_{1/2}Li_{1/2}TiO_3)$ -(1-x)$(Sm_{1/2}Na_{1/2}TiO_3)$ Ukr. Khim. Zhurn. 61: 73-77.

Section 3

Natural Lighting Systems

Chapter 7

Natural Lighting Systems Based on Dielectric Prismatic Film

Daniel Vázquez-Moliní, Antonio Álvarez Fernández-Balbuena and Berta García-Fernández

Additional information is available at the end of the chapter

http://dx.doi.org/10.5772/50352

1. Introduction

Daylight provides high quality lighting, reduces energy use and has numerous beneficial physical and psychological effects on people. Furthermore, natural lighting has many benefits in creating indoor spaces, such as energy saving and better quality of vision; two facts that improve environments and thus, productivity.

The light pipe is a device that can transfer natural light from a building´s roof into the depths of the building, this straight construction consist of a reflective closed walled structure (P.D. Swift & G.B. Smith, 1995). Daylight guidance has been one of the mayor areas of innovation in interior lighting in recent years, with the development of light pipes daylight and electric light are simultaneously delivered into a building where they are combined and distributed via luminaries. As a result, the overall wattage of artificial light is reduced and the consumption of electricity decreases (Mayhoub, et al., 2010). The commonest light pipes are reflective mirror guides which use high reflectance aluminium, also fiber optic guides are widely used for illumination purposes. Optical design with new materials like dielectric ones, with regard to their reflection, transmission and absorption is as important as its geometry study.

In recent works, Vazquez-Moliní et al. introduced an illumination system called ADASY® integrated into a building's façade that consists of a horizontal light guide inside the building. ADASY® comprises a collection system, a light guide, and daylight luminaries (Vazquez-Moliní et al, 2009). Prior developments in solar lighting systems based on micro-replicated light film had been studied showing that prism light guides performance varies with the length of the guide, maintenance conditions, the collecting system, the luminaries, and the direction from which light is directed (Whitehead, L. A., 1982). The objective of previous investigations of the group had been the study and development of efficient dielectric prismatic hollow light pipes that direct natural light into interior spaces applied in office buildings One of the proposed system is a daylight illumination system by vertical transparent prismatic light guide for an office building; this model consist in a hollow tube internally coated with thin polycarbonate prismatic film. In this model, two different prismatic sheets have been used; the light guiding system works with 90º prismatic film and the extraction system is composed of 70º prismatic film perpendicular to the previous one that works extracting the light outside the guide with a specific angular distribution. This design allows us to obtain a transparent simple and beautiful pipe which is integrated in building design; in addition, the spatial distribution of the extraction sheets can be adapted to the requirements of each space. (Alvarez Fernandez-Balbuena A. et al, 2010).

Light color quality is an important issue to evaluate in natural lighting systems. High reflectance aluminum lighting guides are giving bad light quality because the spectral reflectance of the aluminium, changes the color characteristics of the output light at the end of the guide (Vázquez-Moliní et al, 2007). When light guides are made of a dielectric prismatic film, the influence of the spectral reflectance is minimized due to the total internal reflection produced in the surface of the prismatic film, absorptance is not usually considered significant in the literature when the sheets are thin. Color Rendering Index and Correlated Color Temperature are important parameters in order to evaluate lighting quality in Museums, office buildings and production centers to get the normative approval (García-Fernández et al, 2011).

A skylight is a technology for obtaining natural light into a building. Skylights are an opening in a roof that is covered with translucent or transparent material and that is designed to admit light provided a connection to the outdoor environment to occupants. Nowadays the skylight technology is widely used in outstanding buildings (Dubois, 2003). Almost all of the skylights used are just an opening in the ceiling that ensures watertight but without significant optics, just a diffuser on it to prevent direct sunlight.

The Compound Parabolic Concentrators (CPC) is a non-focusing light funnel with specular reflecting surface of a parabolic shape designed to give the maximum concentration ratio for a given acceptance angle θmax. CPCs are relevant for solar energy caption because they achieve good concentration for many acceptance angles. There are many applications where the Compound Parabolic Concentrator is used not only for concentration but also for collimation (in reverse mode) like in natural lighting, thermal collector, LED's optic, car light and optical fiber coupling (Winston, 1975). Winston et al. had explored a growing field that is applicable to areas where the collection, concentration, transport and distribution of light is important. Systems for natural light caption offer very important advantages when use some kind of CPC optics (Winston et al., 2005). CPC passive optics made of dielectric film allows the operation of the collimating system during long periods of time without need to fit its orientation since a CPC in reverse mode is capable of redirecting all the light entering in 2π in the designed angle (Alvarez Fernandez-Balbuena et al., 2009).

2. Prismatic film

A prismatic film is a thin plastic that works with the optical principle of total internal reflection (TIR) through the prism structure. It can be used in several applications for replacing metal guides with best performance. The film geometry has one flat surface and the other one is a textured surface consisting of an array of linear right angle prisms inclined at 45 degrees to the flat surface. This configuration has a light angular acceptance cone determined by the refractive index of the prismatic film material, approximately a 30º semiangle cone. If this angular criterion is not met, the light will instead be rejected out of the light pipe through the prismatic material (OLF).

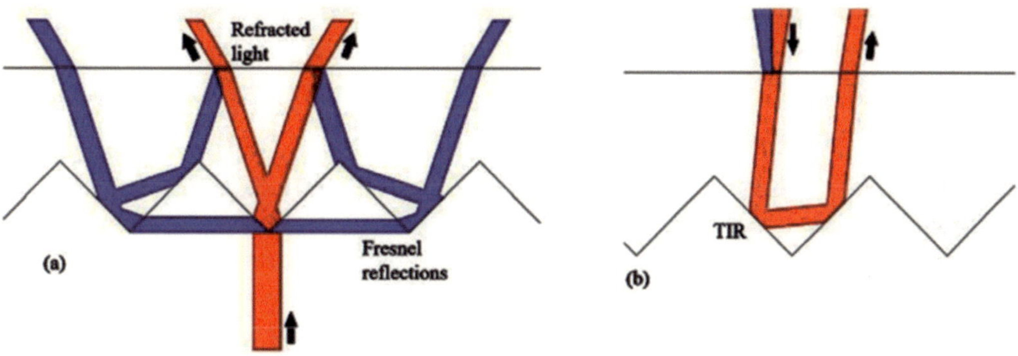

Figure 1. (a) Example of refraction and (b) total internal reflection in prismatic film.

Figure 1 (a) is an example of light entering in a prismatic film through prism apex side, in this case light is divided in two main arms (refracted light). The incidents rays are drawn in red lines and the Fresnel refractions are plotted with blue lines. Most of the rays directly emerge from the prism as displayed in red lines; rest of light than is guided in the prismatic structure is Fresnel light with low flux energy. In figure 1 (b) the prismatic film produces TIR, in this case the light is returned in the direction from which it came, this principle is used for guiding light with a prismatic film.

The degree to which the film's prisms deviate from perfect prisms also affects the efficiency of the total internal reflection process, and therefore, the effectiveness of the film in transporting and distributing light. These imperfections include 90° corners which are not precise, surfaces which are not optically flat or which deviate from the correct angle and optical inhomogeneities in the material (Remillard et al.1992). Absorption is due to bulk absorptivity of the material used to produce the film and transmission can be used to advantage of the application in light distribution. With the typical losses due to absorption and transmission, the reflectance efficiency has been estimated as approaching 99% (Keipp, 1994). Precision micromachining, polymer processing and certain other manufacturing technologies like microreplication have made possible the development of an optimized prismatic film (Wang, 2009). The structure of prism film under different magnifications is shown in Figure 2.

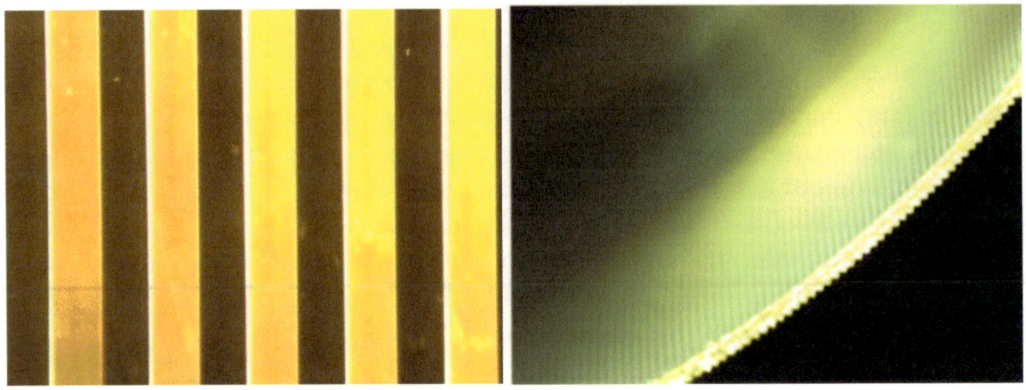

Figure 2. Prismatic film used to distribute light. a) Surface area detail of prisms structure (25X), b) prismatic film (2.5X).

2.1. Image processing algorithms

This section deal with a brief description of the recognition procedure algorithm to analyse the prism profile structure. First, a diagram with a description of the recognition procedure to extract the inclination of the lines from the image and to calculate the radio of the peak prism is given (Fig.3), secondly, a more detailed description will follow.

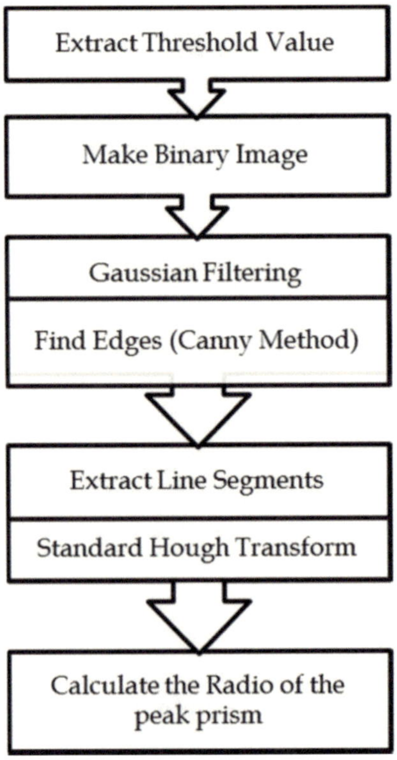

Figure 3. Structure of the recognition procedure and extraction process performed.

In order to investigate the light behavior in the system the effect of accuracy of prismatic structures and peak defects has been examined; computer analyses such as changes of prism angle and plane shape were carried out. By analyzing the image processing prism structure and by using morphological operations measurement of the inclination angle with high accuracy can be achieved. To execute the processing, the prismatic film profile showed in figure 4 is employed (the optical microscope used to obtain the prism image is Motic SMZ-143 equipped with a digital camera Moticam 2000).

Figure 4. Prism structure used to analyse the parameters of the prism (57X)

Firstly, a threshold is applied to the input image in order to make it binary. The threshold value is determined from a grey-level histogram of the image, later the edge is separated from the background (Fig.5.b).

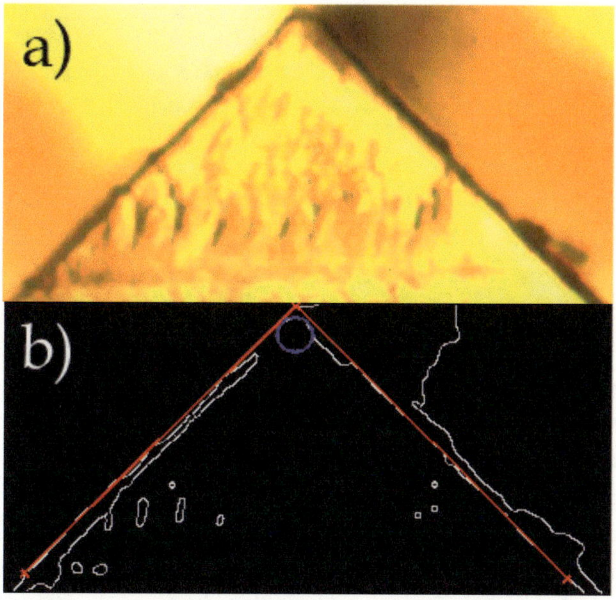

Figure 5. (a) Cropped image before the digital processing used to obtain the dimensions of micro prism structures. (b) Edge detected images resulting from the Canny Method (white), the line after Hough transforms (red), and the estimation to calculate the radius of the prism´s peak (blue).

The edge map is used to determine the existence of lines around the regions. This edge description is obtained from the operator Canny (Canny, 1986) this operator is consider as an optimal edge detector to find boundaries between poorly defined objects as well as hard edges. The Canny method finds edges by looking for local maxima of the gradient of the image; the gradient is calculated using the derivative of a Gaussian filter. The method uses two thresholds, to detect strong and weak edges, and includes the weak edges in the output only if they are connected to strong edges.

Later, we use the Hough transform (Fig.5. (b) and Fig.6) to detect the parameters that controls the accuracy of the right angle at the vertex of the prism (Hough, 1962).

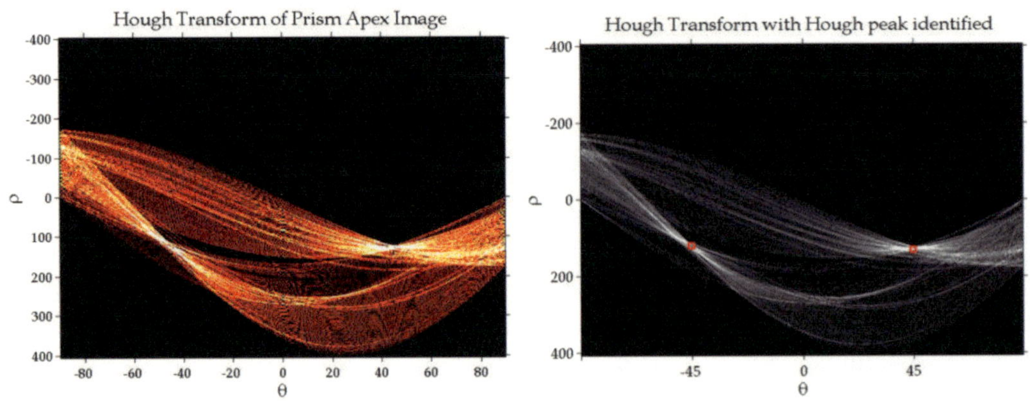

Figure 6. (a) The Hough transform of the prism image. (b)The red square shows the peaks of data in the Hough matrix. ϱ is the distance from the center and θ the angle at which the sum of intensities in the image peaks, it is thus the slope of the line along with the position.

Originally, the Hough Transform was proposed to extract straight lines in the particle tracks recognizing procedure. Nowadays, the Hough transform is a technique which is used to insulate features of a particular shape within an image. In this case, the Hough transform is used to identify the parameters of the line and it uses the parametric representation of a line which is fits to a set of given edge points. It takes as input the grey scale image, and produces as output an image showing the positions of tracked intensity discontinuities. The output of the edge detector defines where features are in the image, and the Hough transform determine what the features are and how many of them exist in the image. The main advantage of the Hough transform technique is that it is tolerant of gaps in feature boundary descriptions and is minimal unaffected by image noise. The result of the Hough transform is stored in a matrix that often is considered an accumulator (Fig. 6 (a)). One dimension of this matrix is the angles θ and the other dimension are the distances ϱ, and each element has a value telling how many points/pixels are positioned on the line with parameters (ϱ, θ). So the element with the highest value shows the line that is most represented in the input image.

After performing probabilistic Hough Transform, two lines are obtained determining and ensuring the right angle prism used to guide the incident light. Figure 5 shows in blue color the plot of the inclination angle of the prism through the digital processing. The main peaks are located in the Hough transform matrix. The left slope is calculated to be an apex angle of -45.38º and the right slope has an apex angle of 44.50º forming a total angle of 89,88º (Fig. 6 (b)), after that, an approximation was made to relate the prism rounding radius with the slope values and the contour of the vertex profile. The circle obtained used to achieve the radius of the prism vertex is showed in red color (Fig. 5 (b)). The prism base obtained has a width of 400 μm and the radius value obtained is 13.78 μm; this result could be affected by the pressure exerted to make the cut of the prism film.

3. Prismatic lightguide analysis

The lightguide structure analysed is a prismatic hollow tube with the thin polycarbonate film with right angle prism sections. The prism light guide transmits light by total internal reflection, which gives higher efficiency and homogeneous light distribution through the guide. Prismatic sheeting developments have provided further improvements in sunlight systems (Fig. 7), boosting efficiency for specific incidence angles with regard to the aluminum guides.

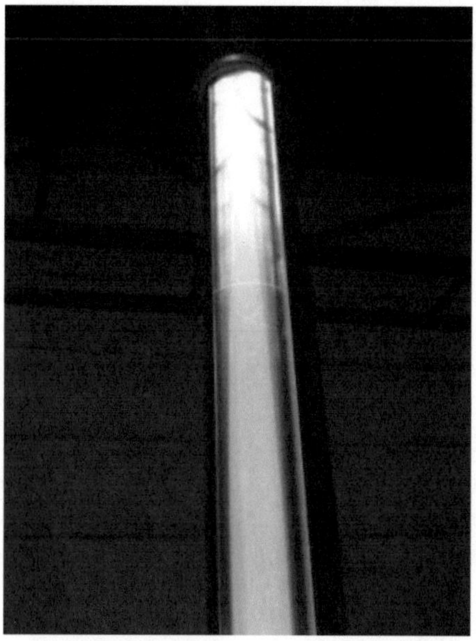

Figure 7. Prismatic light tubes are used for transporting and distributing natural light. Experimental setup in the School of Optics (Complutense University of Madrid)

Light travels mainly in the hollow air space inside the guide and bounces off by total internal reflections (TIR) when the input light is highly collimated. This configuration has an angular acceptance cone, which is not an isotropic distribution in the space determined by the refractive index of the prismatic film, if the refractive index of the dielectric material is

1.5, as is the case of acrylic plastic, then the input light angle must be approximately less than 27.5º from the guide´s axial direction, even though the acceptance cone can be higher in the meridional plane.

Two different analysis have been done: Firstly, a theoretical simulation based on the specular reflection model in order to analyse the color characteristics to the output of two guides, rectangular and cylindrical of different lengths, one of them of aluminum material and the other one, internally coated of prismatic film structure this model have been developed by means of a mathematical software as Matlab. Secondly, the prismatic guides proposal evaluates with a ray tracing software through which we show the efficiency for a wavelength evaluated at the end of the guide, this model have been studied with Monte Carlo ray tracing.

3.1. Theoretical simulation

The analysis presented is based on the specular reflection model. There are two important objectives to evaluate with regard to the light quality as a function of the Spectral Power Distribution; these are Correlated Color Temperature (CCT) and the Color Rendering Index (CRI) of the light source, which we evaluate in this work using the guidelines of CIE 13.3.

Color rendering of an illuminant is the effect of the illuminant on the color appearance of objects by conscious or subconscious comparison with their color appearance under a reference illuminant. Colour Rendering serves to describe the effect of an illuminant on the color appearance of objects. The most fundamental natural source is daylight, thus the primary reference for comparing the rendering of a source should be the Standard Illuminant D65.

The Correlated Color Temperature is the temperature of the Planckian radiator whose perceived color most closely resembles that of a given stimulus at the same brightness and under specified viewing conditions (CIE 17.4, 1989). According to this definition, CCT can be calculated using one of the chromaticity diagrams. CIE still recommends to calculate CCT using the 1960 (u, v) chromaticity diagram (now deprecated). On the (u, v) diagram, find the point on the Planckian locus that is at the shortest distance from the given chromaticity point. CCT is the temperature of the Planck's radiation at that point. CCT can be calculated from CIE 1931 chromaticity coordinates x and y in many different ways. Complicated algorithms and simple equations alike have been proposed and used for several decades.

The simulated construction of the prismatic guides is based on a tube internally coated polycarbonate sheet whose outer face is composed of 90º micro-prisms; rays that enter on the guide with an angle suitable undergo total internal reflections. In this model, in order to include the round in corners and the absorption we estimate the spectral reflectance of the inner face of the prism sheet will be estimated in 0.99 throughout the spectrum analyzed. In this section two different types of light guides: circular and rectangular cross sections have been analyzed (Fig. 8).

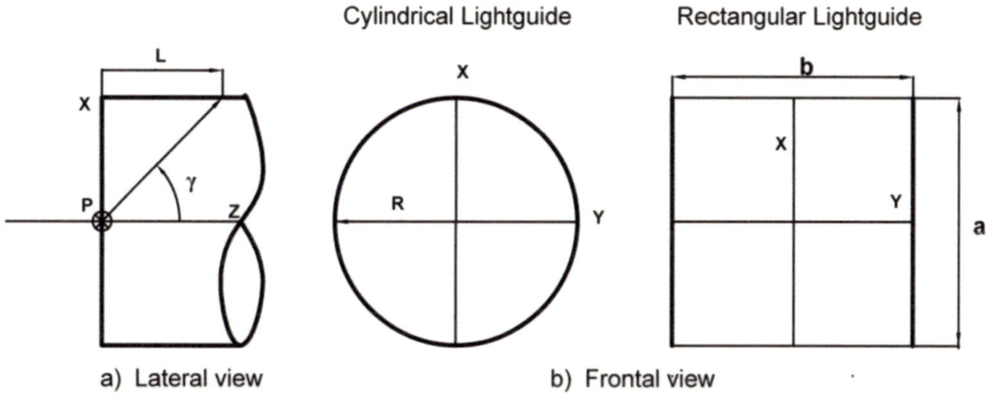

Figure 8. The Coordinate system defines the angles of incidence and observation in the guides. Longitudinal section (a) and transverse section (b).

The photometric profile of the source is an angular distribution adapted to the optimal transmission features of the prismatic film used, and the analysis was restricted to the spectral range between 380 and 780 nm. The CIE Illuminant D65 is the reference light source.

The spectral reflectance of aluminum is estimated as isotropic over the fence and has been experimentally determined using a Hitachi U-3400 spectrophotometer with special accessories for measuring the specular reflectance at 12° incidence. Experimental spectral reflectance measured of aluminum is shown in figure 9.

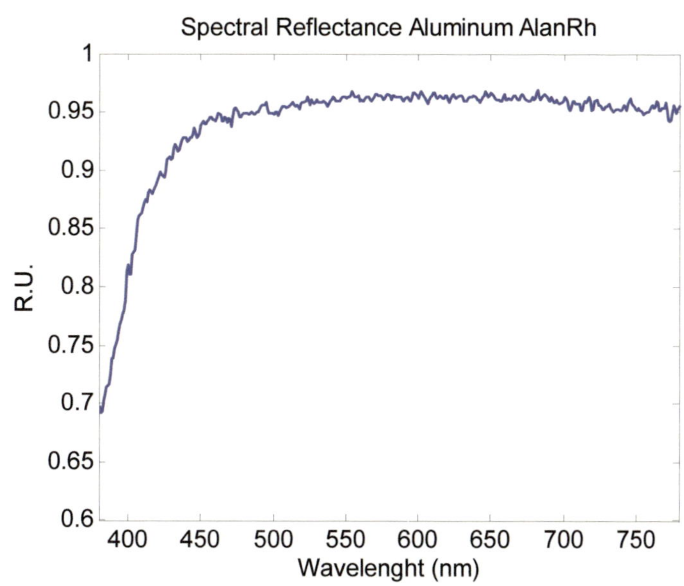

Figure 9. Aluminum spectral reflectance.

To determine the spectral distribution of radiation emerging from the rectangular guide, it is used the model developed by Whitehead (Whitehead, L.A., 1982). According to this model, the number of traversals (Δn) per unit length (Δz) of a ray of light undergoes inside the guide is given by the following expression:

$$\frac{\Delta n}{\Delta z} = \frac{\tan\theta}{Dr}, \tag{1}$$

where Dr is the average cross-sectional distance travelled by a ray in crossing the guide air space and θ is the angle by which any ray deviates from the guide's axial direction.

If the guide air space is rectangular, with dimensions a and b (fig.8), then

$$Dr = (a^{-1} + b^{-1})^{-1}, \tag{2}$$

If roughly circular with radius R

$$Dc = \frac{4R}{\pi}, \tag{3}$$

In the rectangular case, the dimension a is the same as b (0.8862 meters) , the cylindrical guide aperture radius is 0.5 meters in order to maintain the same input area than rectangular guide, and the length of the guides (L) evaluated are 5, 10 and 15 meters.

The dependence of the radiant flux with the angle to the incident beam is given by equation 4

$$\varphi(\theta) = 2\pi I_0 \left[\cos\theta_2 - \cos\theta_1\right], \tag{4}$$

where I_0 is the total intensity power radiated by the source in a specific direction θ. Moreover, we obtain the spectral distribution of radiant flux incident S_λ, where S_λ is the primary illuminant spectral distribution. θ_1 and θ_2 are the angular limits and the intensity has I_0 value.

The spectral and angular distribution at the exit of the guide will be:

$$S' = \int_0^{\pi/2} P(\lambda,\theta)\rho(\lambda,\theta)^n d(\lambda,\theta), \tag{5}$$

ρ is the reflectance of the material for each wavelength studied and n is considering the number of internal reflections obtained for each incidence angle theta θ in the guide.

The prismatic guide has a higher transmission throughout the spectrum (Fig. 10) and the relative spectral power is constant. In the aluminum guide, the energy area of the shorter wavelengths obtained is reduced due to low spectral reflectance of the aluminum.

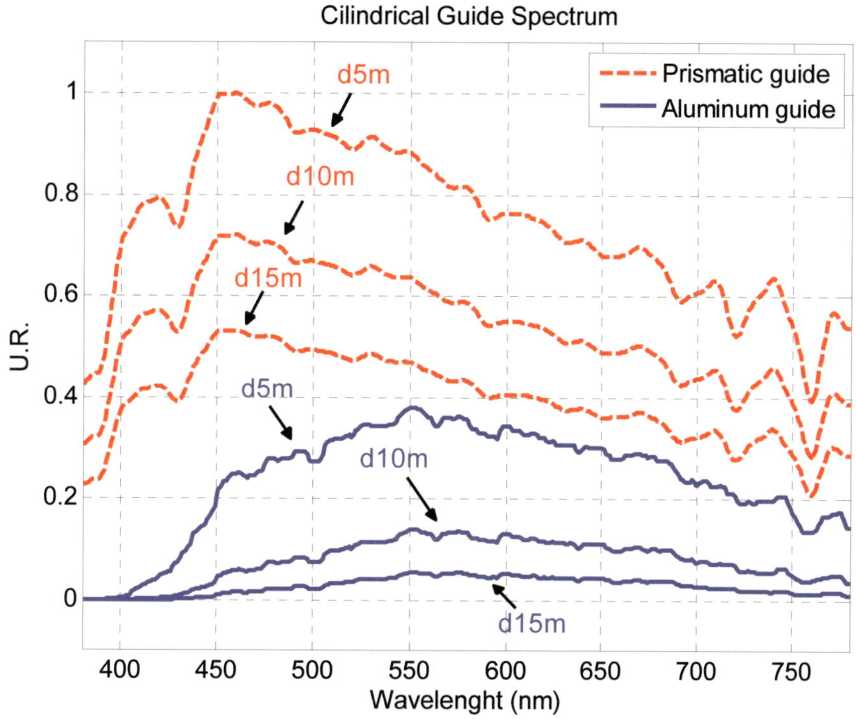

Figure 10. Spectral Power Distribution of the two types of guides for different lengths.

3.2. Color analysis for circular and rectangular lightguides

The spectral power distribution is used for the estimation of the spectral ratio resulting. The spectral ratio is used to compare the spectral distribution at the input and output of the guides. The spectral ratio (η_{λ}.) will be:

$$\eta(\lambda) = \frac{S'(\lambda)}{S(\lambda)} \tag{6}$$

Considering $S(\lambda)$ the input spectral power distribution of the source and $S'(\lambda)$. Spectral power distribution ratio emergence from cylindrical guide is greater from rectangular guides with the same dimensions (figure 11). In addition, the ratio decreases significantly in 10 and 15 meters from aluminum guides. The prismatic guide has high transmission across the spectrum and the spectral power is constant. There is a downward trend in short wavelengths in aluminum guide energy due to the spectral reflectance characteristics of the material.

We define the spectral ratio, η_{λ} as the fraction of the incident spectral power distribution transmitted by the guide:

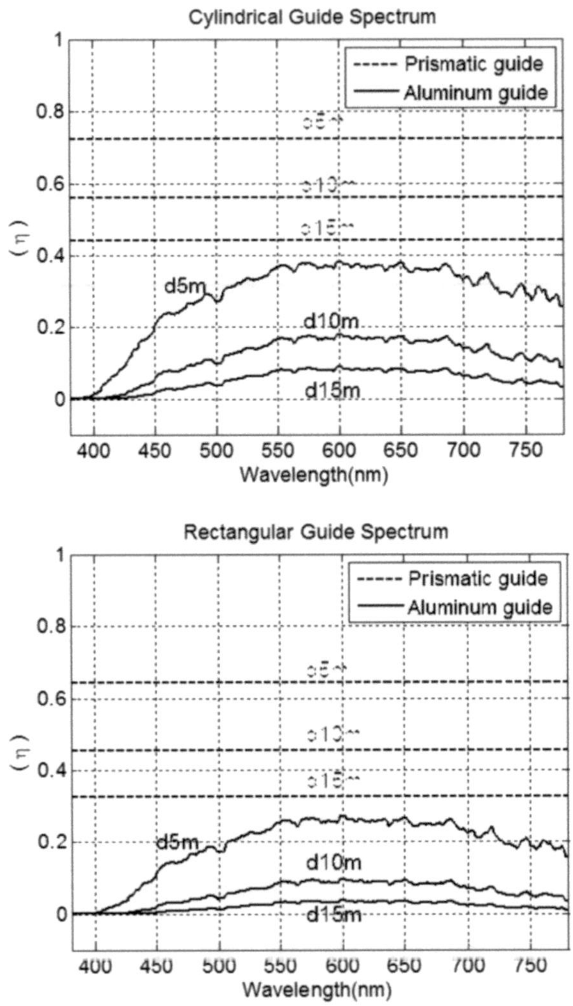

Figure 11. The ratio of spectral power distribution of the two types of guides for different lengths.

For comparing different behavior of light guide regarding color output flux it is used CIE 1931 diagram due to it is more spread on known color representation (CIE 13:3, 1995).

Color coordinates in CIE 1931 chromaticity diagram for studied lengths of aluminum and prismatic guides are show in figure 12. The chromatic coordinates with the light source (D65) chosen for different lengths of guide will be compared. The three consecutive black square points and blue circle points represent the aluminum guide in 5, 10 and 15 meters.

The three consecutive points (d5m, d10m and d15m) represent the aluminum guide. The results of the CIE 1931 chromaticity coordinates of the full spectrum transmitted in the prismatic guides differ from those transmitted in the aluminum guides approaching to the yellow zone. When the length of the aluminum guides increases, there are no changes in the results obtained for the prismatic guide (P), and the result are superposed with the illuminant D65 because the reflectance is maintained for the entire spectrum.

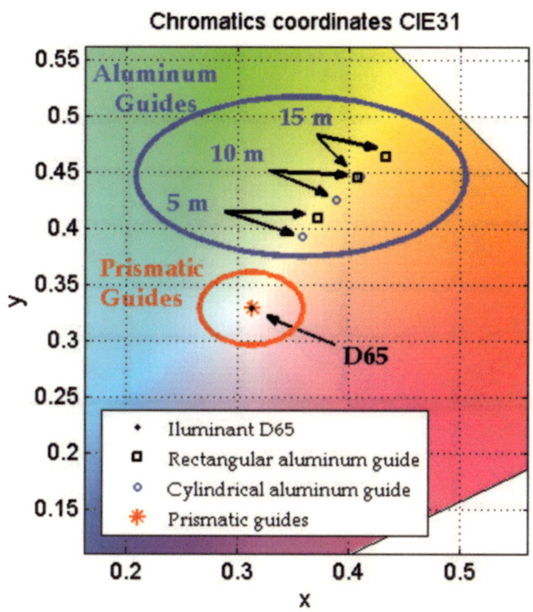

Figure 12. CIE 1931 chromatic coordinates of the studied lightguides.

The CCT is obtained from the spectral distribution at the exit of the guide (Table 1). This calculation is carried out for different lengths of the two types of guides considered in this work. To determine the reference light source for a given test source, we must find the CCT of the test source. Once this data is known, the reference light source is a Plankian black body which has the same temperature.

Moreover, the Color Rendering Index (CRI) of the sources at the end of the guides it is calculated (fig.13). From the result obtained (Fig.10), it is possible to determine the significant color change in the aluminum fence, showing less color change difference for all lengths than the prismatic guides.

	Cylindrical Guide		Rectangular Guide	
	CCT (ºK)	CRI	CCT (ºK)	CRI
Iluminant D65	6503	99.97	6503	99.97
Prismatic guide (5m)	6503	99.97	6503	99.97
Prismatic guide (10m)	6503	99.97	6503	99.97
Prismatic guide (15m)	6503	99.97	6503	99.97
Aluminum guide (5m)	4707	89.81	4407	87.74
Aluminum guide (10m)	4126	85.45	3818	82.39
Aluminum guide (15m)	3781	81.98	3478	78.15

Table 1. Correlated Color Temperature (CCT) and Color Rendering Index of each spectral power.

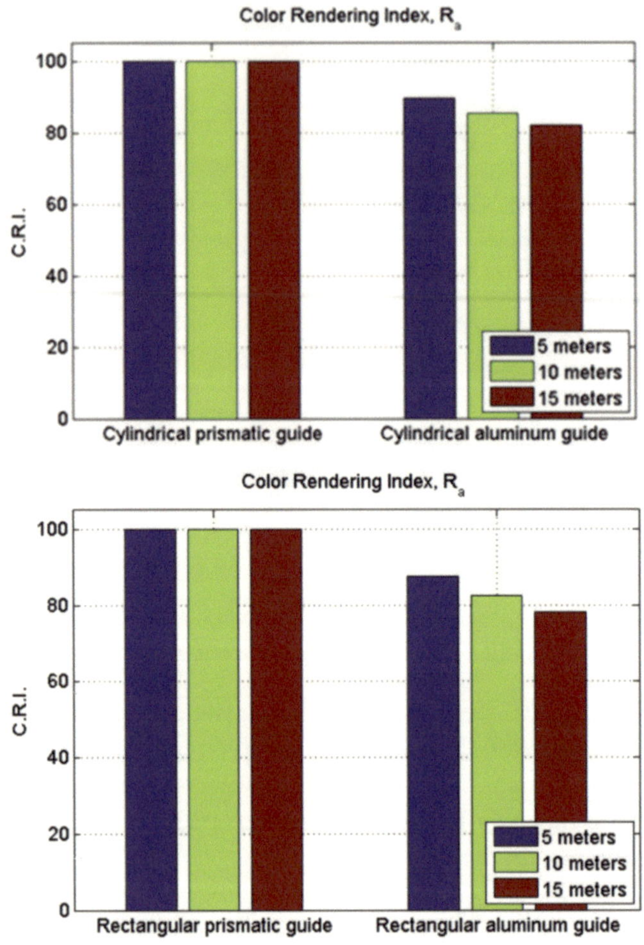

Figure 13. CRI of measured lightguides (cylindrical and rectangular).

3.3. Flux transfer analysis

The flux transmitted of the two types of prismatic guides for a wavelength is studied in order to analyze the efficiency of both systems. For this purpose, the distribution characteristics are calculated using a three-dimensional ray tracing. In this case, the lightguides have a 1.49 refractive index for the studied wavelength without absorption losses. Generally, the associated absorption loss in prismatic film is rather low, and taking into account that it is considered the model of light reflection, in this study this parameter will not be studied on the color quantity estimated.

The dimensions of the guides analyzed in the ray tracing study (Fig.14) are the same defined in the theoretical simulation (Fig.8) and the enlargement scale factor of the prism shape is on the order of 10. The light cone input is an extensive emitter with a random distribution of 30º semi angle (θ) which has the same size as the section of the light guide.

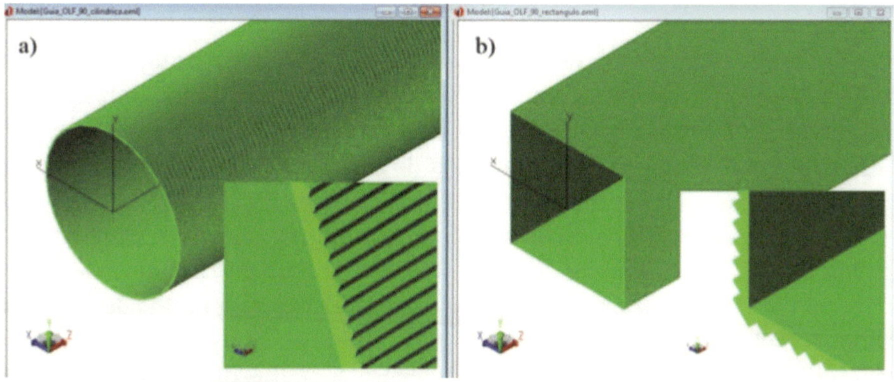

Figure 14. Perspective view of Ray trace simulation model of lightguides with a detail of the prismatic structure: cylindrical (a) and rectangular (b).

The comparison of efficiency in rectangular and cylindrical prismatic guides is shown in Figure 15; the graphics indicates the influence of the shape in maintaining the transported flux and the efficiency for a wavelength (546 nm). The output flux efficiency of 15 meters cylindrical guide is 2% higher than rectangular guide; in addition, in the first meters the flux remains more constant.

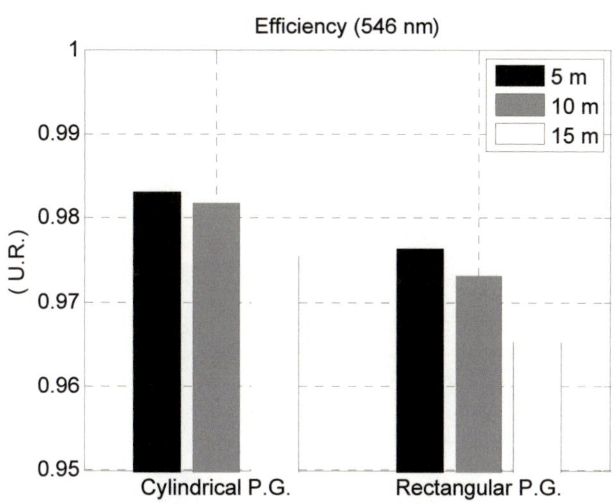

Figure 15. Output efficiency results for wavelength of 546 nm in Prismatic Guides.

4. Hollow prismatic CPC

Compound parabolic concentrator (CPC) is an optical devices used in the solar energy related areas and also in other applications where radiant energy concentration is needed, being defined as one of the first devices that resulted from the practical application of nonimaging optics (Welford and Winston, 1978). Light from a defined range of angles of incidence is reflected by total internal reflection on the parabolic walls of the CPC and concentrated at the exit of the CPC.

The authors propose an innovative 3D hollow prismatic CPC (PCPC) in reverse mode made of a prismatic dielectric material, which has a high efficiency comparing it with aluminium CPC (ACPC). The basic idea is to use a hollow prismatic light guide with CPC shape. In figure 16 (up-left), we can observe the design in 2D geometry in the inverse mode proposal; all the rays entering at the focus of the parabola (F_1 and F_2) emerge through the exit aperture with the design angle θ. This paper reports 2D, 3D design (Fig. 16) and numerical analysis by ray-tracing software, furthermore experimental results are shown. A prototype has been developed and tested showed in figure 16 (down). The hollow PCPC in reverse mode has an entrance pupil that is small compared to the exit pupil depending on the design angle. This CPC design accepts light in 2π entering the entrance pupil and redirecting it in the CPC design angle. This new concept is made of a prismatic film; this dielectric layer accepts light not only in the entrance pupil (Entry 1) but also through the layer itself. This property allows an increase in efficiency compared with the ACPC.

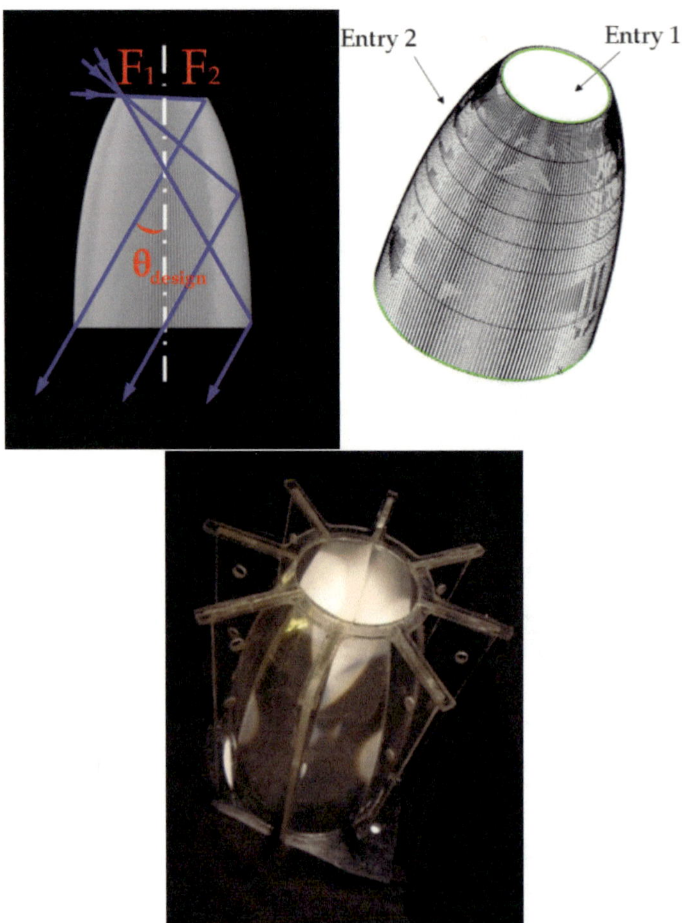

Figure 16. The CPC profile (up-left) with a ray tracing showing the design angle and the maximum input angle of design θ, 3D hollow HCPC 30º software design (up-right) and the experimental prototype (down) in which the reflector surface is a prismatic film supported by eight polycarbonate ribs.

In this section we analyzed the PCPC by raytracing software to determine the angular transmission, optical efficiency and irradiance distribution on the system exit aperture in a 3D system, later a comparison between ACPC and PCPC is presented.

4.1. Ray-tracing: Efficiency of the PCPC compared to ACPC

Ray-tracing is processed for two kinds of CPCs, one designed with standard aluminum with a reflectance of 1.0 and the other one, the prismatic CPC designed with PMMA.

4.1.1. Polar Isocandela Plot of ACPC and hollow PCPC

The polar intensity diagram provides the shape of the light distribution of both parabolic systems. Figure 17 (a) shows isocandela plot representing the ACPC intensity and angular distribution. The 17 (b) illustrations show the isocandela plot for PCPC when light enter through entry 1+2.

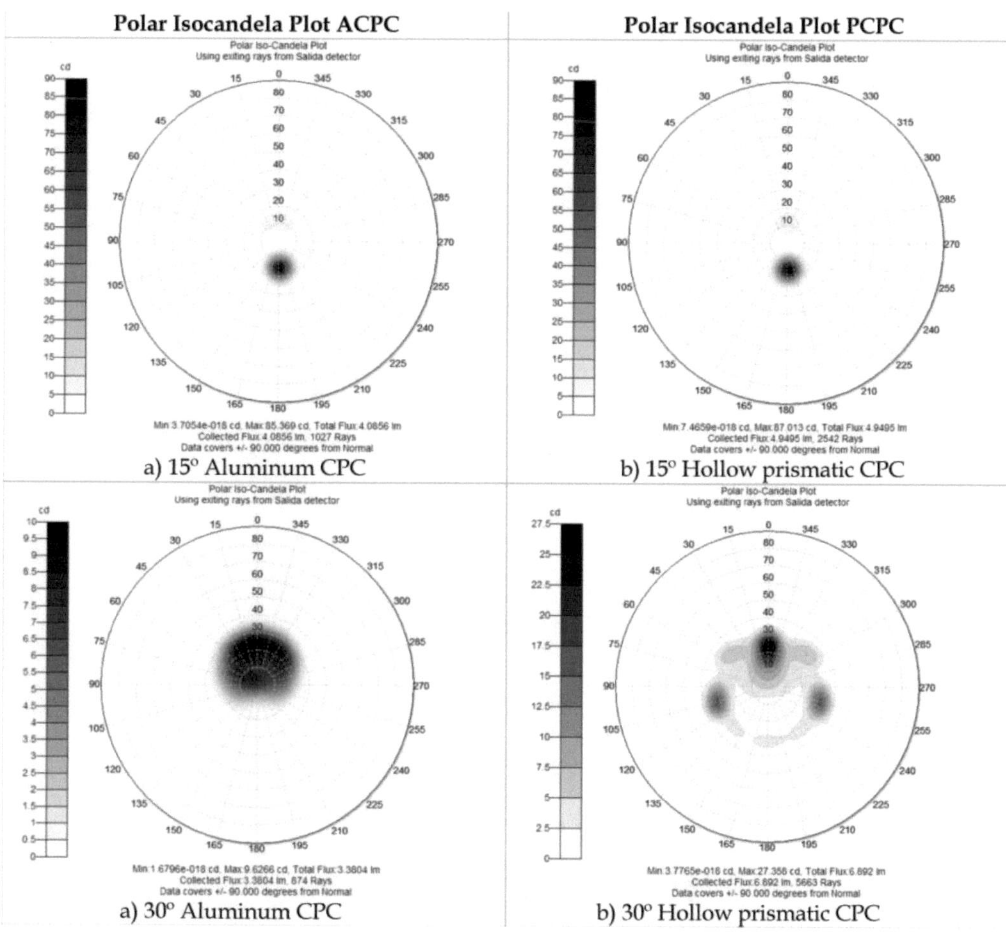

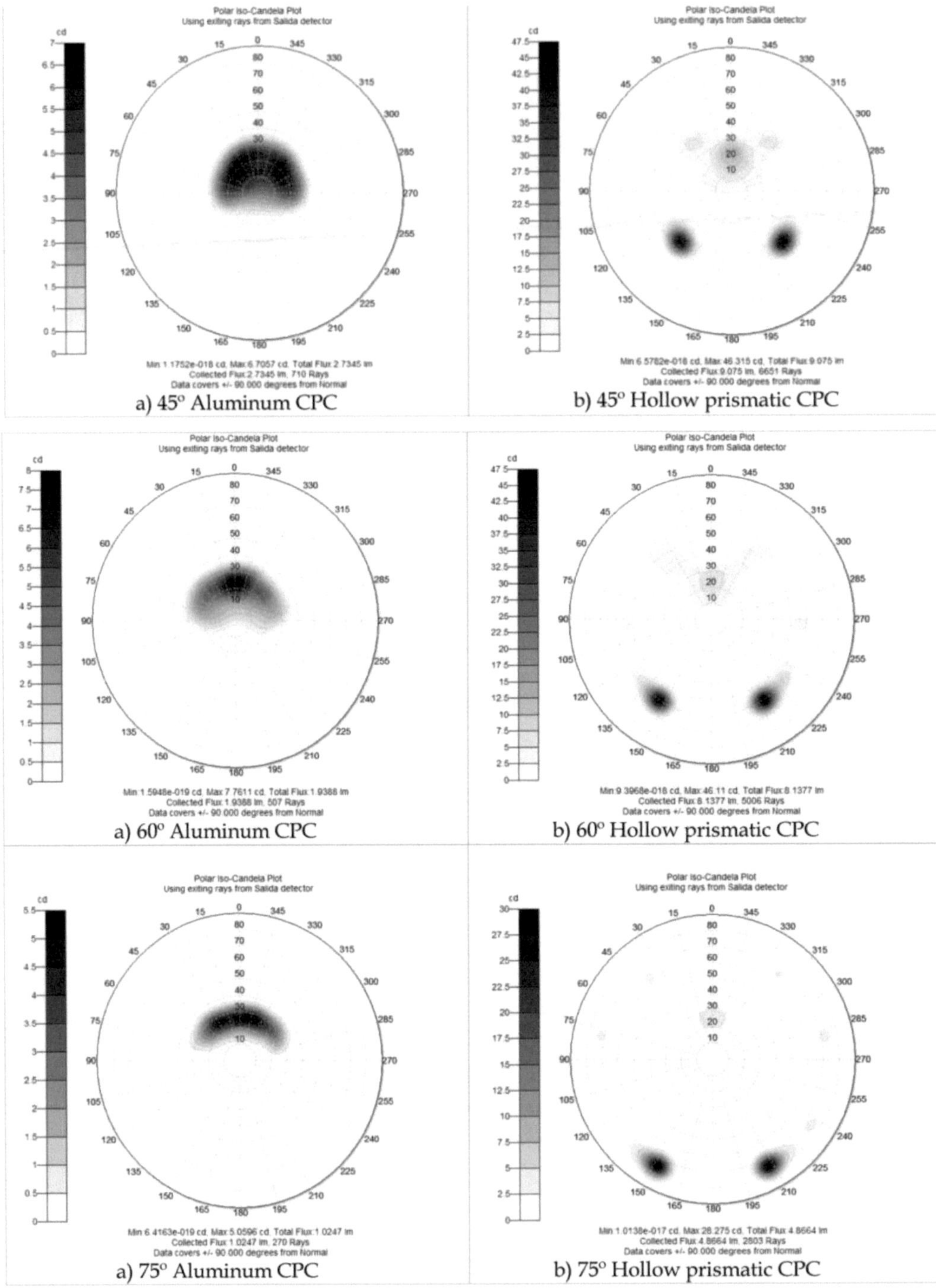

Figure 17. The polar intensity diagram provides the shape of the light distribution to the exit of both parabolic systems

5. Efficiency comparative

Ray-tracing is processed for two kinds of CPCs, one designed with standard aluminum with a reflectance of 100 % and the other one, designed with dielectric prismatic film.

Computed efficiency is obtained by tracing collimated rays in 5º angle intervals for both systems and computing the obtained flux at the exit pupil of the system.

The comparison of PCPC and ACPC collectors is shown in figure 18; it indicates the effects of incidence angles and the efficiency of the PCPC. To compare both systems it is used collimated light at different angles. The flux obtained in the entry 1 (Φ) is used to calculate the final efficiency (η),

$$\eta = \frac{\phi_{entry1}}{\phi_{exitpupil}} \tag{7}$$

The PCPC accepts light out of this entry pupil Entry 1 so η can be higher that 100%.

When it is analysed the efficiency obtained in both systems adding the prismatic surface of the PCPC (entry1+entry2) to 85º, it is observed the improvement with regard to the ACPC, reaching a 600 % higher efficiency flux than an ACPC.

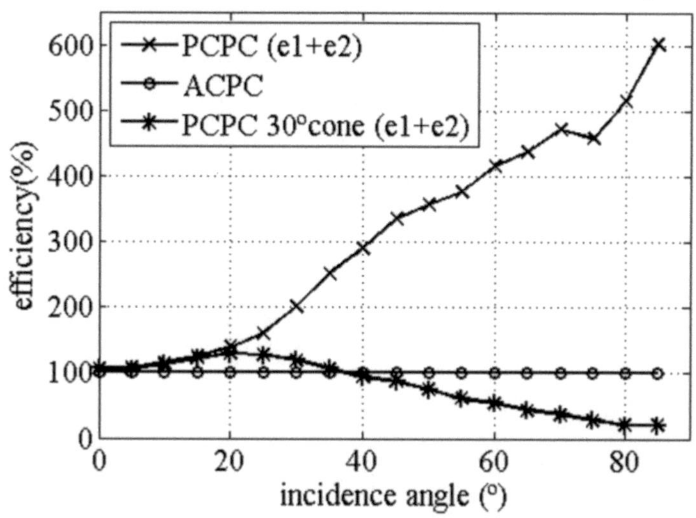

Figure 18. Output flux HPCPC VS output flux HPCPC 30º cone using entry1 and entry2

The efficiency of the PCPC to compare the outflow of 30º cone (*) (entry1+entry2) with regard to the efficiency of entry 1 it is evaluated. There is a clear profit for incidence angles ranging from 0º to 35º, though it is necessary to improve the efficiency for the higher incidence angles.

It is necessary to investigate how the PCPC is working when light enters in the entry pupil (entry 1) as it is done in ACPC.

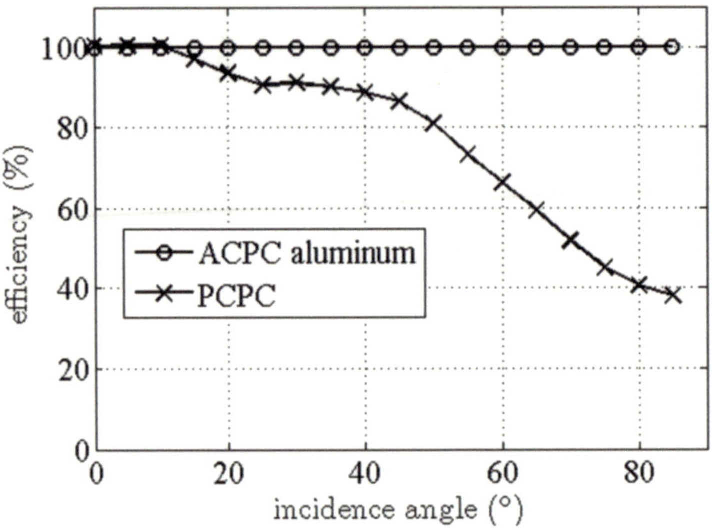

Figure 19. ACPC VS PCPC using entry 1.

The ACPC works better than PCPC as the incidence angle increases if we use only entry 1 in the prismatic PCPC. Figure 19 show this behaviour for different incidence angles. The TIR that suffers light beam in the outer surface of the PCPC is not happening when incidence angle increases, this behavior explains the decrease of efficiency according to the incidence angle increases The division of the beam showed in figure 20 (b) is due to the prismatic effect when light reach the film in the outer side as shown basic raytrace in figure 1a.

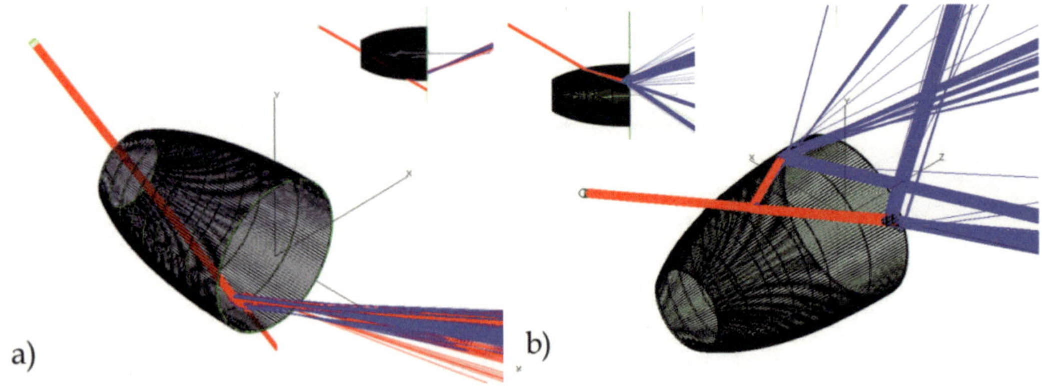

Figure 20. Examples of collimated light entering entry 1 30º (a) and entry 2 30º (b).

The light guiding after PCPC is a good alternative to light far away from the collecting system, the use of a hollow light guide is demonstrated as a good way to transport light, and new bending systems show a good way to reach all the lighting necessities.

6. Experimental measures

A CCD video photometer (Radiant Imaging Prometric 1400) is used to measure output light distribution in exit pupil with a Lambertian screen (Fig.21). We measure output light in two experimental assemblies changing the incidence angle between 0º and 75º increasing the source angle in steps of 15º. Firstly, we evaluate entry 1, and secondly entry 1+2 is evaluated.

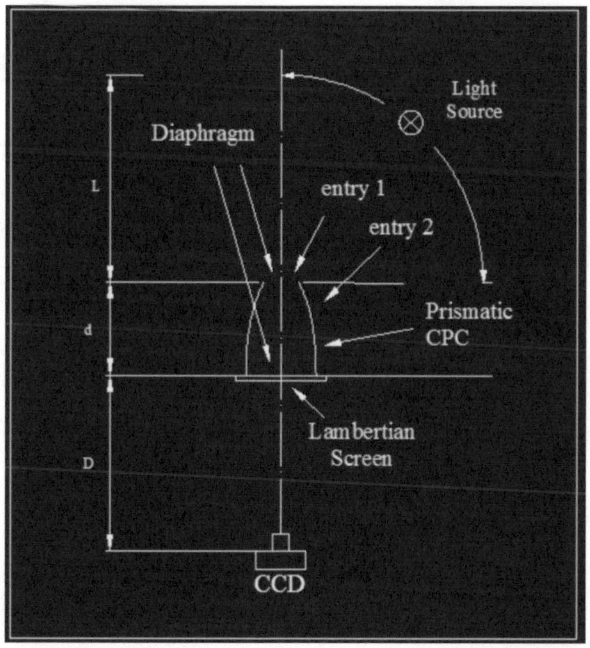

Figure 21. Schematic diagram of experimental setup The HPCPC has the following parameters: input aperture diameter (entry1): 88 mm; output aperture diameter (Lambertian screen): 187 mm, L: 4000 mm, d: 260 mm.

Normalized light distribution map onto exit pupil is shown in the figures 22 to 28. The (a) figure represents the illuminance map obtained with the ray tracing software, the (b) figure represents the map obtained with the experimental setup. Figures 22 to 24 show data for using entry 1 and 25 to 28 show data using entry 1+entry 2 for some light angles.

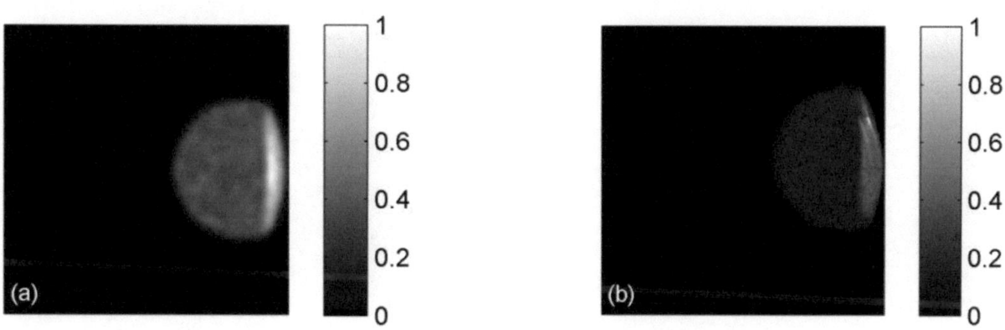

Figure 22. 15º entry1. (a) Raytracing simulation, (b) Experimental measurement.

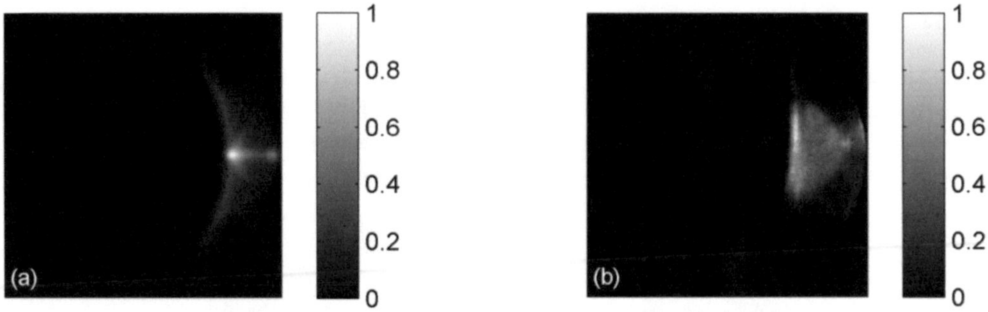

Figure 23. 30º entry1. (a) Raytracing simulation, (b) Experimental measurement.

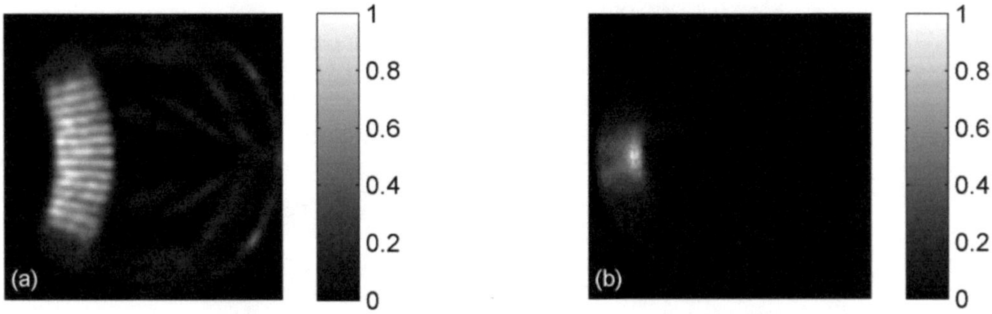

Figure 24. 75º entry1. (a) Raytracing simulation, (b) Experimental measurement.

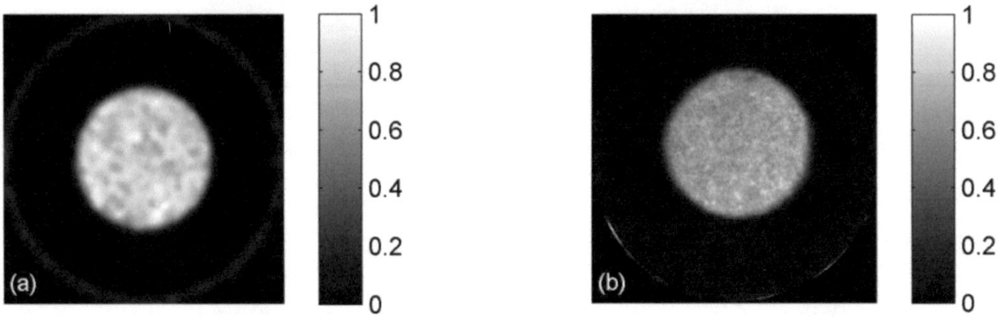

Figure 25. 0º entry1+entry2. (a) Raytracing simulation, (b) Experimental measurement.

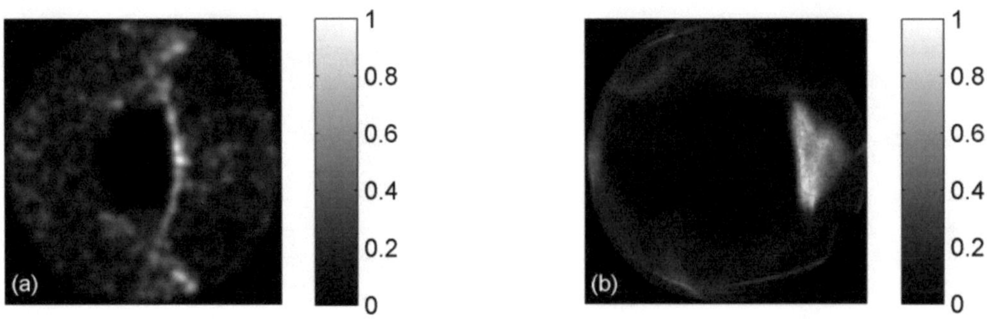

Figure 26. 45º entry1+entry2. (a) Raytracing simulation, (b) Experimental measurement.

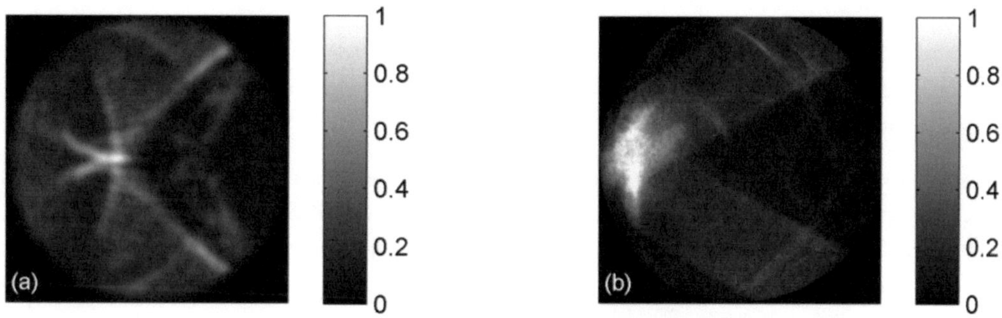

Figure 27. 60º entry1+entry2. (a) Raytracing simulation, (b) Experimental measurement.

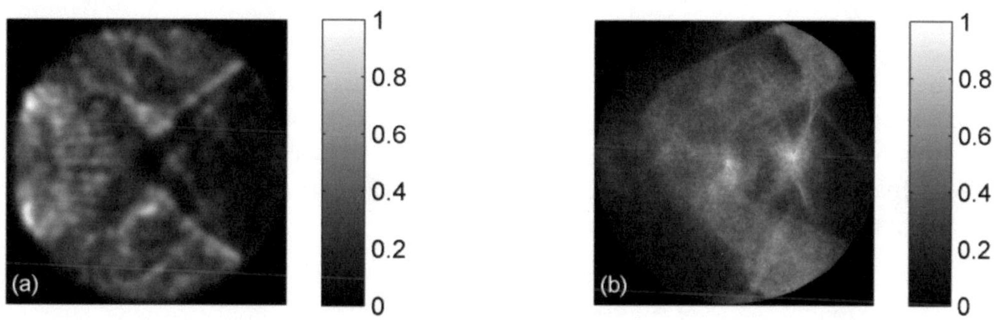

Figure 28. 75º entry1+entry2. (a) Raytracing simulation, (b) Experimental measurement.

The circular peripheral wreath in figure 25 is due to the multiple internal reflections of the light inside the prismatic film. When we increase the angle of incidence, we can observe the division of the beam showed in figure 27 due to the prismatic film structure.

7. Conclusions

The present study shows the interesting combination of light transmission and reflection capabilities of dielectric prismatic film analysed in different guidance geometries and the approach of a collecting system for sunlight applications. Prismatic film can work as a perfect mirror or transparent material depending upon the angle that light strikes the material, it allow us to produce lighting products with unique properties.

It is important to analyse the micro-structure prism imperfections in the form of the surfaces: imperfect corners (apex and valley), surfaces which are not optically flat or which deviate from the expected angle, optical inhomogeneities in the material and the existence of the curved area on the peaks prism which modify the optical behaviour of the prism film; this imperfections modify the optical path and therefore the rays can be directed to other directions instead of undergoing total internal reflections. In order to check the importance of prism rounding, the apex is analysed through Hough Transform by digital image processing. The peak curved radio obtained is despised because of the radius size obtained by image processing is small with regard to the size of the prism structure. The prism are calculated to have an angle of 89.88º and the diameter of the curved region obtained in the prism´s peak is 27.56 μm, this result could be affected by the pressure exerted to make the cut of the film.

Dielectric prismatic guides have a high quality output flux regarding standard light guides based on specular material. Specific shapes such as a circular cross section show a soft increase in flux transmission, with the extent of the improvement dependent on the input position of the light ray.

When input light is in the admitted angle the prismatic light guides have a higher transmission in the entire spectrum than the reflective guides and their spectral efficiency is more constant. There is a downward trend in short wavelength in aluminum guide energy due to the spectral reflectance characteristics of the material.

Prismatic light guides turn out to be more robust in lighting quality maintenance than the aluminum guides, which are efficient but only capable of maintaining light quality distances lower than 5 meters. Cylindrical light guides have a 2% higher efficiency than rectangular although the rectangular shape could be more convenient in office buildings due to the occupied space and construction constraints.

In this study, the authors propose an innovative 3D hollow prismatic CPC (PCPC) working in reverse mode and made of a dielectric material, which has a high efficiency compared with aluminium CPC (ACPC). Transportation of daylight over longer distances requires an optimized collector, the PCPC is an appropriate design for natural light systems like skylights and collector guiding systems since it has properties as collimator to catch the light and to direct it and transport it long distances from a remote source with little attenuation. The hollow PCPC has an entrance pupil that is small compared to the exit pupil depending on the design angle. This CPC design accepts light in 2π entering the entrance pupil and redirecting it in the CPC design angle. This new concept is made of a prismatic structure film; this dielectric layer accepts light not only in the entrance pupil but also through the layer itself.

The results obtained shows that measured PCPC efficiency compared with standard aluminum is 600% higher at 85º incidence angle, a medium value of 300% increase is obtained in the range from 0 to 85º. There is a clear profit for incidence angles ranging from 0 º to 35 º, though it is necessary to improve the efficiency for the higher incidence angles.

The design of big structures in buildings is easier with this new system because of the minor weight of the plastic material which can be conformed in independent parts and development in moulding fabrication can improve the cost of the system.

Author details

Daniel Vázquez-Moliní, Antonio Álvarez Fernández-Balbuena and Berta García-Fernández
Dept. of Optics, School of Optics, University Complutense of Madrid, Spain

8. References

Alvarez Fernandez-Balbuena A., Vázquez-Moliní D., Garcia-Fernandez B., Garcia-Botella A. & Bernabeu E. (2009). Skylight: a hollow prismatic CPC, *Proceedings of the SPIE*, 7423, 74230T.

Alvarez Fernandez-Balbuena A., Vazquez-Moliní D., García-Fernandez B., García-Rodríguez L. & Galán-Cañestro T. (2010) Daylight illumination system by vertical Transparent Prismatic Lightguide for an office building. *Colour and Light in Architecture*. Knemesi, Verona, pp. 360-365.

Canny, J. (1986). A Computational Approach to Edge Detection, *IEEE Trans. Pattern Analysis and Machine Intelligence*, 8(6):679–698.

CIE Publ. No.13.3 (1995). Technical Report- *Method of measuring and specifying colour rendering properties of light source.*

CIE Publ. No.17.4 (1989). IEC Pub. 50(845) *International.Lighting Vocabulary.*

Dubois, M.D. (2003). Shading devices and daylight quality: an evaluation based on simple performance indicators, *Lighting Research and Technology*, 35(1): 61-74

García-Fernández, B., Vázquez-Moliní, D. & Álvarez Fernández-Balbuena, A. (2011). Lighting quality for aluminum and prismatic light guides. *Proceedings of the SPIE* 8170, 81700T.

García-Botella A, A. Fernández-Balbuena A., Vázquez-Moliní D. & Bernabéu E. (2009), Ideal 3D asymmetric concentrator, *Solar Energy*, 83 pp. 113-119. Hough, P.V.C. (1962). Methods and means for recognizing complex patterns, U.S. Patent 3,069,654, Dec.

Hsieh, C. K. (1981). Thermal analysis of CPC collectors *Solar Energy*, 27 (1): 19-29.

Remillard J.T., Everson M.P. & Weber W.H. (1992). Loss mechanisms in optical light pipes, *Applied Optics*, 31(34):7232-7241.

Kneipp, K.G. (1994). Use of prismatic films to control light distribution, *International Lighting in Controlled Environments Workshop*, NASA-CP-95-3309, pp. 307-318.

Mayhoub M. S. & Carter D J. (2010). Towards hybrid lighting systems: A review, *Lighting Research and Technology*, 42:51-71.

Wang M.W. & Tseng C.C. (2009). Analysis and fabrication of a prism film with roll-to-roll fabrication process. *Optics Express*, 17(6):4718-4725.

OLF 2301 Data sheet, www.3M.com/lightingproducts.

Swift P.D. & Smith G.B. (1995). Cilindrical mirror light pipes, *Solar Energy Materials and Solar Cells*, 36(2) 159-168.

TracePro® Opto-Mechanical Design Software, http://www.lambdares.com/

Vázquez-Moliní D., Álvarez Fernández-Balbuena A., González-Montes M., Bernabeu E., García-Botella A., García-Rodríguez L., and Pohl W. (2009), Guiding daylight into a building for energy-saving illumination. SPIE Newsroom, *Proceedings of the SPIE* DOI: 10.1117/2.1200911.1825.

Welford, W.T. & Winston, R (1978). *The Optics of Nonimaging Concentrators: Light and Solar Energy*. Academic Press, London.

Winston, R. (1975). Development of the compound parabolic collector for photo-thermal and photo-voltaic applications *Proceedings of the Society of Photo-Optical Instrumentation Engineers*. Optics in Solar Energy Utilization, 68: 136-44.

Winston, R., Miñano, J. C. & Benitez P. (2005). *Nonimaging Optics*, Elsevier Academic Press, San Diego, CA.

Whitehead, L. A. (1981). U. S. Patent 4,260,220. *Prism Light Guide Having Surfaces which are in Octature.*

Whitehead, L. A. (1982). Simplified Ray Tracing in Cylindrical Systems, *Applied Optics*, 21(19): 3536-3538.

Chapter 8

Effect of Dielectric in a Plasma Annealing System at Atmospheric Pressure

N.D. Tran , N. Harada, T. Sasaki and T. Kikuchi

Additional information is available at the end of the chapter

http://dx.doi.org/10.5772/50594

1. Introduction

Thin metallic wire is widely used in electrical, electronic and automobile technology. It is, however, necessary to anneal and clean the copper wire after drawing. The traditional manufacturing process of fine wire consists of three processes: drawing, annealing and cleaning as shown in Figure 1 and 2 [1]. These processes use Joule heating and chemicals to anneal and clean the wire [2-4]. However, this method has some drawbacks: low efficiency due to the division of the annealing and cleaning processes and environmentally harmful because of using chemical to clean thin wire, for example, Tri-chloethane, it is hazardous to human body and has ozone layer effect. The new annealing system in which the annealing and cleaning processes are simultaneously operated in Atmospheric Pressure Dielectric-Barrier Discharge (APDBD) [5-9] is the potential solution for these drawbacks. In previous studied, it is shown that wire annealing [10, 11] and cleaning [12] using APDBD is totally possible. Also, our previous studies showed that annealing using APDBD is possible for thin copper wire [13], [14] however; the annealing efficiency is low due to choosing the material, size and shape of the dielectric, discharge gas, applied power,... In this study, the efficiency of dielectric material on the annealing and cleaning results was investigated. It is, however, the dielectric material and the dielectric size are the first two important parameters need to be investigated. Moreover, the efficiency of the frequency and the applied voltage of the applied power in the dielectric permittivity were also investigated. To clarify these dependences, an equivalent circuit model is used to analyze the effect of dielectric.

1.1. Annealing

Based on Kalpakjian's Manufacturing Engineering and Technology, annealing is a general term used to describe the restoration of a cold-worked or heat-treated metal or alloy to its original properties, such as to increase elongation rate and reduce hardness and strength, or to modify the microstructure. The traditional thin metallic wire annealing is using external heating source or Joule heating as shown in Figure 2.

The purpose of annealing thin metallic wire is using heat to increase elongation rate in short duration. The temperature and duration of annealing affect the crystal size and crystal texture of thin wire. And elongation rate depends on the size and orientation of crystal. In addition, the annealing temperature is configured by drawing process; generally lower than 2/3 melting point of the wire material. With assuming that the drawing process is stable then deformation degree is constant. Therefore, in order to reach the required elongation rate, the temperature, and duration of annealing need to be properly chosen.

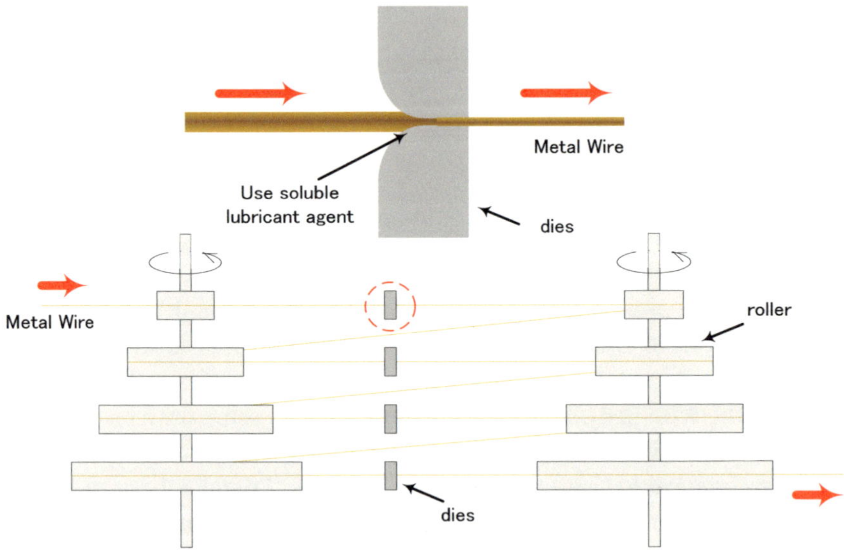

Figure 1. Mechanism of drawing process

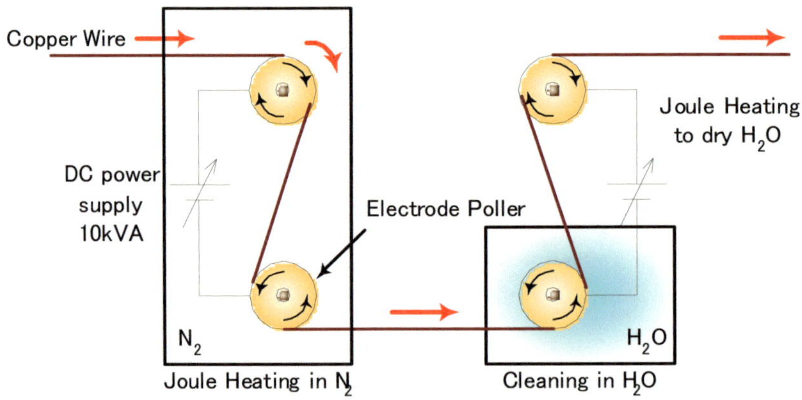

Figure 2. Traditional mechanism of annealing and cleaning process

Heating thin metallic wire at annealing temperature is a thermodynamic process, which is used to rearrange or eliminate of dislocations (recovery), create new crystal (primary recrystallization) and grow crystal (recrystallization). After annealing, if the crystal size is enlarged and in good orientation, the elongation rate increases. At the first state, the recovery process, some the crystals are not fully rearranged therefore the elongation rate is still low. At primary recrystallization process, a number of crystals in the wire are born and this process strongly depends on the temperature but not duration. However, when the temperature is so high, the number of formed crystal increases then crystal size is small. Small size crystal means elongation rate is low. Therefore, choosing temperature in this process is important to get good orientation of crystal. At the recrystallization process, the crystals were born from the previous process will grow up. The crystal growth depends on duration. No noteworthy crystal growth will occur in the short annealing period then crystal size remains the same. Therefore, the duration of annealing in this process is important to get good orientation of crystal. For continuously thin wire annealing, to have good period annealing, applied voltage and the velocity of thin wire moved through or the length of chamber along the thin wire of plasma reactor have to be considered. If annealing duration is long, the annealing temperature at the end of the process is high and it is the cause of some disadvantages like increasing surface roughness. Moreover, during annealing at high temperature, the presence of hydrogen in air can increase the roughening effect of thin wire.

1.2. Plasma annealing

In order to solve the environmental harm of traditional annealing method, the plasma annealing is replaced. At low pressure such as in vacuum, the temperature generated by the ion bombarding on a target was studied for using in plasma immersion ion implantation (PIII) [15, 16]. This phenomenon was generally applied in modifying the target surface [16, 17] or heating the target [18, 19]. However, these systems operate under expensive vacuum systems. Recently, the atmospheric pressure dielectric barrier discharge (APDBD) for wire annealing has become greatly interesting [20, 21] because of its low-cost system and being environmentally friendly.

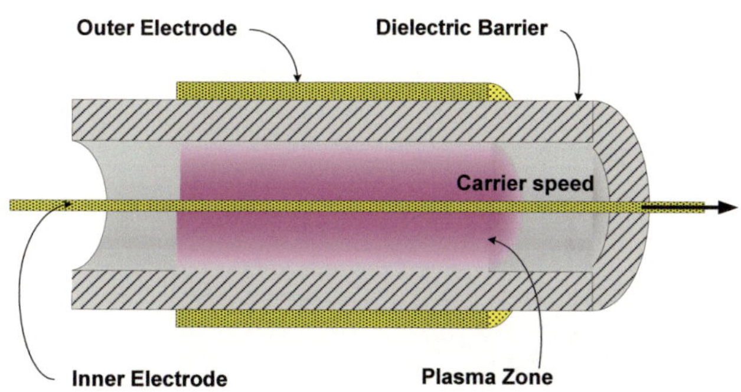

Figure 3. Annealing thin wire at atmospheric pressure dielectric barrier discharge

At atmospheric pressure dielectric barrier discharge, the thin metallic wire is annealed by moved through the plasma reactor as shown in Figure 3. The dielectric is used to prevent the arc and the gas is fed into reactor to assist plasma discharge. The conceptual layout of plasma annealing is shown in Figure 4. In the plasma reactor, the discharge gas is ionized into electron and ion. Under a strong electric field, ion and electron bombard the thin wire surface. The essence of the generated temperature is the impact energy of electrons and ions on thin wire surface as shown in Figure 5. Furthermore, the electron-neutral particle collisions in streamers also assist generating temperature. The total temperature generated by the charged particle bombarding (ions and electrons) and the electron-neutral collision continuously heats the wire surface to annealing temperature.

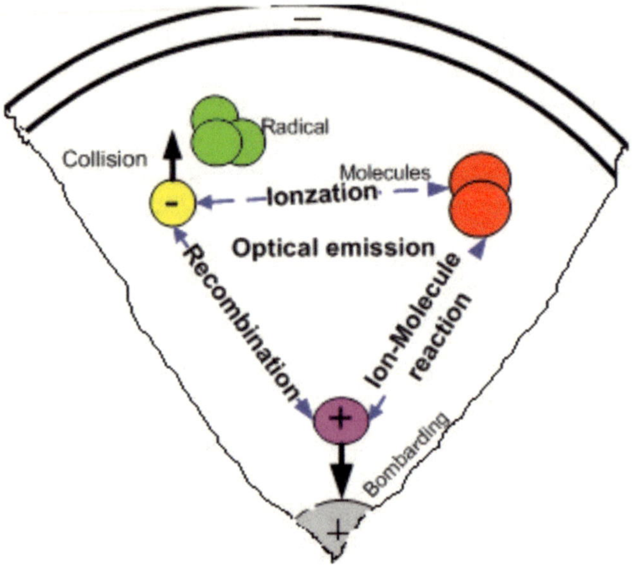

Figure 4. Plasma annealing phenomenon

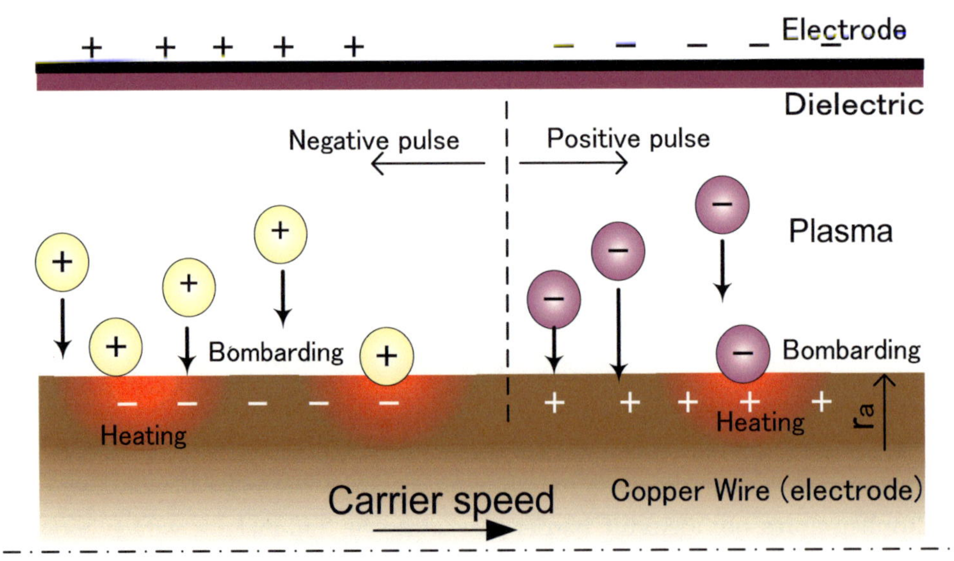

Figure 5. Annealing mechanism

1.2.1. Electron bombarding

Electron bombarding on the target was studied for melting target [22-25]. The electrons bombard the anode (thin wire) and then convert their kinetic energy into thermal energy. Heat transfer occurs from the plasma to the anode in several ways. Firstly, the electrons have thermal energy, which they release upon contact with the anode. Secondly, the electrons also have kinetic energy, which partially gets converted to thermal energy as it passes through the anode. It is important to note that the transfer kinetic energy from electron to anode depends on the current density.

1.2.2. Ion bombarding

The ion bombarding on a target was studied for using in plasma immersion ion implantation (PIII) [15, 16]. When the negative high-voltage pulse is applied to the target, the electrons near the cathode are driven away on the time scale of inverse electron plasma frequency, which is relatively shorter than the time scale of inverse ion plasma frequency, leaving the ions behind to form an ionic space charge sheath. On the time scale of inverse ion plasma frequency, the ions within this sheath are accelerated and then bombard the target surface under the sheath electric field. This phenomenon was generally applied in modifying the target surface [16, 17] or heating the target [17, 18] using low-pressure plasma. It is known that the treatment effect depends on ion bombarding energy, i.e., sheath thickness, ion current, and sheath electric field. On the basis of the fluid model [26] or kinetic model [27-29], it is possible to calculate dynamic sheath thickness, ion current, or ion bombarding energy. Recently, the atmospheric pressure dielectric barrier discharge (APDBD) for annealing thin wire has become greatly interesting [21, 30]. Also, our previous studies showed that annealing using APDBD is possible for thin copper wire [31]. However, the generated temperature by ion bombarding was not estimated. In this study, an analysis model using helium, argon or nitrogen gases and low-frequency (35 – 45 kHz) applied voltage is proposed to analyze the annealing result in APDBD.

2. Experiment setup

The system consists of a gas tank, a power supply, a plasma reactor, a spectrometer sensor and a temperature sensor, as shown in Figure 6. The connecting power to the cylindrical reactor is shown in Figure 7. The design parameters of the reactor shown in Figure 8 are shown in Table 1. The thin cylindrical aluminum electrode (outer electrode) is covered with a dielectric to prevent arcing and it is connected to a power supply (20 kVp-p, 2 Ap-p, and 45 kHz). The thin copper wire (inner electrode) is driven through the reactor by carriers and connected to the ground. Before annealing, the reactor is filled up with discharge gas; helium, argon or nitrogen (purity > 99.9%). During annealing, purified discharge gas is continuously fed into the plasma reactor under a control flow rate to assist the plasma discharge. The applied voltage and current waveform are recorded using a digital oscilloscope (Yokogawa SL1000) with a high voltage probe (Iwatsu HV-P30), and a Rogowski coil (PEARSONTM current monitor 4997), respectively. The elongation rate of three samples is measured by SAIKAWA ET-100. The industrial required elongation rate is higher 20%.

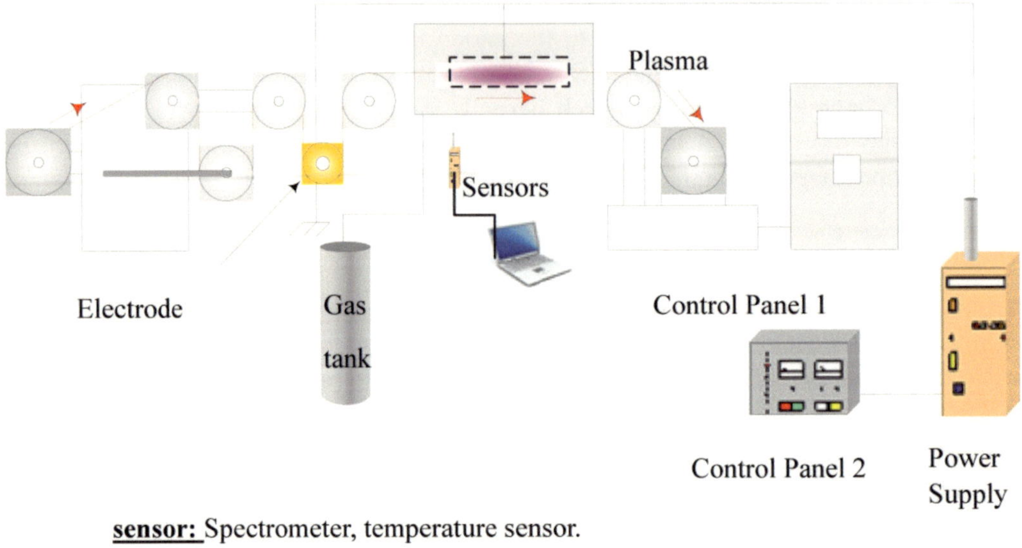

Figure 6. The schematic diagram of the thin wire plasma annealing system

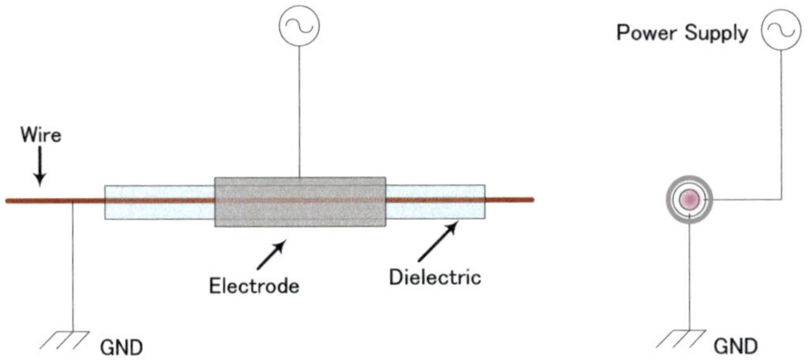

Figure 7. Cylindrical DBD reactor

Diameter of copper wire (mm)	0.2
Dielectric thickness (mm)	2.5
Dielectric material	Glass
Gap length (mm)	4.9
Reactor length (mm)	100
Gas	Ar/He/N2
Gas mass low (l/min)	5
Applied voltage frequency (kHz)	45
Velocity of copper wire (m/min)	30

Table 1. Experimental setup parameters

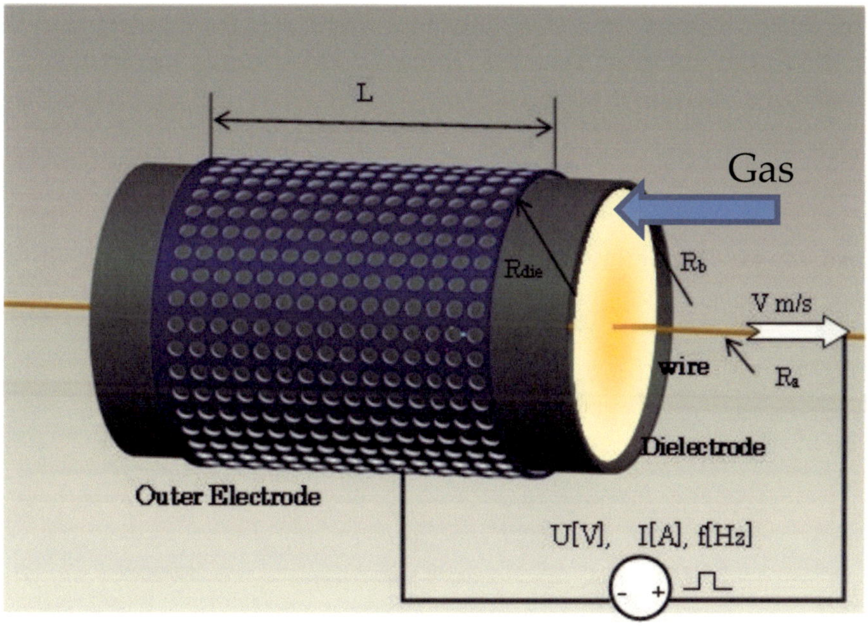

Figure 8. Cylindrical DBD reactor with design parameters

Temperature range	300 °C~1000 °C
Accuracy	0.5% of measurement value +1°
Responding time	2ms
Distance and area measurement	88mm/ϕ0.45~4500mm/ϕ22

Table 2. IGAR12-LO MB10's specifications parameters.

3. Results and discussions

3.1. Dielectric permittivity measurement

The dependence of the dielectric permittivity on the frequency in APDBD annealing system is measured by electrode contact and CLR as shown in Figure 9. Dielectric material is placed sandwich in two aluminum electrodes, high voltage, high frequency function generator is used as power supply. Table 3 shows the experiment equipment be used in this study.

As shown in Figure 9, the dependence of dielectric permittivity on the dielectric voltage V is expressed via capacitance C as

$$V = \frac{I}{j \cdot \omega \cdot C} \tag{1}$$

where parallel capacitor C is expressed by

$$C = \frac{\varepsilon_r \cdot \varepsilon_0 \cdot S}{d}, \tag{2}$$

where S is area of dielectric, d is discharge gap, ε_r and ε_0 is the dielectric and free-space permittivity, respectively. From equation (1) and (2) the dielectric permittivity (3) can be expressed as

$$\varepsilon_r = \frac{d}{\varepsilon_0 \cdot S} \cdot \frac{I}{\omega} \cdot \frac{1}{V}. \tag{3}$$

For comparison, the CRL meter also is used to measure the dielectric permittivity as shown in Figure 9.

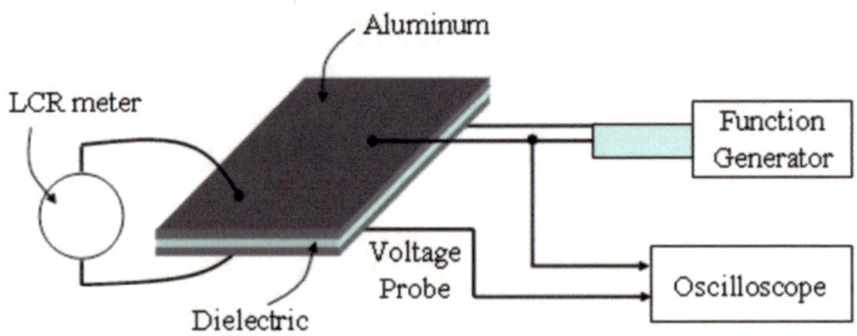

Figure 9. The schematic diagram of the dielectric permittivity measurement

From equation (3), the dependence of dielectric permittivity on frequency was shown in Figure 10. The results show that the dielectric permittivity in the APDBD decreases when frequency increases. At 45kHz, the dielectric permittivity of BN and Al_2O_3 are 12.9 and 6.79, respectively. Moreover, Table 4 shows the comparison of the dielectric permittivity at three measurement methods; our analysis, LCR, and literature. The results also show that at low frequency, the dielectric permittivity is higher than that literature however, at high frequency they are nearly the same.

Function generation	NF Corporation model WF1973
Oscilloscope	Agilent Technologies DSO 1024A
High voltage probe	Agilent Technologies N 2863A
LCR meter	Agilent Technologies U1732A

Table 3. Equipment used

Frequency f [kHz]	Dielectric permittivity ε_r					
	Electrode contact		LCR		Literature [20, 21]	
	BN	Al_2O_3	BN	Al_2O_3	BN	Al_2O_3
1			5.95	4.82	6.8*	8.1
10	31.7	22.9	4.95	4.76		
45	12.9	6.79				
100	7.19	5.46	7.83	5.12		

Table 4. Comparison dielectric permittivity between measurement and literature

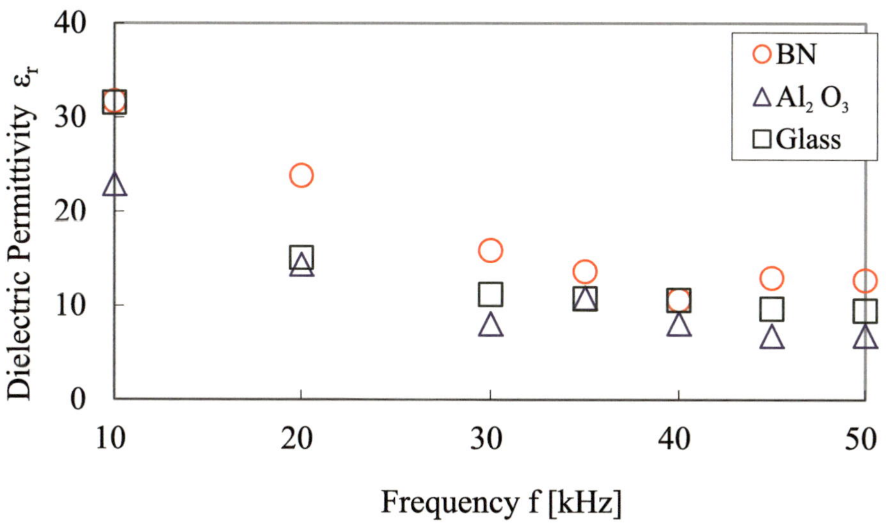

Figure 10. Dependency of dielectric permittivity on frequency

3.2. Dependence of annealing temperature on frequency

Figure 11 shows the experiment result, the dependence of elongation rate on frequency. The result shows the weakly positive relation between the elongation rate and the input frequency. When the frequency increases from 30 kHz to 40 kHz, the elongation rate slightly increases. Thus, we estimate that frequency has no effect on the annealing temperature.

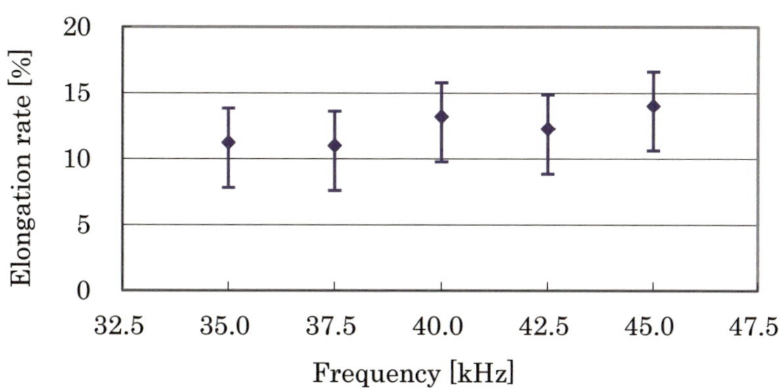

Figure 11. The dependence of elongation rate on frequency

3.3. Dependence of annealing temperature on dielectric material

In this part, the effects of dielectric's material on annealing condition are observed. Figure 12 shows the comparison of the elongation rate with the dielectric substances. This result shows that the elongation rate strongly depends on the dielectric materials. The elongation rate using

BN is higher rate than using SiO_2. According to a comparison of properties of some dielectric materials (SiO_2, Al_2O_3, BN and glass), we acknowledge that dielectric with higher dielectric constants is more effective to reach the annealing temperature as shown in Table 5. Physical characteristics of the dielectric, such as thermal expansion and melting point, are also important. For example, from our experiment the alumina can be suddenly broken with long duration annealing using a water-cooled electrode due to its large thermal expansion. For the best annealing result, boron nitride is an excellent choice for the dielectric material.

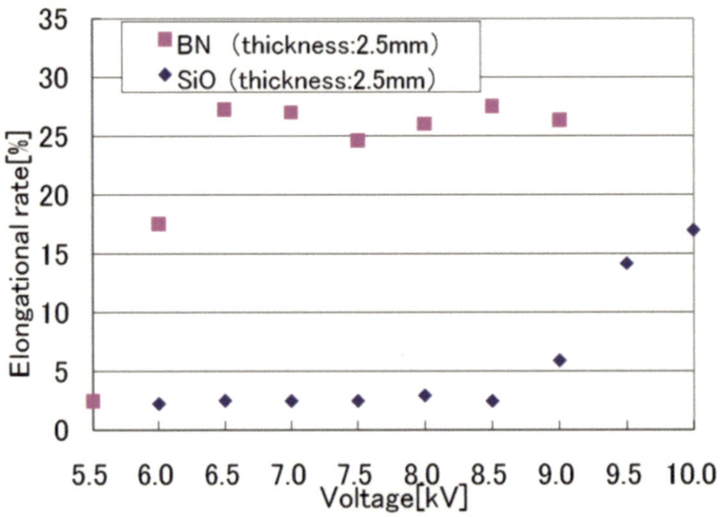

Figure 12. The dependence of elongation rate on dielectric material

Material properties	SiO_2	BN	Al_2O_3	Pyrex Glass
Maximum Temperature (°C)	1713	850	1500	821
Permittivity (F/m)	4.0	6.7	9.5	4.7
Bending strength (Mpa)	105	60.2	310	59.3
Thermal conductivity (W/mK)	1.9	30.98	24	1.005
Thermal expansion (K-1)	8×10^{-6}	0.87×10^{-6}	6.4×10^{-6}	3×10^{-6}

Table 5. Material properties of dielectric

3.4. Dependence of annealing temperature on the thickness of dielectric

Figure 13 shows the dependence of elongation rate on dielectric thickness. This result shows that the annealing effect strongly depends on the dielectric thickness. Compared to the elongation rate, the elongation rate using thin dielectric is higher rate than that using thick ones.

3.5. Dependence of annealing temperature on dielectric size

Figure 14 shows the relation of the elongation rate and the diameter of dielectric. The size of dielectric is weakly affects elongation rate. It is noteworthy that when the copper wire is driven through the reactor, the chatter motion changes the discharge gap length and heating point on the wire surface and consequently, the annealing result. Moreover, when the thin copper wire is annealed in a wide reactor, discoloration and unevenness occur on the wire surface due to the streamer length density reduction. To obtain a steady discharge state, it is usually necessary to reduce discharge gap length. However, decreasing the reactor gap reduces the discharge volume and consequently, the annealing temperature.

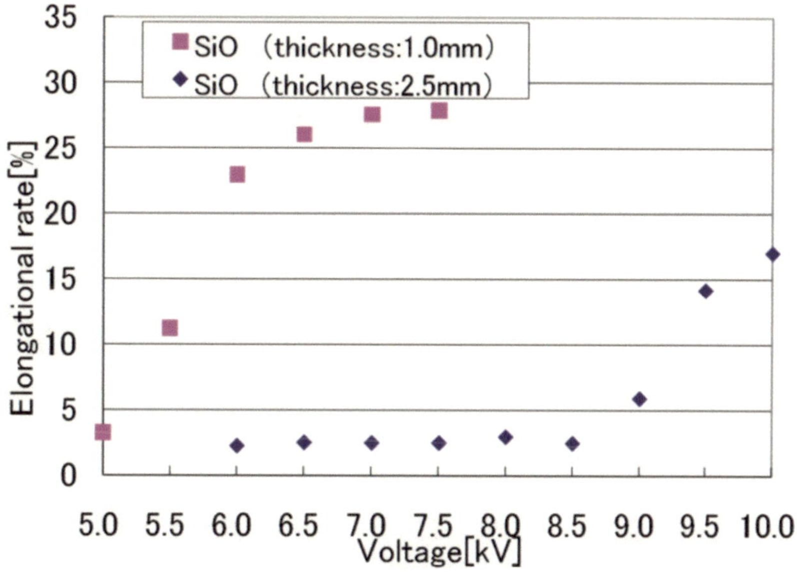

Figure 13. The dependence of elongation rate on dielectric thickness

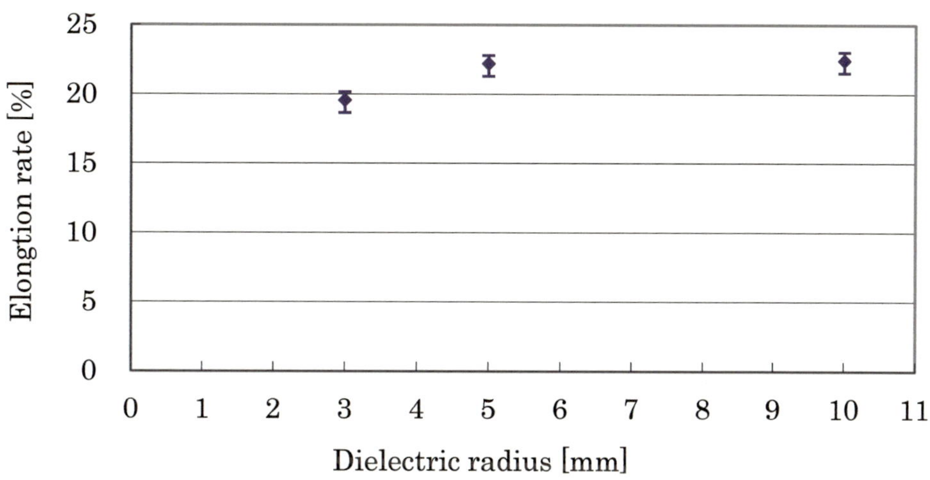

Figure 14. The dependence of elongation rate on dielectric size

4. Equivalent circuit model

In this part, an equivalent circuit model is used to analyze the dependence of the thin copper wire annealing temperature on the dielectric in APDBD. From analysis model, the main factor that determines annealing temperature is the ion bombardment on the wire surface. The average temperature of the thin copper wire in APDBD reactor is calculated as a function of dielectric diameter, dielectric material, applied voltage, ion mass, and gas thermal conductivity. The effect of dielectric on annealing from analysis model is used to compare with that from experiment.

4.1. Discharge mechanism

The moving of electron and ion in the reactor leads to the discharge current behaviors. Figure 15 shows the reflection of discharge current on the applied voltage 9 kV of Al_2O_3 dielectric. At the beginning of the discharge, when increasing the increment of the applied voltage, $dV/dt > 0$, the discharge current crossing the discharge gap increases. The electrons and ions run instantaneously toward anode and cathode and deposit on them to form two space charges. The internal electric field generated by separation of electron and ion space charge. This electric field increases with the increment of applied voltage and is reversely proportional to external electric field. When the incremental applied voltage reaches the peak, $dV/dt = 0$, the discharge current (displacement current) crossing the discharge gap is zero and the internal electric field reaches maximum, the reverse current is formed. When $dV/dt < 0$, the reverse current exists until the applied voltage reaches the local bottom value ($dV/dt = 0$). The process starts again with the same situation when the instant voltage increases again. From experiment, the currents profile of discharge gases is an invariant sine profile and the total current becomes very broad due to the dominant of displacement current. The behavior of voltage-current in the cylindrical APDBD discharge is the same as that of a series RLC circuit.

The discharge characteristics of the annealing system are represented by the total discharge current, which includes the total of displacement current and the conduction current. From Figure 15, the negative current cycle has higher and shaper peak current than that positive one, that leads to the conclusion that the conduction current at negative cycle is higher than that at positive cycle. It is also shown that the phase different between total current waveform and voltage waveform is almost 90°, which indicate that the displacement current takes up large scale value in the total current. The small conduction current in APDBD is because of high resistivity of dielectric. Figure 15 also shows the changing of displacement current (sine shaped profile) and conduction current (narrow peaks) when increasing applied voltage. As shown from the current waveform, in every haft cycle, the waveform has multi peaks and every peak has a sine shape profile during discharge. This characteristic is different from micro-discharge model (contain many very small, short-lived (nanoseconds), many current filaments). In our experiment, the currents profile of all three discharge gases is an invariant sine profile and total current becomes very broad due to the dominant of displacement current. Therefore, the cylindrical APDBD discharge with helium, argon or nitrogen can be modeled by an equivalent series RLC circuit.

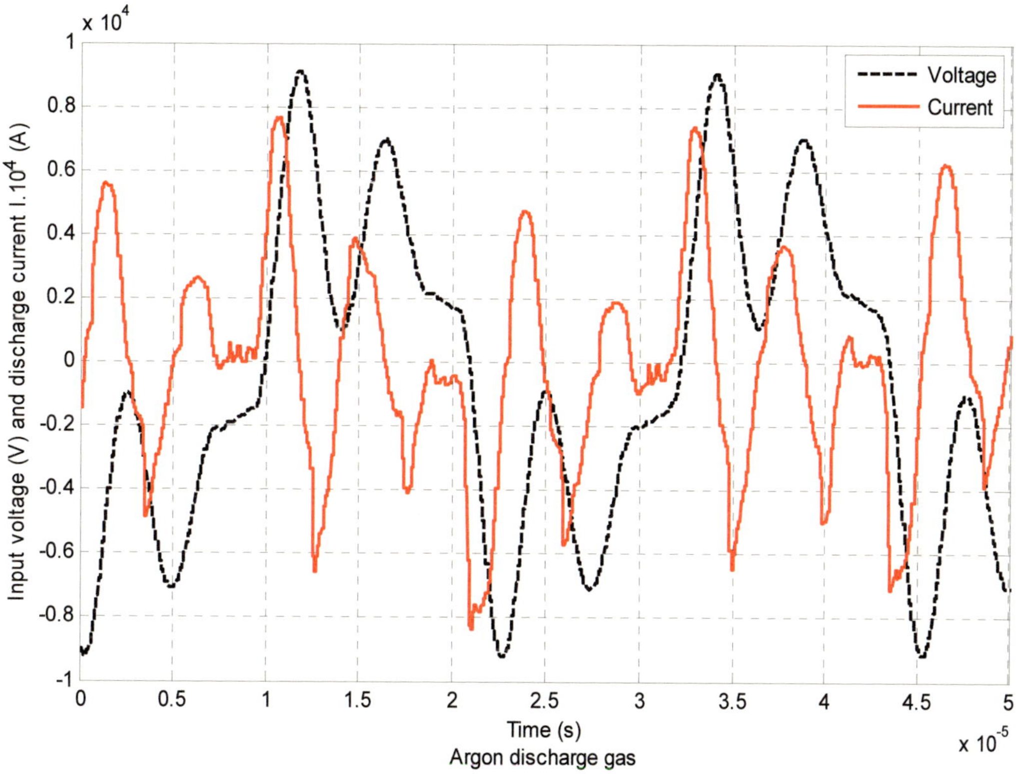

Figure 15. Voltage and discharge current waveform of argon at 8.5 kV

4.2. Equivalent circuit model

Combination between the discharge characteristics of plasma annealing and the reflected physical structure of the plasma reactor, the equivalent circuit model is formed as a RLC circuit in which the total impedance is the combination between series and parallel circuit model. The corresponding main physical part of discharge mechanism shows that plasma reactor can be divided into three parts: (1) dielectric wall, (2) dynamic sheathes and (3) plasma bulk as shown in Figure 16(a). The impedances of dielectric, ionization (Z_p), and non ionization gas capacitance C_g are shown in Figure 16(b). The equivalent circuit model with impedances of dielectric, sheath and plasma bulk is shown in Figure 16(c). In that the impedance of the dielectric is the parallel combination of dielectric capacitance and dielectric heating resistance [32], the impedance of the sheath is the parallel combination of sheath capacitance and ion heating resistance which is presented by ion current in sheath, and the impedance of the plasma is the parallel combination of the R_p (only the ohm heating by streamer is considered) with the cylindrical space capacitance C_p [33]. The diodes, D_a and D_b, are used to specify the sign of the input voltage. The gas capacitance of the reactor before discharging is also connected to the parallel C_g.

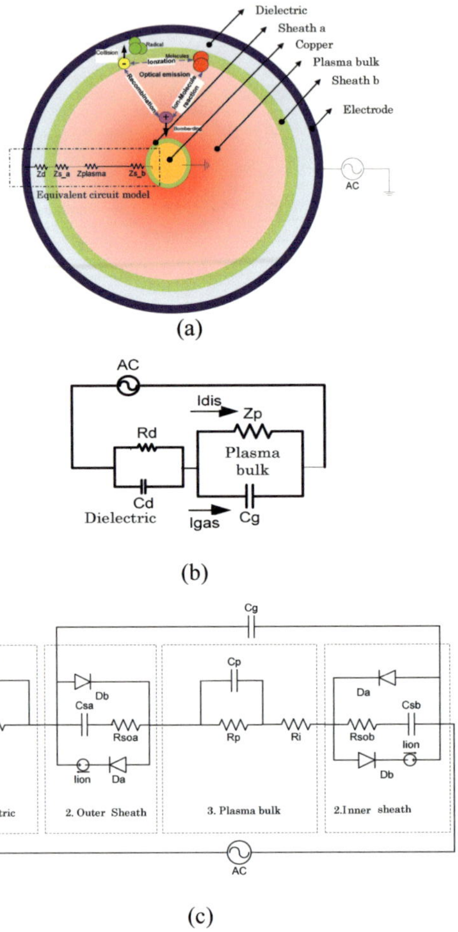

Figure 16. Equivalent circuit model

4.3. The dependence of annealing result on dielectric material

Based on the assumption of the ion heating power, the thin wire temperature from model is [34]

$$T_w = \frac{8 r_a l \varepsilon_0 \sqrt{\frac{2q}{M}} f \int_0^{\frac{\tau}{2}} \left(\frac{0.84\left(V_{in}(t) - Z_{die} I_{in}(t)\right)}{1 + \left(\frac{A_b}{A_a}\right)^i} \left(\frac{A_b}{A_a}\right)^i \right)^{\frac{5}{2}} dt}{9 r_s \beta^2 \left(\rho C_v v r_a^2 + \frac{4\pi\lambda}{\ln\left(\frac{r_c}{r_b}\right)} \right)} + T_0 \tag{4}$$

where $V_{in}(t)$ is the applied voltage, $I_{in}(t)$ is the total discharge current, q is the charge of the ion, M is the ion mass, l is the length of the reactor, and r_a, r_b, r_c and r_s are thin wire radius, the dielectric inner and outer radii and sheath radius, β is a tabulated function of the time-varying ratio r_a/r_s, λ is the discharge gas thermal conductivity, f is the applied voltage frequency, v is the copper wire velocity, T_0 is the copper wire temperature before coming into the plasma reactor, C_v is the specific heat capacity, ϱ is the density, A_a and A_b is the cross-section of thin wire and dielectric, respectivity and T_w is the wire surface temperature. Z_{die} is the dielectric admittance and is a parallel combination of the admittance of the dielectric capacitance Z_d and the dielectric heating R_d,

$$Z_{die} = \frac{R_d Z_d}{R_d + Z_d}. \tag{5}$$

The cylindrical geometry dielectric capacitance Cd is

$$C_d = 2\pi\varepsilon_0 K_d \frac{1}{\ln\left(\frac{r_c}{r_b}\right)}, \tag{6}$$

where k_d is the dielectric permittivity. The dielectric heating R_d is obtained from the power loss in dielectric heating P_d [35],

$$R_d = \frac{dk_d\varepsilon_0 \tan\delta}{2\pi fA_d}, \tag{7}$$

where A_d is the dielectric area, d is the dielectric thickness, tanδ is the loss tangent, and f is the applied voltage frequency.

4.4. Modeling results and discussions

For a typical experimental test, the numerical result is calculated with the same input parameters as in the experiment. Equation (4) is used to calculate the copper wire average temperature in the DBD reactor as a function of the dielectric material parameters. The data in Figure 16 is used as the input current parameter to analyze the effect of the dielectric material and the frequency on the wire temperature.

The plasma annealing system shown in Figure 8 is modeled by an equivalent circuit, as shown in Figure 16 with carrier speed 20m/min, dielectric Al_2O_3. Following Child's law for a collision-less sheath in atmospheric pressure plasma annealing with a high plasma density, the area ratio scaling exponent I = 3. The wire temperature outside of the plasma reactor is assumed to be equal to room temperature (T_0 = 20 °C).

4.4.1. Dependence of annealing temperature on frequency

Figure 17 shows the weakly positive relation between the annealing temperature and the input frequency. When the frequency is increasing from 25 kHz to 40 kHz, the annealing temperature incrementally increases only 10°C. The comparison this result with experiment result we can see that there are the same. Thus, we estimate that frequency has no effect on the annealing temperature.

4.4.2. Dependence of annealing temperature on dielectric material

Figure 18 shows the dependence of annealing temperature on dielectrics material, glass and aluminum oxide. Simulation result shows that the aluminum oxide dielectric is more effective at reaching the annealing temperature than the glass dielectric. This result also is the same as that from experiment. We can conclude that the dielectrics material affects to annealing and cleaning result. The dielectric material that has higher dielectric constants is more effective to reach the annealing temperature.

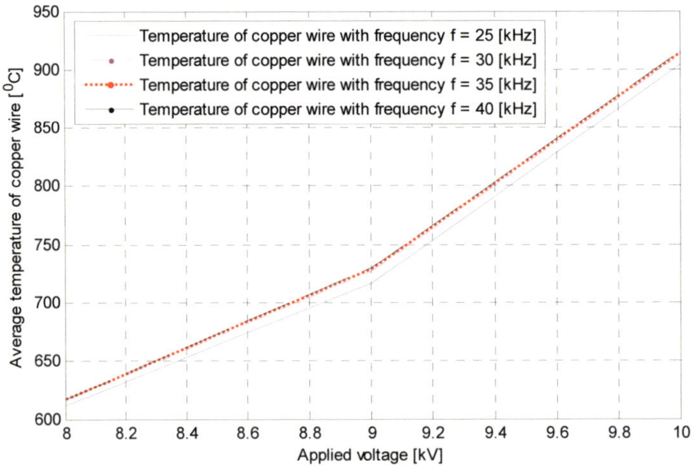

Figure 17. Temperature of copper wire as a function of frequency

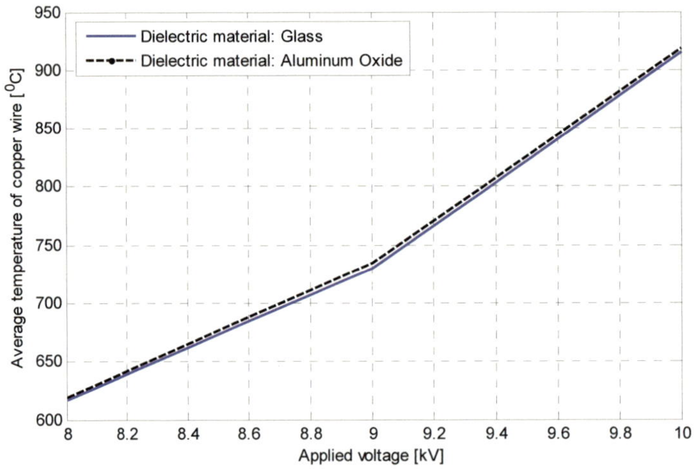

Figure 18. Temperature of copper wire depending on the dielectric material

5. Conclusions

In this chapter, the authors succeeded in setting up the experiment and examining the potential application of dielectric in atmospheric pressure dielectric barrier discharge (APDBD) for annealing and cleaning thin copper wire. The new model of annealing thin copper wire in APDBD also was modeled to RLC equivalent circuit. In conclusion, the dielectric material and dielectric thickness are two parameters which strongly affect to annealing and cleaning thin metallic wire. For optimal annealing conditions, the dielectric with the thinner thickness and the higher dielectric constants is more effective to reach the elongation rate and cleaning surface. Moreover, the dependence of dielectric on frequency of applied power also were considered. The results show that the frequency is slightly effect to dielectric and annealing result.

Author details

N.D. Tran
Energy and Environment Lab., Department of Mechanical Engineering, University of Technical Education Hochiminh City, Thu Duc District, Hochiminh City, Vietnam

T. Kikuchi, N. Harada and T. Sasaki
Plasmadynamics Lab., Department of Electrical Engineering, Nagaoka University of Technology, Kamitomioka, Nagaoka, Niigata, Japan

Acknowledgement

First and foremost, I would like to thank Professor Nobbuhoro HARADA for introducing me to the world of plasmasdynamics. I also would like to thank Associate Professor Takashi Kikuchi and Assistant Professor Toru Sasaki for their generous guidance with worthy advices carrying out this work. I sincerely thank Maeda, Kubo, Kawaii and other colleagues from Plasmadaynamics Lab. for assisting me in experiments and helping me improve my work with their invaluable suggestions.

6. References

[1] K. Maeda, S. Furuya, T. Kikuchi, and N. Harada: 47th AIAA Aerospace Sciences Meeting, Orlando, Florida, (2009), p. 1538.
[2] V. D. Sype: U.S. Patent 6270597B1 (2001).
[3] J. V. O'Gradi: U.S. Patent 2589283 (1951).
[4] Consonen: U.S. Patent 3717745 (1973).
[5] S. Ono, S. Teii, Y. Suzuki, and T. Suganuma: Thin solid films, 518 (2009) 981.
[6] T. Nakamura, C. Buttapeng, S. Furuya, N. Harada: presented at IEEE Int. Conf. Power and Energy, 2008.
[7] M. M. Santillan, A. Vincent, E. Santirso, and J. Amouroux: journal of Cleaner Production, 16 (2008) 198.

[8] C.H. Yi, Y.H. Lee, and G.Y. Yeom: Surf. Coat. Technol. 171 (2003) 237.

[9] T. Nakamura, C. Buttapeng, S. Furuya, N. Harada: Applied Surface Science (2009), doi:10.1016/j.apsusc.2009.05.121.

[10] T. Nozaki, K. Okazaki: Pure Appl. Chem., 78-6(2006) 1157–1172.

[11] Goossens, E. Dekempeneer, D. Vangeneugden, R. Van de Leest and C. Leys: Surface and Coatings Technology , 142-144(2001) 474-481.

[12] G. Borcia C. A. Anderson, N. M. D. Brown: Applied Surface Science, 225- 1-4(2004) 186-197.

[13] J. Park, S. Kyung, G. Yeom: JJAP, 46-38(2007) 942–944.

[14] M. Laroussi: IEEE Trans. Plasma Sci. 30-4(2002)1409-1415.

[15] J. R. Conrad and C. Forest: presented at IEEE Int. Conf. Plasma Science, 1986.

[16] J. R. Conrad, J. Radtke, R. A. Dodd, F. Worzala, and N. C. Tran: J. Appl. Phys. 62 (1987) 4591.

[17] J. L. Shohet: IEEE Trans. Plasma Sci. 19 (1991) 725.

[18] H. J. Hovel and T. F. Kuech: U.S. Patent 4708883 (1987).

[19] R. R. Wei, J. N. Matossian, P. Mikula, and D. Clark: U.S. Patent 5859404 (1999).

[20] S. Choi, S. W. Hwang, and J. C. Park: Surf. Coat. Technol. 142-144 (2001) 300.

[21] D. Wandke, M. Schulze, S. Klingner, A. Helmke, and W. Viol: Surf. Coat. Technol. 200 (2005) 700.

[22] M. Davis, A. Calverley, and R. F. Lever; Appl. Phys., 27 (1956) 195.

[23] K. Marshall and R. Wickham; J. Sci. Instrum., 35 (1958) 121.

[24] J. C. Kelly; J. Sci. Instrum., 36 (1959) 89.

[25] R. E. Haigh and P. H. Dawson; J. Appl. Phys. 12 (1961) 609 doi: 10.1088/0508-3443/12/11/305

[26] G. A. Emmett and M. A. Henry: J. Appl. Phys. 71 (1992) 113.

[27] S. Qin, Z. Jin, and C. Chan: Nucl. Instrum. Methods Phys. Res., Sect. B 114 (1996) 288.

[28] S. Masamune and K. Yukimura: Nucl. Instrum. Methods Phys. Res., Sect. B 206 (2003) 682.

[29] J. O. Rossi, M. Ueda, and J. J. Barroso: IEEE Trans. Plasma. Sci. 28 (2000) 1392.

[30] I. S. Choi, S. W. Hwang, and J. C. Park: Surf. Coat. Technol. 142-144 (2001) 300.

[31] T. Nakamura, C. Buttapeng, S. Furuya, and N. Harada: presented at 2nd IEEE Int. Conf. Power and Energy (PECon 08), 2008, p. 1498.

[32] J. R. Roth and X. Dai: AIAA J. 1203 (2006) 6.

[33] M. A. Lieberman and A. J. Lichtenberg, Principles of Plasma Discharges and Materials Processing (Wiley,New York, 1994), Chap. 4.

[34] Dam N. Tran, Vinh P. Nguyen, Toru Sasaki, Takashi Kikuchi, and Nobuhiro Harada: JJAP, 50(2011) 036202.

[35] E. M. Bazelyan, YU. P. Raizer: Spark Discharge (CRS Press, New Yourk, 1998).

Chapter 9

Empirical Mixing Model for the Electromagnetic Compatibility Analysis of On-Chip Interconnects

Sonia M. Holik

Additional information is available at the end of the chapter

http://dx.doi.org/10.5772/50435

1. Introduction

For over four decades the evolution of electronic technology has followed Moore's low where the number of transistors in an integrated circuit (IC) approximately doubles every two years. This naturally increases the number of internal interconnections needed to complete the system. The increase in chip complexity is achieved by a combination of dimensional scaling and technology advances. A variety of chip types exist, including memory, microprocessors and application specific circuits such as System-on-Chip (SoC). Since it is expensive to fabricate the large and simple passive components such as on-chip capacitors and inductors on the same die as the active circuits, it is desirable to fabricate these on separate dies then combine them in System-in-Package (SiP). The main advantage of SiP technology is the ability to combine ICs with other components, including passive lumped elements already mentioned but also antennas, high speed chips for radio frequency communication etc., into one fully functional package. The high complexity of SiP brings many challenges to the design process and physical verification of the system. In many cases the design process relies on detailed 3-D numerical electromagnetic simulations that tend to be slow and computationally demanding [1,2,3] in many cases limited by the available computer memory capacity and computational speed. Therefore, directly including the detail of the dense interconnect networks into the numerical model is demanding due to the amount of memory required to hold the detailed mesh, and numerical penalties associated with small mesh cell sizes relative to the wavelengths of the signals being modelled.

From a package-level point of view, the on-chip interconnects can be seen as a mixture of metal inclusions located in a host dielectric [4-8]. Most of the current studies based on numerical analysis of 2-D or 3-D structures with two constituents show that the effective properties of the mixture strongly depend on the volume fraction, its geometrical profile and spatial orientation in periodic or random arrangements [9,10]. It has been shown that the macroscopic properties of dielectric-only mixtures can be represented by a homogenous dielectric with an effective permittivity that is determined using an empirical mixing model [11]. Metal-dielectric mixtures have been less thoroughly explored, with work limited to treating spherical or ellipsoidal metal inclusions [12,13]. The approach was extended [4-8] to cope with rectangular cuboid metal inclusions representative of on-chip interconnect structures. The use of a single fitting parameter was retained, and is calculated for a wide range of aspect ratios (0.6 – 3), dielectric host materials (1 – 11.7), metal fill factors (0.2 – 0.6) and signal frequencies (1 – 10 GHz) that are likely to be of interest to System-in-Package designers. Here a simplified empirical mixing model defined for a narrowed down range of aspect ratios (1.4 – 3) which accounts for the interconnect geometries is presented. The model is at the same time straightforward and more accessible as well as more accurate. The accuracy improvement is related to the neglected range of low aspect ratios where the scaling factor Ψ has more rapid increment [4, 5].

2. Methodology

Interconnects often form regular gratings (e.g. bus structures), hence an infinite metallic grating in a homogenous dielectric host, as shown in Fig. 1., represents a straightforward but broadly applicable model.

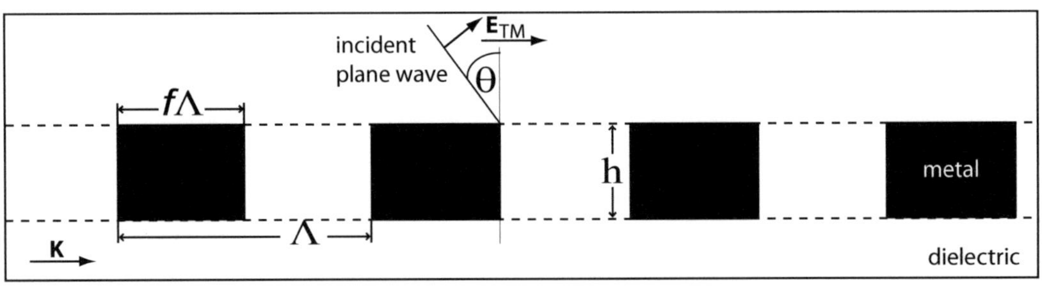

Figure 1. Diagram of 2-D grating structure studied here. The dashed line represents the boundary of the homogenised equivalent (not shown).

Wide parameter space for the dimensions is considered, as follows. The interconnect pattern density is initially limited by the design rules within 20% - 80% metal fill [14]. A statistical analysis of the pattern density of a real chip showed that, typically, the maximum pattern density in actual metal layers does not exceed 60% [15]. Thus metal fill factors f in the range 0.3 - 0.6 is considered. While metal layer height is fixed for any given layer in any given process, track width is less restricted. Aspect ratios (x_{AR}) are continuing to increase as technology develops [16], hence structures with narrowed down [4,5] values within $1.4 \leq x_{AR} \leq 3$ are studied. Due to the growing use of low-k dielectrics a host materials with permittivity ε_e in the range $1 \leq \varepsilon_e \leq 11.7$ are considered. The interconnect pitch Λ is often measured in micrometres or nanometres, whereas the wavelength λ of the clock signal is typically measured in centimetres. Thus, we can expect to successfully apply an appropriate effective medium approximation because the condition $\Lambda << \lambda/4$ is met [17]. Here, the interconnect pitch is fixed $\Lambda = 100$ μm for the sake of clarity in illustrating the method. While we include the effect of the interlevel dielectric, we deliberately neglect the bulk substrate and all but the top layer of interconnects.

The modified Maxwell Garnett mixing rule is used

$$\varepsilon_{eff} = \varepsilon_e + \Psi f \varepsilon_e \frac{\varepsilon_i - \varepsilon_e}{\varepsilon_i + 2\varepsilon_e - f(\varepsilon_i - \varepsilon_e)} \tag{1}$$

where ε_i and ε_e are the dielectric functions of the inclusion and host material respectively (here, a metal and a dielectric), Ψ is a constant relating the fields inside and outside the inclusions (typically $\Psi = 3$ for spherical inclusions), f is the filling factor or ratio of the volume of the inclusion to the total size of the unit cell [18]. In earlier work a limited example of such an approach for a fixed value of Ψ applicable to a single structure was presented [6-8]. Further, the approach was expanded by developing a compact equation to calculate the appropriate value of Ψ for a broad range of parameters [4,5]. Here a new empirical model is presented where the considered values of aspect ratio are within 1.4 – 3. The frequency dependent dielectric function of a metal inclusion $\varepsilon_i(\omega)$ can be expressed by a Drude model [13,19]

$$\varepsilon_i(\omega) = 1 - \frac{\omega_p^2}{\omega(\omega + j\gamma)} \tag{2}$$

where ω is the frequency of interest, ω_p is the plasma frequency and γ is a damping term representing energy dissipation. Despite wide spread use of copper interconnects for the intermediate levels of the interconnect stack, aluminium is often used for the global wiring with which we are concerned, and has $\omega_p = 15$ eV and $\gamma = 0.1$ eV [19]. Note that the energy is related to the free space wavelength λ_0 by $\omega = 1.24\text{x}10\text{-}6(\lambda_0)^{-1}$. This model was used in both the analytical and numerical calculations.

3. Empirical model

Rigorous coupled wave analysis (RCWA) [20] was performed for the TM polarisation with the electric field vector $\mathbf{E}_{TM}$ coplanar with the grating vector $\mathbf{K}$ as shown in Fig. 1. The structure was illuminated by a plane wave with incident angle $-89 \leq \theta \leq 89^{\circ}$ and free-space wavelengths $\lambda = 30$ cm, $\lambda = 10$ cm, $\lambda = 6$ cm, and $\lambda = 3$ cm. Hence, the adjusted height h of the homogenised layer does not simplify the model the height of the homogenised equivalent layer was kept the same as the grating. The reflection and transmission coefficients for the homogenised structure were calculated using an analytical formula defined for a stratified medium comprising a stack of thin homogenous films [21].

It is not necessary to make Ψ dependent on the host dielectric as this is already accounted for explicitly in Eq. (1). It was verified by numerical experiment. The scaling factor Ψ in the narrowed down range of aspect ratios was observed to have a linear dependence on this parameter, therefore the general form of the empirical model was chosen:

$$\Psi(x_{AR}) = \alpha \cdot x_{AR} + \beta \tag{3}$$

where the coefficients α, β are determined by linear regression from data obtained from nearly 5000 simulations spanning a four dimensional parameter space. The coefficients are represented as a linear function of metal fill factor by

$$k(f) = k_1 \cdot f + k_2, \qquad k = \{\alpha, \beta\} \tag{4}$$

where k_1, k_2 are well approximated by

$$\begin{aligned} k_1(\nu) &= k_{11} \cdot \nu + k_{12}, \\ k_2(\nu) &= k_{21} \cdot \nu + k_{22}, \qquad k = \{\alpha, \beta\} \end{aligned} \tag{5}$$

where ν is the frequency (in units of GHz), and factors k_{11}, k_{12}, k_{21}, k_{22} are presented in Table 1.

α_{11}	0.0064	β_{11}	-0.0341
α_{12}	0.2309	β_{12}	-1.7494
α_{21}	-0.0037	β_{21}	0.0213
α_{22}	-0.2346	β_{22}	2.8473

Table 1. Coefficients for calculation of the scaling factor Ψ.

The fit of the model was assessed using a linear least square method. Figure 2 illustrates the good agreement between Ψ obtained from a 'brute force' fitting algorithm and that from our linear approximation for an example grating structure. The grating has $f = 0.5$, and the illumination frequency is 5 GHz.

In Fig. 3 the reflection coefficient obtained using RCWA and homogenised model for a sample of 27 different structures are depicted. For frequencies in the range $1\text{ GHz} \leq \nu \leq 10\text{ GHz}$ the error between RCWA results for the detailed structure and those obtained for the homogenised structure is less than 2.5% for reflection coefficient and 0.2% for the transmission coefficient (not shown here) when $0.3 \leq f \leq 0.6$ and $\theta \leq \pm 30°$. The accuracy of the model for reflection coefficient calculations tends to improve with an increase of the metal fill factor. For structures with metal fills $0.4 \leq f \leq 0.6$ the error varies between 0 – 1.5%. When the model is applied, without modification, to interconnects with a trapezoidal cross section, sometimes found in fabricated structures, the error remains similar as for the empirical model presented in [4,5] and is below 5% for sidewalls with angles of up to 5° and incident angle up to 30°.

The calculated effective permittivity varies according to the particular mixing rules used to analyse a given mixture. However, there are theoretical bounds to the range of calculated effective permittivieties. For the compound of the two dielectrics the effective permittivity calculated from the Maxwell-Garnett mixing rule has to fall in between the following bounds [13]

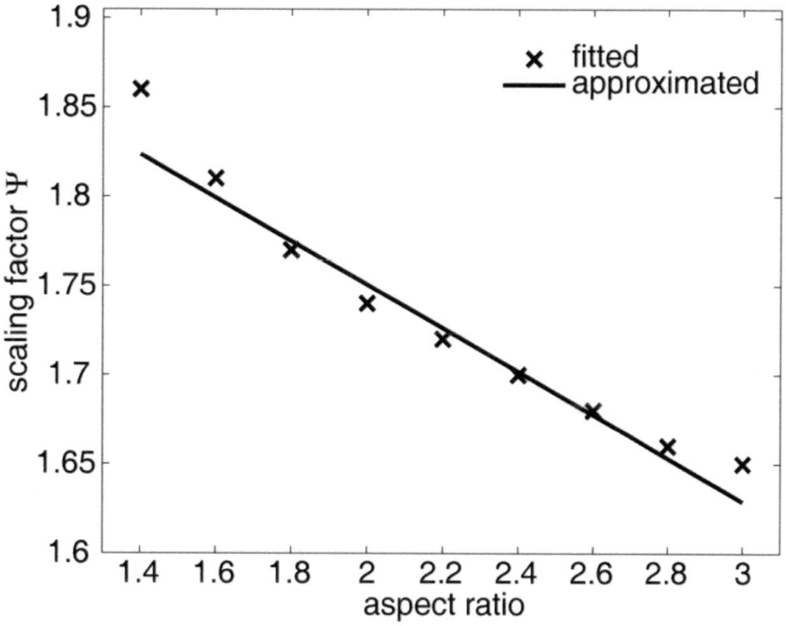

Figure 2. Plot of the scaling factor Ψ obtained for an example grating structure. The fitted values (crosses) show good agreement with approximated data (lines). Grating parameters: $f = 0.5$, $\Lambda = 100$ μm, $\nu = 5$ GHz, $\varepsilon = 6.25$, $1.4 \leq x_{AR} \leq 3$.

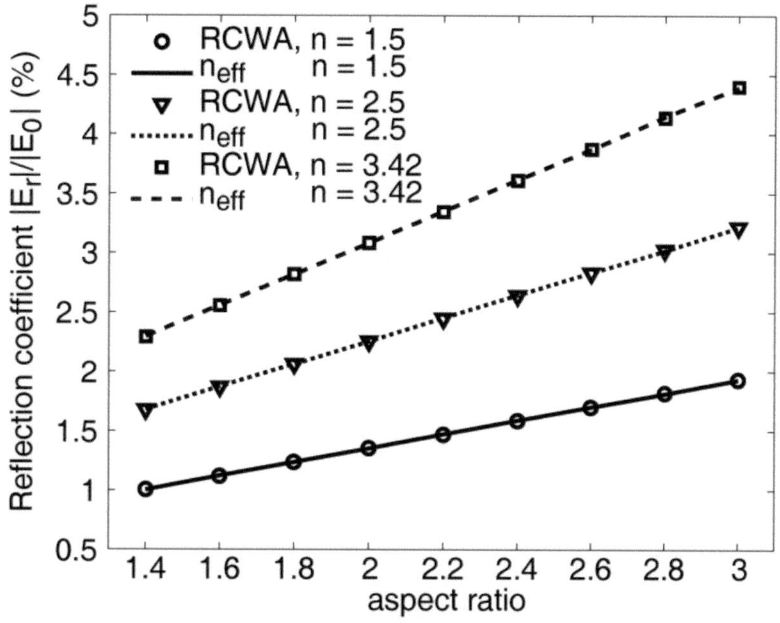

Figure 3. Plot of the reflection coefficient obtained from a subset of the gratings studied, as a function of aspect ratio and host permittivity. Results from the homogenised model are drawn as lines, while those from the detailed structure simulated with RCWA are plotted as markers. Fixed parameters: $\nu = 5$ GHz, $f = 0.5$.

$$\varepsilon_{eff,\max} = f\varepsilon_i + (1-f)\varepsilon_e,$$

$$\varepsilon_{eff,\min} = \frac{\varepsilon_i\varepsilon_e}{f\varepsilon_e + (1-f)\varepsilon_i}.$$

These bounds are also called Wiener bounds. The upper limit for the effective permittivity $\varepsilon_{eff,max}$ is defined for a layered material with boundaries between inclusions and host dielectric parallel to the field vector. The lower bound $\varepsilon_{eff,min}$ is obtained for the case where the field vector is perpendicular to the boundaries between inclusions and host. Since the Wiener bounds are defined for anisotropic mixtures, stricter bounds, Hashin-Shtrikman bounds, have been defined for the statistically homogenous, isotropic and three dimensional mixtures. The upper and the lower bounds are as follows

$$\varepsilon_{eff,\max} = \varepsilon_i + \frac{1-f}{\frac{1}{\varepsilon_e - \varepsilon_i} + \frac{f}{3\varepsilon_i}},$$

$$\varepsilon_{eff,\min} = \varepsilon_e + \frac{f}{\frac{1}{\varepsilon_i - \varepsilon_e} + \frac{1-f}{3\varepsilon_e}},$$

where it is assumed that $\varepsilon_i > \varepsilon_e$. The lower limit corresponds to the Maxwell-Garnett mixing rule whereas the upper limit is the Maxwell-Garnett rule for the complementary mixture obtained by transferring the constituents: $\varepsilon_i \rightarrow \varepsilon_e$, $\varepsilon_e \rightarrow \varepsilon_i$, $f \rightarrow 1 - f$.

It was verified that in a set of about 5000 simulations of the grating structure run to define the empirical model all effective refractive indices ($n_{eff} = \sqrt{\varepsilon_{eff}}$) are well within the Wiener bounds. Nevertheless, the predicted n_{eff} has values close to the lower limit. This is related to the specific alignment of the grating structure (single layer of interconnects) and the angle of incidence wave. Such regular and linearly distributed arrangement of the inclusions with the field vector perpendicular to the grating surface results in an effective permittvity from the bottom range of the possible values defined by Wiener bounds. The upper limit is several orders higher in magnitude, hence even if satisfied, for the purpose of the analyses of this particular grating structure it can be lowered by replacing it with the Hashin-Shtrikman lower limit. It is illustrated in Fig. 4, for a random structure, that the real parts of n_{eff} obtained from the empirical model are within the lower limits of the Wiener and Hashin-Shtrikman bounds.

The more strict Hashin-Shtrikman bounds overestimate the obtained values of n_{eff}. Hence these limits are based on the Maxwell-Garnett mixing rule for the complementary mixtures and the lower limit is just the classical Maxwell-Garnett rule with $\varepsilon_i > \varepsilon_e$. Therefore, for the analysed interconnect grating structure it can be assumed that the upper bound for the effective refractive index is the classical Maxwell-Garnett rule whereas the lower bound is the Wiener lower limit.

4. Experimental validation

Experimental validation was carried out by the free space measurement of S-parameters of an air-copper grating structure ($\Lambda = 500\mu m$, $f = 0.3$, AR = 1) attached to a Rogers 4350 dielectric plate (thickness $762\mu m$, $\varepsilon_r = 3.66$) illuminated by a plain wave. A pair of horn antennas operated at the X-band (8.2 – 12.4 GHz) frequencies with Teflon's hemispherical lenses connected to the network analyzer was used. The plane wave illumination focused on a relatively small area was achieved by the special equipment arrangement. A free space calibration method along with smoothing procedure was implemented in order to eliminate systematic errors occurring in the measurement data [22,23]. A 2-D finite difference time domain (FDTD) was defined as shown in Fig. 5, with the case of detailed grating structure in Fig. 5(a) and its homogenised equivalent in Fig. 5(b).

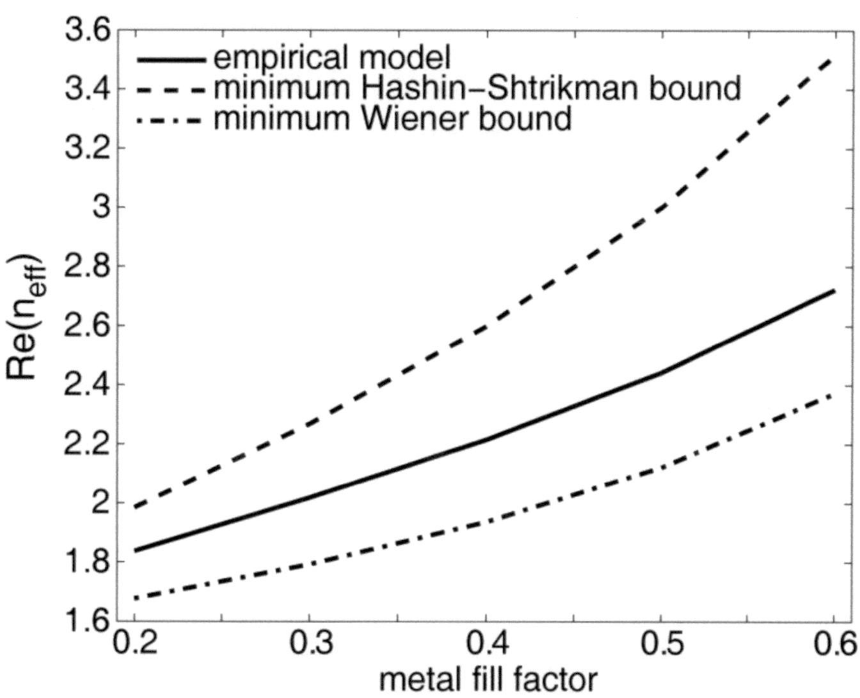

Figure 4. Plot of the real part of the effective refractive index of the grating structure compared with theoretical bounds. Grating parameters: $\chi_{AR} = 1$, $\varepsilon = 11.7$, $\Lambda = 100$ μm, $\nu = 5$ GHz, $0.2 \le f \le 0.6$.

The domain size was 60 cells in x by 1 cell in y and 8203 cells in z direction. The space increment in both directions was set to $5\mu m$ and it was ensured that the domain size in the z-direction was at least a half wavelength from each of the absorbing Perfect Matched Layer (PML) boundaries as the behaviour of these boundaries is not reliable in the presence of evanescent fields. The grating structure is periodic in x-direction, with one period of the grating defined in the domain. The structure was illuminated by a wave with frequencies within 8 – 12 GHz in steps of 1 GHz applied at the top of the domain and propagated in the negative z-direction.

The homogenised equivalent structure was obtained by replacing the grating layer with a solid dielectric. The dielectric properties were calculated from the modified Maxwell-Garnett mixing rule. The value of scaling factor Ψ was empirically found as due to the structural difference between experimental settings and the structure studied in order to define the empirical model, the straightforward application of the empirical model underestimated factor Ψ. It was verified that Ψ, when equal to 2.5, gives good approximation of the calculated effective permittivity ε_{eff}.

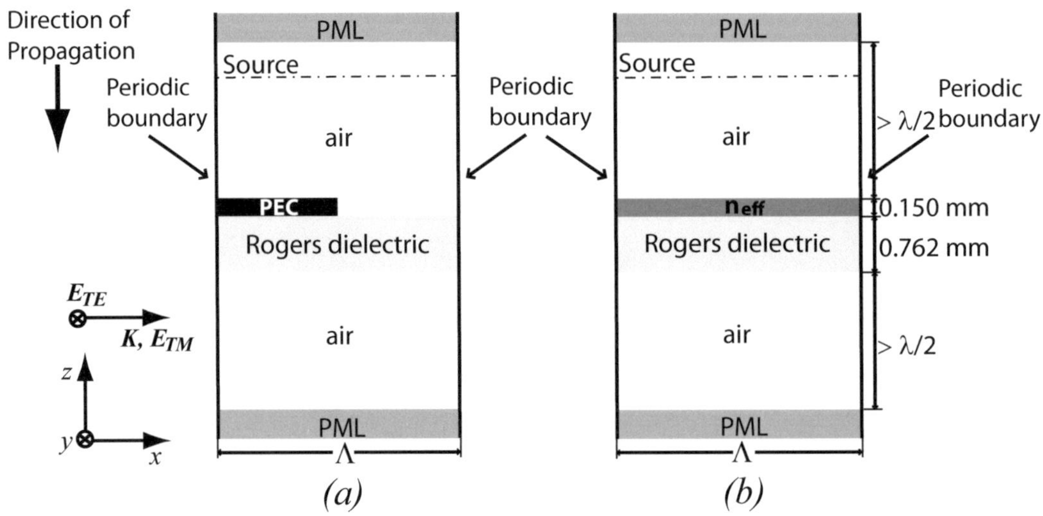

Figure 5. Diagram of the 2-D FDTD simulation domain. The domain is one period wide and periodic in (a) x-direction, with its (b) homogenised equivalent

The FDTD and analytical calculations, based on characteristic matrix method, of reflection and transmission coefficients for the gratings with structural period 500μm and its homogenised equivalent are plotted in Fig. 6 and Fig. 7 along with measured return and transmission losses respectively. In order to validate the proposed approach for a wider range of frequencies, 1 - 18 GHz, numerical calculations were performed using finite element method (FEM) [24]. This analysis shows that the results of the two numerical techniques and measured results follow the same trend over a wide range of frequencies and allowed extrapolation of the measured reflection and transmission coefficients outwith the measured domain. Simulated and measured results are in good qualitative agreement and the observed deviation tends to increase simultaneously with frequency increase.

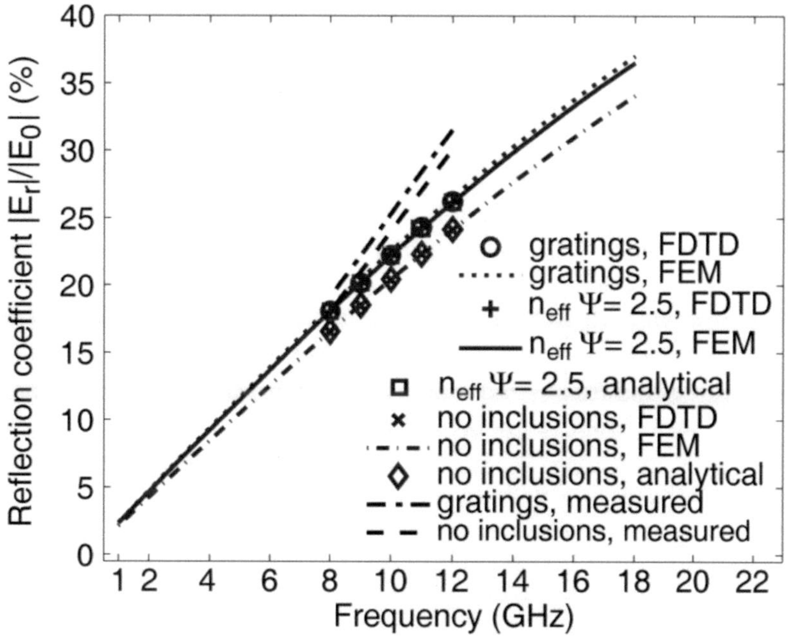

Figure 6. Plot of the reflection coefficient for a grating structure. The experimental data follows the same trend as the numerical FDTD and FEM calculations obtained for grating, homogenised and reference structure. Grating parameters: $f = 0.3$, $\Lambda = 500$ μm, AR = 1; substrate: thickness 762μm, $\varepsilon_r = 3.66$.

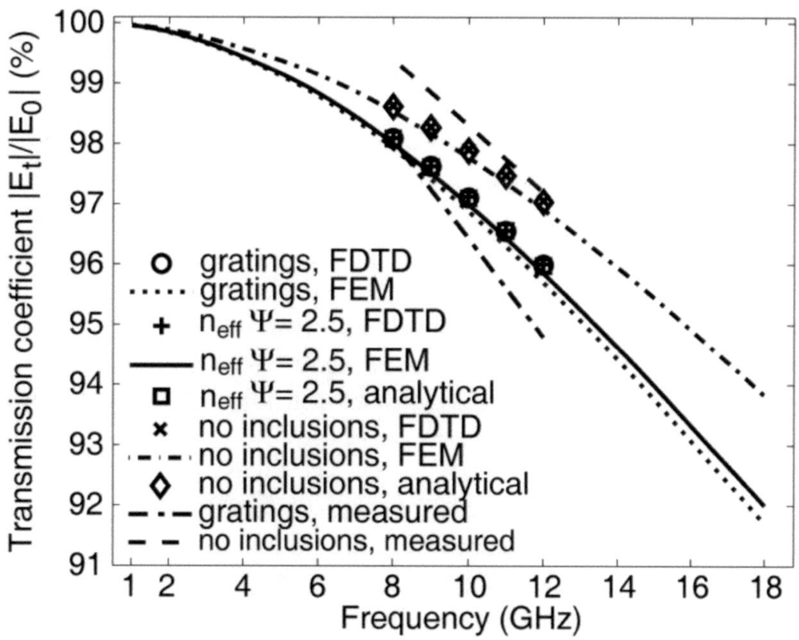

Figure 7. Plot of the transmission coefficient for a grating structure. The experimental data follows the same trend as the numerical FDTD and FEM calculations obtained for grating, homogenised and reference structure. Grating parameters: $f = 0.3$, $\Lambda = 500$ μm, AR = 1; substrate: thickness 762μm, $\varepsilon_r = 3.66$.

5. Conclusion

The empirical model which allows a single layer of on-chip interconnects to be accurately replaced by homogeneous material slab in electromagnetic simulations containing integrated circuits was presented. The model is applicable to a wide range of interconnect dimensions (metal fill 20 - 60%, aspect ratio 1.4 – 3 and host permittivity 1 – 11.7) and is accurate to better than 2.5% (0.2%) error for reflection (transmission) when illuminated by plane waves with frequency 1 - 10 GHz, incident at up to 30° off the normal. The error does not increase with respect to the change in the grating profile and it can be applied to trapezoidal gratings with sidewall angles up to 5°. Our approach allows the behaviour of on-chip interconnects to be accurately captured in full vector electromagnetic simulations without incurring the significant computational penalties associated with a finely detailed mesh. The experimental data supports the conception that the metal-dielectric grating structure specified for interconnects can be homogenised. The validation was carried out by comparing the numerical results with experimental data.

Author details

Sonia M. Holik
University of Glasgow, United Kingdom

6. References

[1] Fontanelli, A.: 'System-in-Package technology: opportunities and challenges', Proc. IEEE 9th Int. Sym. Quality Electronic Design (ISQED'08), San Jose, CA, USA, 2008, pp. 589-593

[2] Trigas, C.: 'Design challenges for System-in-Package vs System-on-Chip', Proc. IEEE Custom Integrated Circuits Conference (CICC'03), Munich, Germany, 2003, pp. 663-666

[3] Sham, M. L., Chen, Y. C., Leung, J. R., Chung, T.: 'Challenges and opportunities in System-in-Package business', Proc. IEEE 7th Int. Conf. Electronics Packaging Technology (ICEPT'06), Shanghai, China, 2006, pp. 1-5

[4] Holik, S. M., Drysdale, T. D.: 'Simplified model for on-chip interconnects in electromagnetic modelling of System-in-Package', Proc. 12th International Conference on Electromagnetics in Advanced Applications (ICEAA'10), Sydney, Australia, 2010, pp. 541-544.

[5] Holik, S. M, Arnold, J. M, Drysdale, D. D.: 'Empirical mixing model for the electromagnetic modelling of on-chip interconnects', PIER Journal of Electromagnetic Waves and Applications, 2011, 26, pp. 1-9.

[6] Holik, S. M, Drysdale, D. D.: 'Effective Medium Approximation for Electromagnetic Compatibility Analysis of Integrated Circuits', Proc. 2nd Int. Congress Advanced Electromagnetic Materials in Microwaves and Optics, Pamplona, Spain, 2008, pp. 413-415

[7] S. M. Holik, T. D. Drysdale.: 'Simplified model of a layer of interconnects under a spiral inductor,' Journal of Electromagnetic Analysis and Applications, 2011, 3, (6), pp. 187-190.

[8] S. M. Holik, T. D. Drysdale.: 'Simplified model of interconnect layers under a spiral inductor', Journal of Microwaves, Optoelectronics and Electromagnetic Applications, 2011, 10, (2), pp. 337-342.

[9] Sereni, B., Krahenbuhl, L., Beroual, A., Brosseau, C.: 'Effective dielectric constant of periodic composite materials', J. Appl. Phys., 1996, 80, (3), pp. 1688-1696

[10] Sereni, B., Krahenbuhl, L., Beroual, A., Brosseau, C.: 'Effective dielectric constant of random composite materials', J. Appl. Phys., 1997, 81, (5), pp. 2375-2383

[11] Kärkkäinen, K., Sihvola, A., and Nikoskinen, K.: 'Analysis of three-dimensional dielectric mixture with finite difference method', IEEE Trans. Geosci. Remote Sensing, 2001, 39, (5), pp. 1013-1018

[12] Chylek, P., and Strivastava, V.: 'Effective dielectric constant of a metal-dielectric composite', Phys. Rev. B, 1984, 30, (2), pp. 1008-1009

[13] Sihvola, A.: 'Electromagnetic mixing formulas and applications', IEE Publishing, London, 1999

[14] Lakshminarayann, S., Wright, P. J., and Pallinti, J.: 'Design rule methodology to improve the manufacturability of the copper CMP process', Proc. IEEE Int. Interconnect Technology Conf. (IITC'02), Burlingame, CA, USA, 2002, pp. 99-102

[15] Zarkesh-Ha, P., Lakshminarayann, S., Doniger, K., Loch, W., and Wright, P.: 'Impact of interconnect pattern density information on a 90nm technology ASIC design flow', Proc. IEEE 4th Int. Sym. Quality Electronic Design (ISQED'03), San Jose, CA, USA, 2003, pp. 405-409

[16] International Technology Roadmap for Semiconductors, Interconnect, 2007 Edition, http://www.itrs.net, accessed March 2010

[17] Rytov, S. M.: 'Electromagnetic properties of a finely stratified medium', Sov. Phys. JETP, 1956, 2, pp. 466-475

[18] Maxwell Garnett, J. C., 'Colours in metal glasses and in metallic films, Phil. Trans. R. Soc. London, 1904, 203, pp. 385-420

[19] Pendry, J. B., Holden, A. J., Stewart, W. J., and Youngs, I.: 'Extremely low frequency plasmons in metallic mesostructures', Physical Review Letters, 1996, 76, pp. 4773-4776

[20] GSolver5.1, Grating Solver Development Company, Allen, TX 75013, USA, http://www.gsolver.com, accessed September 2010

[21] Born, M., Wolf, E.: 'Principles of optics: electromagnetic theory of propagation, interference and diffraction of light', 7th Edition, Cambridge University Press, 1999

[22] Ghodgaonkar, D. K., Varadan, V. V., Varadan, V. V.: 'Free-space measurement of complex permittivity and complex permeability of magnetic materials at microwave frequencies', IEEE Trans. Instrum. Meas., 1990, 39, (2), pp. 387-394

[23] Ghodgaonkar, D. K., Varadan, V. V., Varadan, V. V.: 'A free-space method for measurement of dielectric constants and loss tangents at microwave frequencies', IEEE Trans. Instrum. Meas., 1989, 37, (3), pp. 789-793

[24] Ansoft High Frequency Structure Simulator (HFSS), http://www.ansoft.com, accessed September 2010

Section 4

Electrical Characterization of Microelectronic Devices

Chapter 10

Electrical Characterization of High-K Dielectric Gates for Microelectronic Devices

Salvador Dueñas, Helena Castán, Héctor García and Luis Bailón

Additional information is available at the end of the chapter

http://dx.doi.org/10.5772/50399

1. Introduction

The continuous miniaturization of complementary metal-oxide-semiconductor (CMOS) technologies has led to unacceptable tunneling current leakage levels for conventional thermally grown SiO_2 gate dielectrics [1,2]. During the last years, many efforts have been devoted to investigate alternative high-permittivity (high-k) dielectrics that could replace SiO_2 and SiON as gate insulators in MOS transistors [3]. The higher dielectric constant provides higher gate capacitances with moderated thickness layers; however, other requirements such as lower leakage currents, high breakdown fields, prevention of dopant diffusion, and good thermodynamic stability must also be fulfilled. A number of high-k materials have been investigated as candidates to replace the SiO_2 as gate dielectric, being Al_2O_3 and HfO_2 among the most studied ones [3–5], since both have a larger permittivity than SiO2 and are thermodynamically stable in contact with silicon. The electrical characteristics of the as-deposited layers of these materials, however, exhibit large negative fixed charge and interface state densities and charge trapping as compared to SiO_2, although these characteristics can be improved by including an intermediate oxide between the high-k layer and the silicon substrate [6–8] or by high-temperature post-deposition processes.[9–13]. In addition to binary oxides, laminates of them show an improvement of the electrical characteristics as compared to the single oxide layers [14]. In particular, Al_2O_3–HfO_2 laminates and alloys benefit from the higher k of HfO_2 and the higher crystallization temperature of Al_2O_3 [15,16]

In this chapter we review the standard techniques as well as the new ones which we have developed for the electrical characterization of very thin insulating films of high k dielectrics for metal-insulator-semiconductor (MIS) gate and metal-insulator-metal (MIM) capacitor applications. These techniques have been conceived to provide detailed information of defects existing in the insulator bulk itself and interface traps appearing at the insulator-semiconductor substrate interface. Several methods exist to obtain defect densities at insulator/semiconductor interface, such as deep level transient technique (DLTS), high and low (quasi static) frequency capacitance-voltage measurements and admittance spectroscopy.

However, the study of defects existing inside the gate dielectric bulk is not so widely established. Two techniques have been developed by us to accomplish it: conductance transient technique (GTT) and Flat-Band Voltage transients (FBT) measurements.

GTT is very useful when exploring disordered-induced gap states (DIGS) defects distributed inside the dielectric. This technique has been successfully applied to many high-k dielectric films on silicon. From conductance transient measurements we have obtained 3D profiles or contour maps showing the spatial and energetic distribution of electrically active defects inside the dielectric, preferentially located at regions close to the dielectric/semiconductor interface.

The FBT approach consists of a systematic study of flat-band voltage transients occurring in high-k dielectric-based metal-insulator-semiconductor (MIS) structures. While high-k material can help to solve gate leakage problems with leading-edge processes, there are still some remaining challenges. There are, indeed, several technical hurdles such as threshold voltage instability, carrier channel mobility degradation, and long-term device reliability. One important factor attributed to these issues is charge trapping in the pre-existing traps inside the high-k gate dielectrics. Dependencies of the flat-band voltage transients on the dielectric material, the bias history, and the hysteresis sign of the capacitance-voltage (C-V) curves are demonstrated. Flat bat voltage transients provide the soft optical phonon energy of dielectric thin-films. This energy usually requires chemical-physical techniques in bulk material. In contrast, FBT provides this magnitude for thin film materials and from electrical measurements, so adding an extra value to our experimental facilities.

Throughout the chapter we will give detailed information about the theoretical basis, experimental set-up and how to interpret the experimental results for all the above techniques.

Another topic widely covered will be the current mechanisms observed on high k materials. The above-mentioned methods allow determining the density and location of defects on the dielectric. These defects are usually responsible for the conduction mechanisms. The correlation between conduction mechanisms, defect location and preferential energy values provides very relevant information about the very nature of defects and how these defects can be removed or diminished.

We have studied many high-k materials during the last years, covering all proposed around the world as gate dielectric on silicon. These dielectrics consist of single layers of metal oxides and silicates (e.g.: HfO_2, ZrO_2, $HfSiO_x$, Gd_2O_3, Al_2O_3, TiO_2, and much more) directly deposited on n- and p- type silicon, combination of them in the form of multilayers, and gate stacks with silicon oxide or silicon nitride acting as interface layers which prevent from thermodynamic instabilities of directly deposited high-k films on silicon substrates. Figure 1 summarizes the atomic elements used as precursors of the high-k dielectrics we have studied in our laboratory. An extended summary of the more relevant results obtained will be also included in the chapter. Another topics covered in this chapter include: high-k fabrication methods: ALD, CVD, High Pressure Sputtering, etc., influence of the process parameters on the quality of as-grown and thermally annealed materials, or charge trapping at the inner interface layers on gate-stacks and multilayer films.

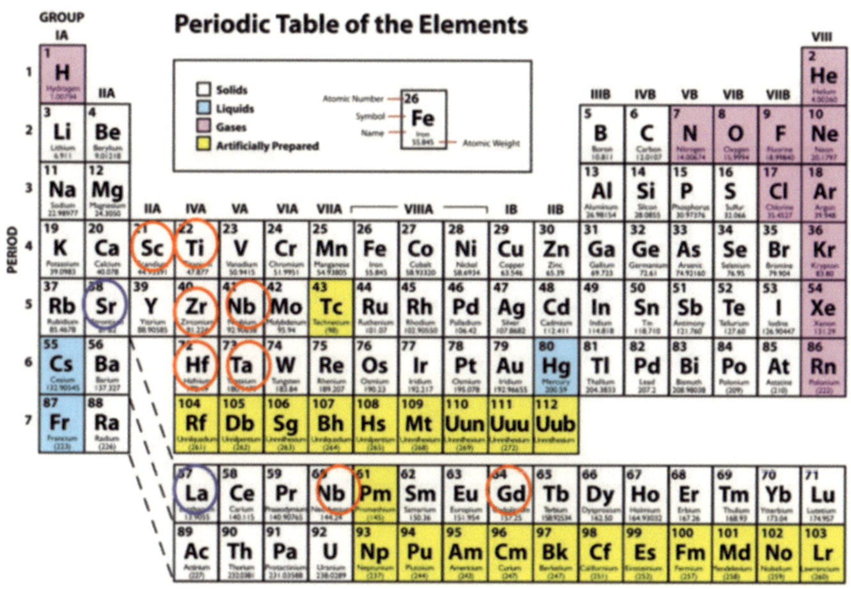

Figure 1. Periodic table with marks on the atomic elements from which high-k materials (mainly oxides and silicates) have been fabricated.

2. Standard characterization methods

2.1. Capacitance-Votage measurements

Capacitance-Voltage is the most frequently used electrical technique to assess the properties of both the thin oxide layer and its interface with the semiconductor substrate. In thicker oxide layers (more than 4-5 nm) C–V curves can be fitted satisfactorily with classical models, described in textbooks. The C–V technique can be used to determine flatband and threshold voltage, fixed charge, and interface state density. It is also often used to calculate the oxide thickness.

The ideal expression of a MIS structure in accumulation regime is: $C_{ac} = \frac{k\varepsilon_0 A}{t_{ox}}$. Non-ideal effects in MOS capacitors include fixed charge, mobile charge and surface states. Performing a capacitance-voltage measurement allows identifying all three types of charge. Charge existing in the dielectric film shifts the measured curve. Trapping and detrapping of defects inside the insulator produce hysteresis in the high frequency capacitance curve when sweeping the gate voltage back and forth.

Finally, surface states at the semiconductor-insulator interface also modify the CV curves. As the applied voltage varies the Fermi level at the interface changes and affects the occupancy of the surface states. The interface states cause the transition in the capacitance measurement to be stretched out In Figure 2 we show experimental high frequency C-V results for hafnium oxide MIS structures measured at room temperature. Atomic Layer Deposition technique was used to grow these 20 nm-thick HfO_2 films

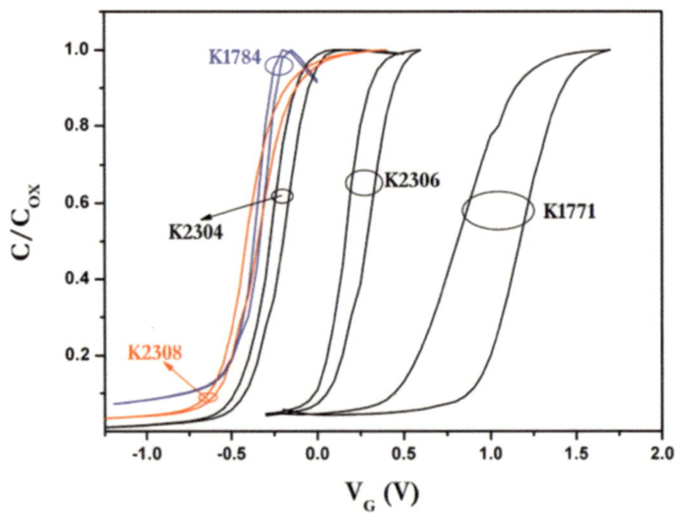

Figure 2. 1MHz C-V curves of Al/HfO_2/n-Si capacitors obtained by Atomic Layer Deposition

The combination of the low and high frequency capacitance (HLCV) [17] allows calculating the surface state density. This method provides the surface state density over a limited (but highly relevant) range of energies within the bandgap. Measurements on n-type and p-type capacitors at different temperatures provide the surface state density throughout the bandgap. A capacitance meter is usually employed to measure the high-frequency capacitance, C_{HF}. The quasi-static measurement of the low frequency capacitance, C_{LF}, consists on recording the gate current whereas a ramp-voltage is applied to the gate terminal. Interface state density is obtained according the following expression:

$$D_{it}=\frac{C_{ox}}{q}\left(\frac{C_{LF}}{C_{ox}-C_{LF}}-\frac{C_{HF}}{C_{ox}-C_{HF}}\right) \quad (1)$$

In sub-4 nm oxide layers, C–V measurements provide the same information, but the interpretation of the data requires considerable caution. The assumptions needed to construct the ''classical model'' are no longer valid, and quantum mechanical corrections become mandatory, thus increasing the complexity of the analytical treatment: Maxwell–Boltzman statistics no longer describe the charge density in the inversion and accumulation layers satisfactorily, and should be replaced by Fermi–Dirac statistics. In addition, band bending in the inversion layer near the semiconductor–insulator interface becomes very strong, and a potential well is formed by the interface barrier and the electrostatic potential in the semiconductor. The correct analytical treatment requires solving the complex coupled effective mass Schroedinger and Poisson equations self-consistently.

2.2. Current measurements and conduction mechanisms

The performance of MOS devices strongly depends on the breakdown properties and the current transport behaviors of the gate dielectric films. The conduction mechanisms are very sensitive to the film composition, film processing, film thickness, and energy levels and densities of trap in the insulator films. Therefore, the analysis of the dominant conduction mechanisms may provide relevant information on the physical nature of the dielectric film and complements other characterization techniques when optimizing fabrication process. The most commonly found mechanisms as well as the voltage and temperature laws for each one are summarized in Figure 3.

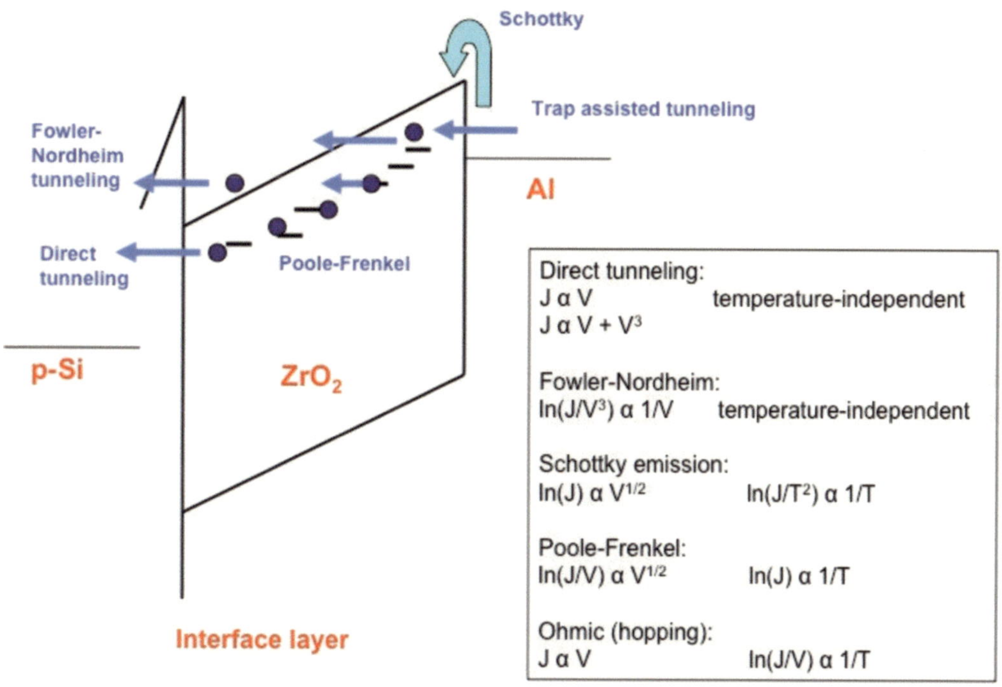

Figure 3. Main conduction mechanisms on Metal-Insulator- Semiconductor devices

- **Electrode-limited mechanisms**: When the dielectric has high bandgap, high energy barrier with electrodes and low trap density, conduction is more electrode-limited than bulk limited. For a large applied bias, the silicon surface is n-type degenerated regardless of the bulk doping. Hence, for a large applied voltage the current is limited by tunneling (independent of the temperature) from the vicinity of the silicon conduction band edge through the triangular barrier into the oxide conduction band

(*Fowler-Nordheim effect*). When barriers are no so high, conduction may occur when electrons or holes are promoted from the corresponding band to the insulator bands (*Schottky effect*). That occurs at lower voltages than Fowler-Nordheim mechanisms.

- **Bulk limited mechanisms:** As the insulators become more defective, as is the case of practically all high-k dielectrics, bulk-limited conduction predominates due to traps inside the insulator. Sometimes current density is due to field enhanced thermal excitation of trapped electrons into the conduction band. This process is known as the *Internal Schottky or Poole Frenkel effect*. The *hopping* of thermally excited between isolated states gives an ohmic I–V characteristic, exponentially dependent on temperature.
- **Tunnel limited mechanisms**: As dielectric films become thinner, tunneling conduction gradually dominates the conduction mechanisms. It may occur via defects in a two-step (or trap-assisted) tunneling or by direct tunneling from one electrode to the other.

In Figure 4 we draw the I-V characteristics at different temperatures of an Al_2O_3–based MIS sample fabricated by Atomic Layer Deposition. Leakage current clearly increases with temperature at lower gate voltages. I-V curves of all the samples were fitted according to the Poole-Frenkel emission, so indicating that the main conduction mechanism is bulk related.

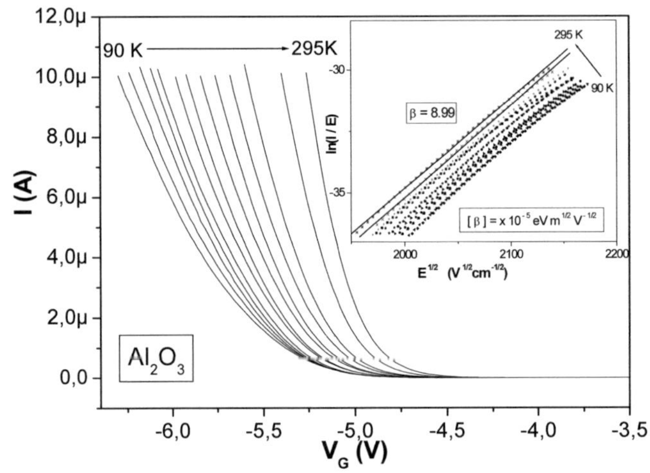

Figure 4. I-V curves at several temperatures and Poole-Frenkel fitting of an ALD Al_2O_3–based MIS sample

2.3. Admittance spectroscopy

The admittance spectroscopy or conductance method, proposed by Nicollian and Goetzberger in 1967, is one of the most sensitive methods to determine D_{it} [18]. Interface trap densities of 10^9 cm^{-2} eV^{-1} and lower can be measured. It is also the most complete method, because it yields *Dit* in the depletion and weak inversion portion of the bandgap, the capture cross-sections for majority carriers, and information about surface potential fluctuations. The technique is based on measuring the equivalent parallel conductance of an MIS capacitor as a function of bias voltage and frequency. The conductance, representing the loss mechanism due to interface trap capture and emission of carriers, is a measure of the interface trap density. Interface traps at the insulator-Si interface, however, are continuously distributed in energy throughout the Si band gap. Capture and emission occurs primarily by traps located within a few kT/q above and below the Fermi level, leading to a time constant dispersion and giving the normalized conductance as

$$\frac{G}{\omega}=\frac{q\omega\tau_{it}D_{it}}{1+(\omega\tau_{it})^2} \tag{2}$$

where $\tau_{it}=\left[v_{th}\sigma_p N_A \, exp\left(-\frac{q\Phi_S}{kT}\right)\right]^{-1}$ is the emission time constant of interface traps with energy Φ_S.

The conductance is measured as a function of frequency and plotted as G/ω versus ω. G/ω has a maximum at $\omega = 1/\tau_{it}$ and at that maximum $D_{it} = 2G/q\omega$. For equation (2) we find $\omega \approx 2/\tau_{it}$ and $Dit = 2.5\ G/q\omega$ at the maximum.

It is also possible to make measurements by varying the temperature and keeping the frequency constant [19], instead of changing the frequency at constant temperature. This has the advantage of not requiring measurements over a wide frequency range and one can chose a frequency for which series resistance is negligible. Elevated temperature measurements enhance the sensitivity near mid-gap allowing the detection of trap energy levels and capture cross sections [20]. It also is possible to use transistors instead of capacitors and measure the transconductance instead of the conductance but still use the concepts of the conductance method [21]. This allows interface trap density determination on devices with the small gate areas associated with transistors without the need for capacitance test structures.

2.4. Other thecniques

In this section we include several electrical characterization techniques that are useful for probing microscopic bonding structures, defects, and impurities in high-k dielectrics, as described in [22].

2.4.1. Inelastic electron tunneling spectroscopy (IETS)

IETS is a novel technique that can probe phonons, traps, microscopic bonding structures, and impurities in high-k gate dielectrics with a superior versatility and sensitivity when compared with other techniques. This technique basically takes the second derivative of the tunneling I–V characteristic of an ultrathin MOS structure. The basic principle of the IETS technique is illustrated in Figure 5. Without any inelastic interaction, the I–V characteristic is smooth and its second derivative is zero. When the applied voltage causes the Fermi-level separation to be equal to the characteristic interaction energy of an inelastic energy loss event for the tunneling electron, then an additional conduction channel (due to inelastic tunneling) is established, causing the slop of the I–V characteristic to increase at that voltage, and a peak in its second derivative plot, where the voltage location of the peak corresponds to the characteristic energy of the inelastic interaction, and the area under the peak is proportional to the strength of the interaction.

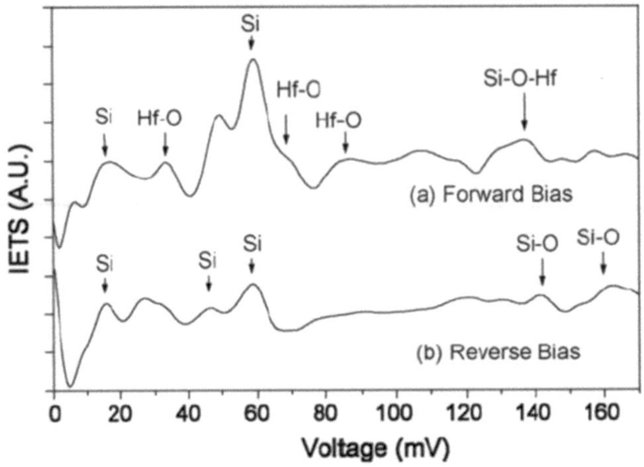

Figure 5. Principles of IETS technique [22].

In a typical MOS sample, there are more than one inelastic mode, as a wide variety of inelastic interactions may take place, including interactions with phonons, various bonding vibrations, bonding defects, and impurities. Figure 6 shows an actual IETS spectrum taken on an Al/HfO2/Si sample, where the features below 80 meV correspond to Si phonons and Hf–O phonons, and the features above 120 meV correspond to Hf–Si–O and Si–O phonons. The significance of this IETS spectrum is that it confirms the strong electron–phonon interactions involving optical phonons in HfO_2, and that the Hf–O phonons have very similar energy range as Si phonons which we know are a source of scattering centers that degrade the channel mobility.

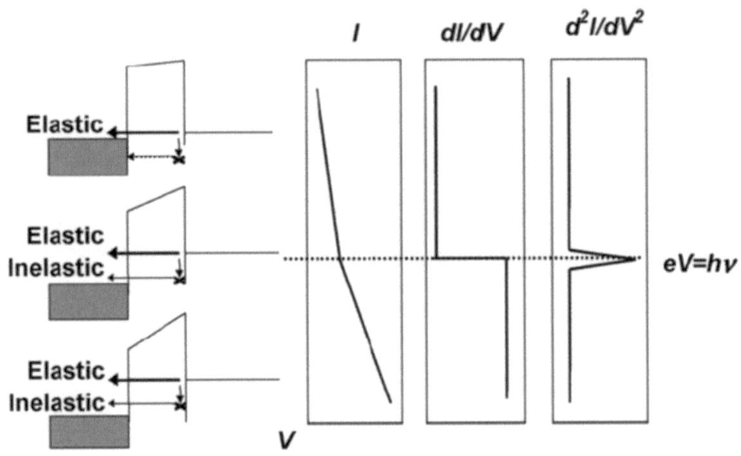

Figure 6. IETS for HfO2 on Si under different bias polarities: (a) forward bias (gate electrode positive), (b) reverse bias (gate electrode negative) [22].

2.5. Lateral profiling of threshold voltages, interface traps, and oxide trapped charge

Lateral profiling is a charge-pumping technique that enables one to profile the lateral distribution of threshold voltages of a MOSFET, and the lateral distributions of interface traps and oxide trapped charge generated by hot-carrier damage [23-24]. Figure 7(a) shows the I_{cp}–V_h curves for the source (curve 1) and the drain junction (curve 2) prior to hot-carrier damage, from which one can obtain the threshold voltage distributions near the two junctions (Figure 7(b)) using the V_h-$V_t(x)$ relationship as described in [25, 26]. Then a channel hot-carrier (CHC) stressing for 300 s to damage the device is used. Comparing curves 2 and 1 in Figure 8(a), one can see that the CHC stressing is not only generated N_{it} but also caused by positive charge inside the insulator gate, Q_{ot}. Therefore, one must neutralize this Q_{ot} before proceeding, and this was accomplished by a light hot electron injection as shown by curve 3 in Fig. 8. Note that this step did not cause any increase in N_{it} as evidenced by the unchanged $I_{cp,max}$.

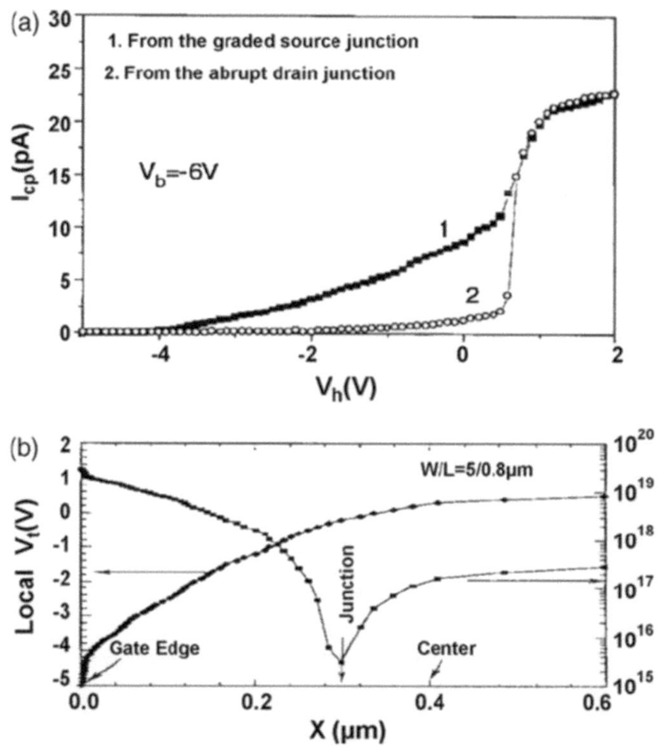

Figure 7. (a) Single-junction charge pumping curves measured either with the source floating (curve 1) or with the drain floating (curve 2). (b) Local Vt distribution across the channel as deduced from the data in (a).(From Reference [22])

These three I_{cp} curves were then used to extract the $N_{it}(x)$ from the difference between curves 3 and 1 at a given V_h, and $Q_{ot}(x)$, from the voltage shift between 2 and 3 at a given I_{cp} (Figure 8(b)).

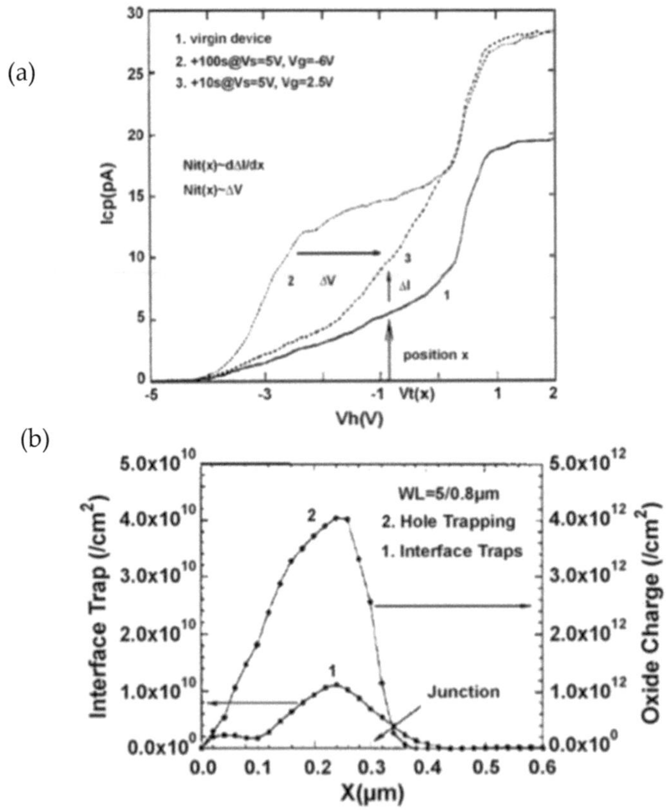

Figure 8. (a) Three charge pumping curves measured for the purpose of directly profiling the erase-induced damage, and graphically illustrating the direct lateral profiling principle. (b) Lateral profiles of both positive oxide charge and interface traps near the source junction, transformed from the three charge pumping curves in (a).(From Reference [22]).

2.6. Pulse agitated substrate hot electron injection (PASHEI) technique for studying trapping parameters

PASHEI technique [27, 28] can be used to study charge trapping in the gate dielectric of an MOSFET under low gate biases. The commonly used carrier injection techniques, such as Fowler–Nordheim (FN) tunneling, and CHC techniques require high gate field to obtain high injection flux, which makes it impractical to study trapping effects under low gate fields when the injection flux is extremely low. Another technique, the substrate hot-electron injection (SHEI) technique, does allow high flux injection at low gate fields, but it requires a separate p–n junction injector in the vicinity of the MOSFET being tested, which rules out most of the devices available for test. In contrast, the PASHEI technique, which will be described below, allows substrate hot-electron injection with just an ordinary MOSFET without a separate injector. The PASHEI technique relies on properly timed pulse sequences to achieve SHEI, as illustrated schematically in Figure 9 for an n-MOSFET. As shown in Figure 9(b), during the electron-emitting phase, the S/D junction is forward biased, and electrons are injected into the substrate. Subsequently, the S/D is reverse biased to create a deep depletion region, which will cause the previously injected electrons in the substrate (those that have not recombined away) to be accelerated across the depletion region and injected into the gate dielectric. This period is called the collecting phase, during which the emitting voltage can control the gate voltage, and large injection current can be achieved with low V_g. Figure 10 illustrates the use of the ΔV_{th} vs. N_{inj} curve, obtained by the PASHEI technique, to extract trap parameters. For this particular sample, we obtained a trap density of 2.7 $x10^{12}$ cm^{-2}, and capture cross-section of 7.7 10^{-19} cm^2, by fitting the trapping theory presented by Zafar [29].

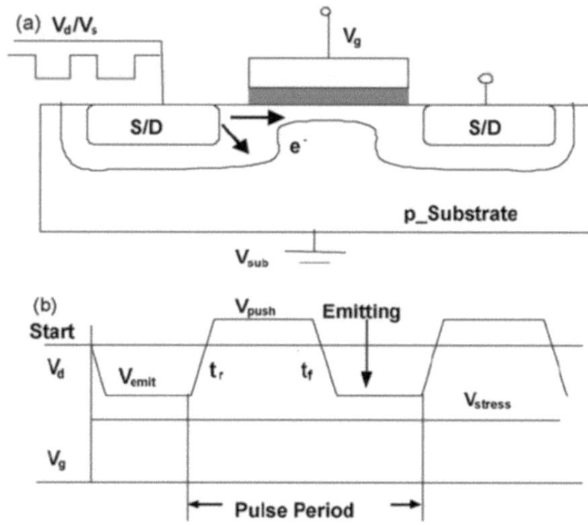

Figure 9. (a) Schematic description of PASHEI. (b) Pulse sequence for PASHEI. (From Reference [22])

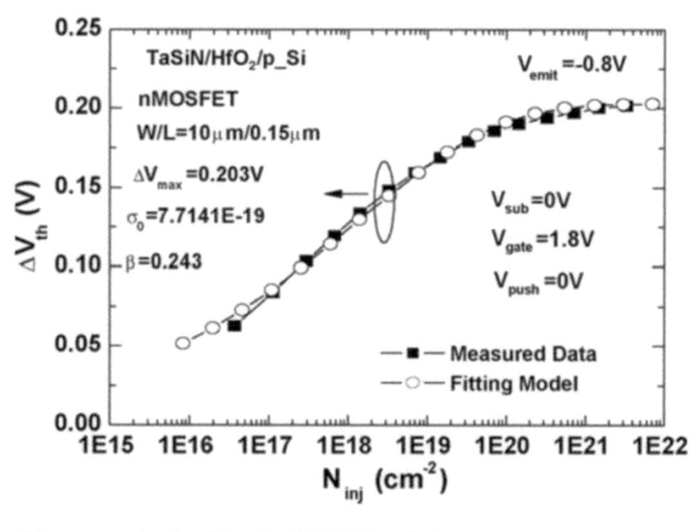

Figure 10. ΔV_{th} vs. N_{inj} curve obtained by the PASHEI technique, to extract trap parameters. (From Reference [22])

3. Advanced techniques

In this section we show three techniques set up in our laboratory: Single shot DLTS , which provides interface state densities), Conductance transient technique used to profile disorder induced gap states in the insulator zones close to the interface, and Flat-band voltage transient technique from which slow traps distribution inside the insulator is obtained.

3.1. Single shot deep-level transient spectroscopy

Deep-level transient spectroscopy (DLTS) has been widely used to characterize localized deep levels in semiconductor junctions. This technique is also useful to measure interface traps in the insulator-semiconductor interface. The instrumentation for interface trapped charge DLTS is identical to that for bulk deep level DLTS. However, the data interpretation is different because interface traps are continuously distributed in energy through the band gap, whereas bulk traps have discrete energy levels.

Single-shot DLTS measurements consist on recording and processing 1-MHz isothermal capacitance transients at temperatures from 77 K to room temperature. A programmable source is used together with a pulse generator to introduce the quiescent bias and the filling pulse, respectively. D_{it} is obtained by first applying a pulse which drives the MIS capacitor to accumulation, in order to fill the interface traps. Afterwards, the bias quickly returns to the limit between depletion and weak inversion, then traps formerly filled are emptied yielding the capacitance transients which are recorded for the DLTS processing. The isothermal capacitance transients are captured by a 1 MHz capacitance meter and a digital oscilloscope. The digital oscilloscope allows us to record the entire capacitance transient and, in this way, we can process the entire energy spectrum with only one temperature scan.

Once the capacitance transients have been captured, we process them as follows: we chose two times t_1 and t_2 (the window rate). The difference in the capacitance value at these times is the DLTS correlation signal which is given by [30, 31]:

$$\Delta C = -\frac{C(t_1)^3}{\varepsilon_S N_D}\frac{1}{C_{ox}}\int_{E_F^{t1}}^{E_F^{t2}}\left[\exp\left(-e_n t_1\right)-\exp\left(-e_n t_2\right)\right]D_{it} \quad (3)$$

The emission rate, e_n, depends on temperature and on energy, E_T, according the well-known Arrhenius law:

$$e_n=\sigma_n v_n N_c \exp\left[\frac{E_T-E_C}{kT}\right] \quad (4)$$

Where σ_n is the capture cross section, v_n is the electron thermal velocity and N_C is the efective state density at the silicon conduction band. According equation (3), all the interface states contribute to the correlation function, but only those with emission rates in the range of the window rate have non negligible contribution. Indeed, the correlation function has a maximum for:

$$e_n^{max} = \frac{\ln\left(\frac{t_2}{t_1}\right)}{t_2 - t_1} \tag{5}$$

If we assume that capture cross section has not strong variations with energy, we can find the energy of interface traps which have the maximum contribution to the correlation function:

$$E_T^{max} = E_C - kT\ln\left[\frac{\sigma_n v_n N_C (t_2 - t_1)}{\ln\left(\frac{t_2}{t_1}\right)}\right] \tag{6}$$

ΔC(E$_T$*)* has a maximum at the energy given by equation (6) and decays very sharply when energy varies from the maximum. Only interface traps with energies close to the maximum contribute to the DLTS signal, and a more simple form equation (3) can be obtained:

$$\Delta C = -\frac{C(t_1)^3}{\varepsilon_S N_D}\frac{kT}{C_{OX}} D_{it}\left(E_T^{max}\right)\ln\left(\frac{t_2}{t_1}\right) \tag{7}$$

And the interface state density at the energy of the maximum:

$$D_{it}\left(E_T^{max}\right) = -\frac{\varepsilon_S N_D}{kT\ln\left(t_2/t_1\right)}\frac{C_{OX}}{C(t_1)^3}\Delta C \tag{8}$$

Equation (6) indicates that for a given window rate the energy is proportional to temperature. Therefore, low temperature transients provide D_{it} for states close to the majority carriers semiconductor band (conduction band for n-type or valence band for p-type). As temperature increases deeper states densities are obtained. Equation (8) says that D_{it} is proporcional to $\Delta C/T$, that is, the sensitivity is lower for deeper states. Since SS-DLTS is a differential technique, its sensitivity is much higher than Capacitance-Voltage or Conductance-Voltage Techniques. Typical sensitivities are in the range of 10^9 eV^{-1} cm^{-2}, which are lower than the state-of-the-art of thermal silicon oxide with silicon interface. Figure 11 is an example of SS-DLTS applied to the case of a hafnium silicate/silicon oxide on n-type silicon. The silicate was deposited by atomic layer deposition. In this case, we studied the effect of post deposition thermal annealing on the quality of the interface.

3.2. Conductance transient technique

All gate dielectrics exhibit conductance transients in MIS structures when are driven from deep to weak inversion [32]. This behavior is explained in terms of disorder-induced gap states (DIGS) continuum model suggested by Hasegawa et al.[33]. These authors proposed that lattice breaking at semiconductor/insulator interface causes defects with a continuous distribution both in energy and in space. Conductance transient phenomena are due to charge and discharge of DIGS states assisted by majority carriers coming from the corresponding semiconductor band by means of a tunneling assisted mechanism. Transients can be understood looking at Figure 12 which is referred to a MIS structure over an n-type semiconductor substrate. When the bias pulse is applied, empty DIGS trap electrons coming from the conduction band (n-MIS structure). E_F and E' are the locations of the Fermi level before and after the pulse. Capture process is assisted by tunneling and is, thus, time consuming, so empty states near the interface capture electrons before the states deep in the dielectric. x_C is the distance covered by the front of tunneling electrons during the time t . It is important to note here that only those states with emission and capture rates of the same order of magnitude than the frequency have non-zero contributions to the conductance [34]. If an experimental frequency ω_0 is assumed, only those states with emission rates in the range $\omega_0 \pm \Delta\omega$ can contribute to the conductance (those located over equiemission line $e_n = \omega_0$), so only when the front of tunneling electrons reaches point A conductance increases. Then, when point B is reached, conductance transient follows the DIGS states distribution which is typically decreasing as we move away from interface, in agreement with Hasegawa's model [33]. Finally, conductance returns to its initial value when the front reaches point C, since after this point DIGS states susceptible to contribute to the conductance signal have energies strongly apart of the Fermi level and, then, they remain empty. Figure 12 is a schematic of the conductance transient principle.

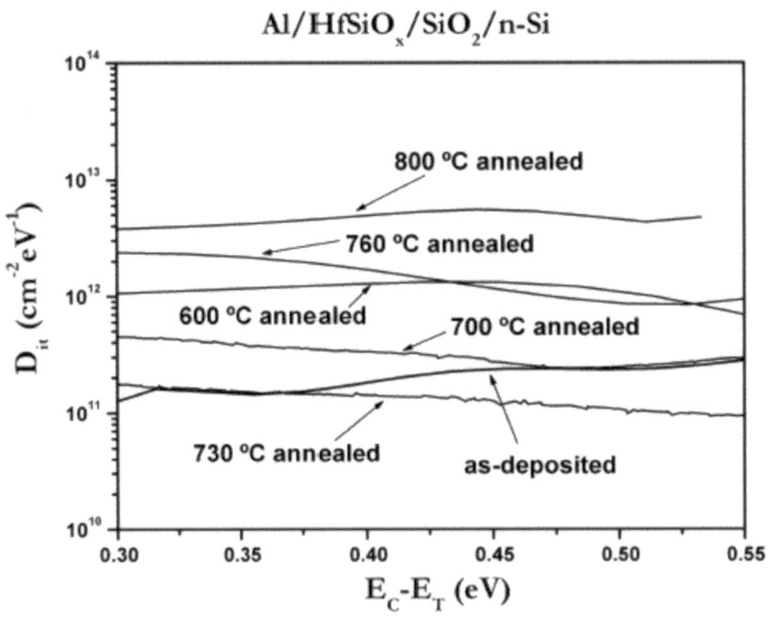

Figure 11. Interface state profiles for *$Al/HfSi_xO_y/SiO_2/n$-Si* capacitors.

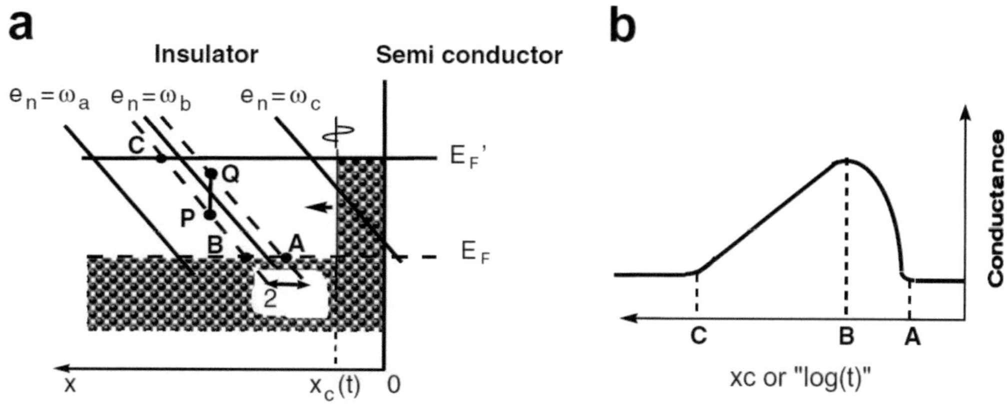

Figure 12. (a) Schematic band diagram of an I–S interface illustrating the capture electrons by DIGS continuum states during a conductance transient. (b) General shape of the conductance transient.

In the following, we show the model developed by us [35] to obtain DIGS states as a function of the spatial distance to the interface and the energy position by measuring conductance transients at different frequencies and temperatures. The calculation details presented here are for the case of an n-MIS structure. Similar equations can be derived for p-MIS devices. Our model departs from the conductance method typically used to obtain the interface state density, D_{it} , in MIS devices. For an angular frequency, ω, D_{it} is related to conductance by $D_{it}=\frac{G_{SS}}{0.4qA\omega}$ [36] where G_{SS} is the stationary value of the conductance. Variations of this value are due to the DIGS contribution to the conductance:

$$N_{DIGS}(E(t),x_C(t))=\frac{\Delta G_{SS}(t)}{0.4qA\omega} \tag{9}$$

where *E(t)* is the energy of the DIGS states which a given time *t* during the transient contribute to the conductance variation. *xc(t)* is the distance covered by the front of tunneling electrons during the time *t, and* is given by $x_c(t)=x_{on}\ln(\sigma_o v_{th} n_s t)$, where $x_{on}=\frac{h}{2\sqrt{2m_{eff}H_{eff}}}$ is the tunneling decay length, σ_0 is the carrier capture cross-section value for x = 0, v_{th} is the carrier thermal velocity in the semiconductor, and n_s is the free carrier density at the interface. Finally, m_{eff} is the electron effective mass at the dielectric and H_{eff} is the insulator semiconductor energy barrier for majority carriers, that is, the dielectric to semiconductor conduction band offset. Figure 13 shows x_{on} for some high-k dielectrics (electron effective mass and barrier height values have been obtained from References [3] and [37] respectively). One can see that x_{on} is higher for dielectrics in which H_{eff} and m_{eff} are low. In these cases, the tunneling front xc is faster and, consequently, transients reach deeper locations in the dielectric. An important trend can be derived from this figure: as permittivity increases, tunneling decay length increases providing deeper DIGS profiles.

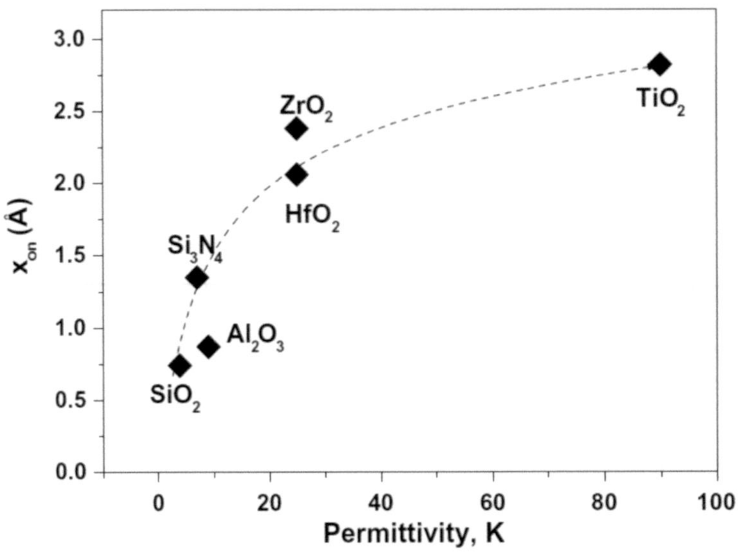

Figure 13. Tunneling decay length versus permittivity for several dielectrics.

Finally, to obtain the energy position of DIGS states in the band gap of the dielectric, we use equi-emission line equations [33] , and considering that the measurement frequency is related to emission rate by $e_n = \omega / 1.98$ [36] , we obtain the following equation:

$$E' - E(x_C, t) = H_{eff} + kT \ln \frac{\sigma_0 v_{th} N_C}{\omega / 1.98} - \frac{kT}{x_{on}} x_C(t) \quad (10)$$

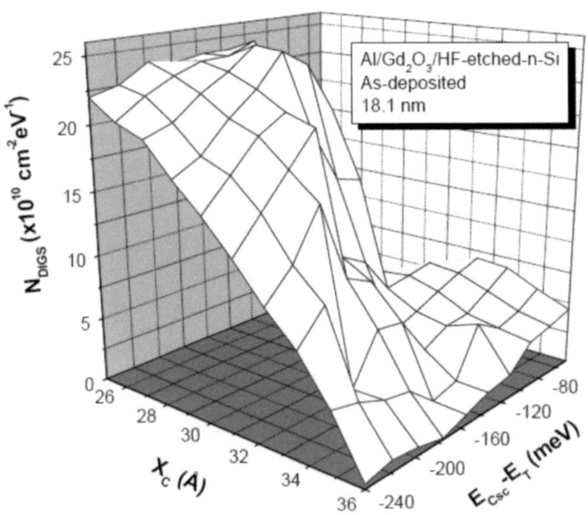

Figure 14. Example of DIGS profile: atomic layer deposited Gadolinium oxide films.

When temperature decreases the emission rates of all interface states exponentially decrease, and the equi-emission lines shift approaching the interface. Thus, transients are modified in a similar way as when frequency is increased while keeping constant the temperature. DIGS three-dimensional profile or contour line maps can be obtained using Equations (9) and (10). As for the experimental sensitivity, temperature measurement involves an error of 0.1 K. Estimated errors of energy and defect concentration values on DIGS profiles are of about 10 meV and 5×10^9 $eV^{-1}cm^{-2}$, respectively. Estimated precision on DIGS depth is of about 2 Å.

The experimental set-up consists of a pulse generator to apply bias pulses, a lock-in analyzer to measure the conductance, and a digitizing oscilloscope to record conductance transients. Samples are cooled in darkness from room temperature to 77 K in a cryostat. Figure 14 is an example of DIGS profiles obtained from conductance transients on MIS structures fabricated with ALD Gadolinium oxide as dielectric.

In section IV.C we review results obtained for several high-k dielectrics grown by atomic layer deposition (ALD) under different processing conditions.

3.3. Flat-Band Transient Technique (FBT)

Several problems must be fixed before the high-k dielectric materials could be extensively used in fabrication. One of them is the instability caused by charge trapping and detrapping inside the dielectric. Fixed and trapped charges cause serious performance degradation by shifting the threshold voltage, limiting transistor mobility and reducing device lifetimes. Threshold voltage shifts are observed under positive bias, negative bias and hot-carrier stressing in high-gate stacks. Charge trapping under positive bias stressing is known to be more severe compared to conventional SiO_2-based gate dielectrics [38]. It is believed to happen due to filling of pre-existing bulk traps. Charge trapping causes threshold voltage shifts and drive current degradation over device operation time. It also precludes accurate mobility (inversion charge) measurements due to a distortion of C-V curves. Negative bias temperature instability (NBTI) induced threshold voltage shifts in high-k devices are also observed and are comparable to those observed for silicon-based oxide devices.

In a previous work [39], we showed the existence of flat band voltage transients in ultra-thin high-k dielectrics on silicon. To obtain these transients, we recorded the gate voltage while keeping the capacitance constant at the initial flat band condition (C_{FB}). Therefore, samples were kept under no external stress conditions: zero electric field in the substrate, darkness conditions and no external charge injection. Under these conditions, the only mechanism for defect trapping or detrapping is thermal activation, that is, phonons. We proved that the energy of soft-optical phonons in high k dielectric is obtained with this experimental approach.

The flat-band voltage, V_{FB}, of a MIS capacitance is given by:

$$V_{FB}(t) = \Phi_{MS} - \frac{Q_i}{C_{ox}} - \frac{1}{\varepsilon_{ox}} \int_0^{tox} \rho_{ox}(x,t) x dx \qquad (11)$$

When the charge density inside the insulator film, $\rho_{ox}(t)$, varied with time, t, or with the distance from the interface, x, the flat band voltage varies. In particular, trapping and detrapping on defects existing inside the dielectric will produce transient variations of the flat-band voltage. According equation (11) these variations are oposite in sign to the charge variation. As it has been suggested elsewhere [40] at flat-band voltage conditions there are not electrons or holes directly injected form the gate or semiconductor, i.e., free charges move by hopping from trap to trap. Moreover, since no optical neither electrical external stimulus are applied, free charges must be originated from trapping or detrapping mechanisms of defects existing inside the dielectric and the energy needed to activate this mechanisms only can be provided as thermal energy, that is, phonons.

The experimental setup of this technique is identical to that used to capacitance-voltage technique. The only difference is that in order to obtain the flat-band voltage transients, a feedback system that varies the applied gate voltage accordingly to keep the flat-band capacitance value was implemented.

The experimental flat band voltage transients become faster when the dielectric thickness diminishes. Time dependences appear to be independent of the temperature. These two facts suggest that there are tunnelling assisted process involved. The amplitude of the transients is thermally activated with energies in the range of soft-optical phonons usually reported for high-k dielectrics. We have proved that the flat-band voltage transients increase or decrease depending on the previous bias history (accumulation or inversion) and the hysteresis sign (clockwise or counter-clockwise) of the capacitance-voltage (C-V) characteristics of MOS structures. In the next section we illustrate all these finger prints.

To illustrate the technique, we have included in Figure 15 some experimental results for the case of a sample of a 20 nm film of hafnium oxide deposited by ALD on silicon. The amplitude of the flat-band voltage transients depends on temperature according an Arrhenius type law:

$$\Delta V_{FB}(T,t)\,\alpha\,\exp-\left({}^{\Phi_{ph}}\!\big/_{kT} \right) \tag{12}$$

where Φ_{ph} is the energy of the soft optical phonons of the dielectric.

4. Some examples

This section includes a selection of different cases to show the applicability of our techniques.

4.1. Effect of interlayer trapping and detrapping on the determination of interface state densities on high-k dielectric stacks

HfO_2 is among the most promising high-k dielectrics, but before qualifying, the nature and formation of electrically active defects existing in these emerging materials should be known. In fact, hafnium based high-k dielectris are already in production [41-43]. While not identified, it is most likely the dielectrics used by these companies are some form of nitrided hafnium silicates (HfSiON). HfO_2 and HfSiO are susceptible to crystallize during dopant activation annealing. However, even HfSiON is susceptible to trap-related leakage currents, which tend to increase with stress over device lifetime. This drawback increases with the hafnium concentration. It is known that defects in SiO_2 are passivated by hydrogen, but this can cause some problems in HfO_2 [44]. Moreover, as most of the high-k materials, when HfO_2 is deposited in direct contact with Si a silicon oxide (SiO_x) interfacial layer (few nanometres thick) is formed [45, 46]. Because of the non-controlled nature of this silicon dioxide layer, its quality is poor and the interfacial state density (D_{it}) and leakage current increase. Moreover, this barrier layer leads to a reduction of the dielectric constant and, hence, to the effective capacitance of the gate dielectric stack. The use of silicon nitride instead of silicon oxide as barrier layer can improve the effective capacitance of the gate dielectric stack, since silicon nitride has a higher permittivity (≈ 7) than silicon oxide (≈ 3.9). Moreover, SiN_x is stable when deposited on Si, preventing the growth of silicon oxides, and the use of nitrides greatly reduces boron diffusion from the heavily doped poly-Si gate electrode to the lightly doped Si channel [3].

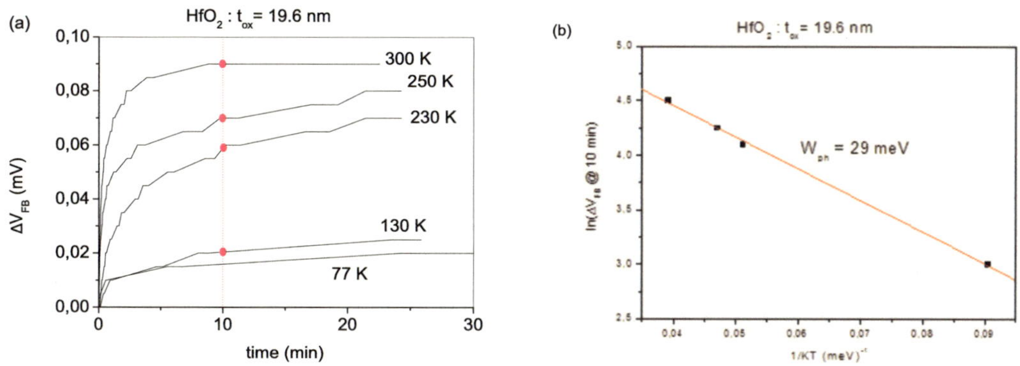

Figure 15. Example of DIGS profile: Atomic Layer Deposited hafnium oxide films.

In a previous work [47] we studied the influences of the silicon nitride blocking-layer thickness on the Interface State densities (D_{it}) of HfO_2/SiN_x:H gate-stacks on n-type silicon. The blocking layer consisted of 3 to 7 nm thick silicon nitride films directly grown on the silicon substrates by electron-cyclotron-resonance assisted chemical-vapour-deposition (ECR-CVD). Afterwards, 12 nm thick hafnium oxide films were deposited by high-pressure reactive sputtering (HPS). Interface state densities were determined by deep-level transient spectroscopy (DLTS) and by the high and low frequency capacitance-voltage (HLCV) method. The HLCV measurements provide interface trap densities in the range of 10^{11} cm^{-2} eV^{-1} for all the samples. However, a significant increase of about two orders of magnitude was obtained by DLTS for the thinnest silicon nitride barrier layers. In this work we probe that this increase is an artefact due to the effect of traps located at the internal interface existing between the HfO_2 and SiN_x:H films. Because charge trapping and discharging are tunnelling assisted, these traps are more easily charged or discharged as lower the distance from this interface to the substrate, that is, as thinner the SiN_x:H blocking layer. The trapping/detrapping mechanisms increase the amplitude of the capacitance transient and, in consequence, the DLTS signal, which have contributions not only from the insulator/substrate interface states but also from the HfO_2/SiN_x:H interlayer traps.

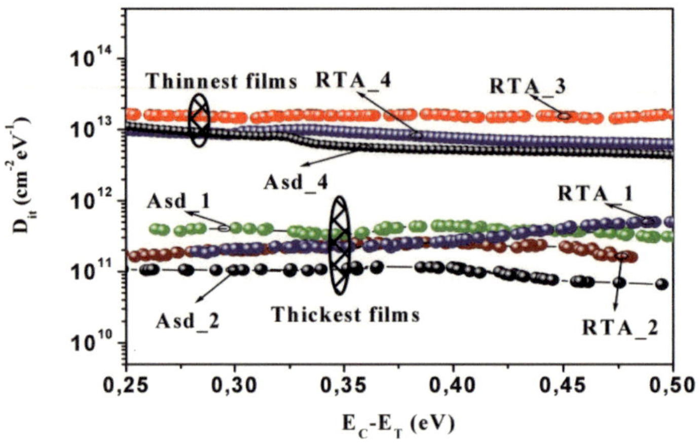

Figure 16. Interface state density measured by DLTS

To determine the interface trap densities we used DLTS and HLCV techniques in order to contrast the results obtained by the two techniques. HLCV measurements are summarized in table 1. This technique provides similar interface density (D_{it}) values (2-4 × 10^{11} cm^{-2}eV^{-1}) for all the samples, regardless the silicon nitride layer thickness. Therefore, interface quality seems not to depend on the blocking layer thickness, as one could expect for these not ultrathin films. In contrast, DLTS results (Figure 16) can be clearly separated in two groups: one corresponding to the thickest samples which has D_{it} densities from 9 × 10^{10} cm^{-2}eV^{-1} to 4 × 10^{11} cm^{-2}eV^{-1}, in good agreement with HLCV results, and the other group corresponding o the thinnest samples wit D_{it} values (from 6×10^{12} cm^{-2}eV^{-1} to 2×10^{13} cm^{-2}eV^{-1}) much higher than those obtained by HLCV. In order to explain these discrepancies we carried out an exhaustive analysis which leads us to conclude that charging and discharging mechanisms of inner traps existing at the HfO_2/SiN_x interface affect the DLTS results.

Figure 17 plots the normalized C-V curves measured at room temperature for the as-deposited samples. The stretch-out is similar for all the samples, meaning a similar trap density, contrary to the DLTS results. Vuillame et al. [48] reported variations in the DLTS signal due to slow traps located inside the insulator, but these changes are only observed for very short filling accumulation pulses times under 50 μs, much lower than the 15 ms used in our experiments. On the other hand, changes were much smaller than those observed in this work. Moreover, slow traps induce hysteresis at the C-V curves and conductance transients. However, a clockwise hysteresis is observed only in the thickest samples and conductance transients have not been detected in any of the thinnest samples. The only difference between the samples is the HfO_2/SiN_x:H interface distance from the substrate, so that we focused our attention in the traps existing at the surface between the SiN_x:H interface layer and the HfO_2 film.

Sample	ECR-CVD time (s)	Silicon nitride Tickness (nm)	RTA	D_{it} from DLTS × 10^{11} ($cm^{-2}eV^{-1}$)	D_{it} from HLCV × 10^{11} ($cm^{-2}eV^{-1}$)
Asd_1	90	6,6 ± 0,4	As-deposited	3 – 5	3.0
RTA_1			600 ºC – 30s	2 - 5	2.2
Asd_2	60	5,9 ± 0,4	As-deposited	0.8 – 1	1.3
RTA_2			600 ºC – 30s	1 - 2	2.7
Asd_3	30	3,9 ± 0,2	As-deposited	Not measured	4.5
RTA_3			600 ºC – 30s	100 - 200	4.4
Asd_4	15	3,0 ± 0,4	As-deposited	50 - 100	2.0
RTA_4			600 ºC - 30s	50 - 100	1.9

Table 1. ECR-CVD deposition time, silicon nitride thickness and interface state densities provided by DLTS and HLCV measurements.

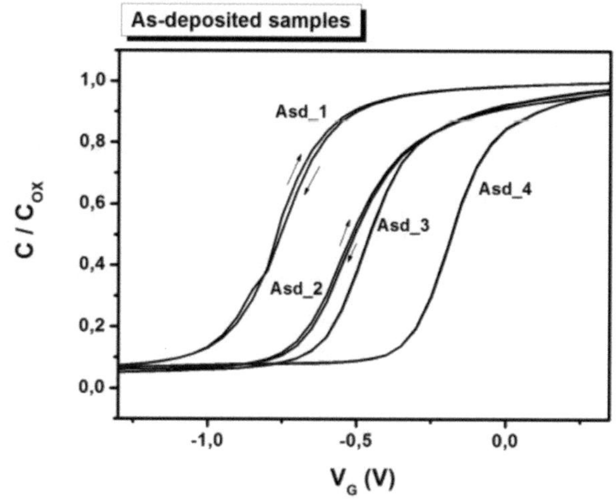

Figure 17. 1 MHz C-V curves measured for the as-deposited samples at room temperature.

To study these discrepancies in depth, we have focused our attention on the sample showing the biggest discrepancies on the D_{it} values measured by HLCV and DLTS. The one selected was the Asd_4 sample, which has the lowest barrier layer thickness (3 nm). First, we recorded the interface state density profiles obtained by DLTS when varying the bias conditions. Figure 18(a) shows important variations in the D_{it} profiles when the accumulation filling pulse voltage is varied while keeping constant the reverse voltage. On the contrary, no significant differences are obtained when varying the reverse voltage (Figure 18(b). Therefore, the mechanisms responsible for these variations must occur during the trap-filling pulse but not under reverse (detrapping) bias conditions, when the capacitance transients are recorded.

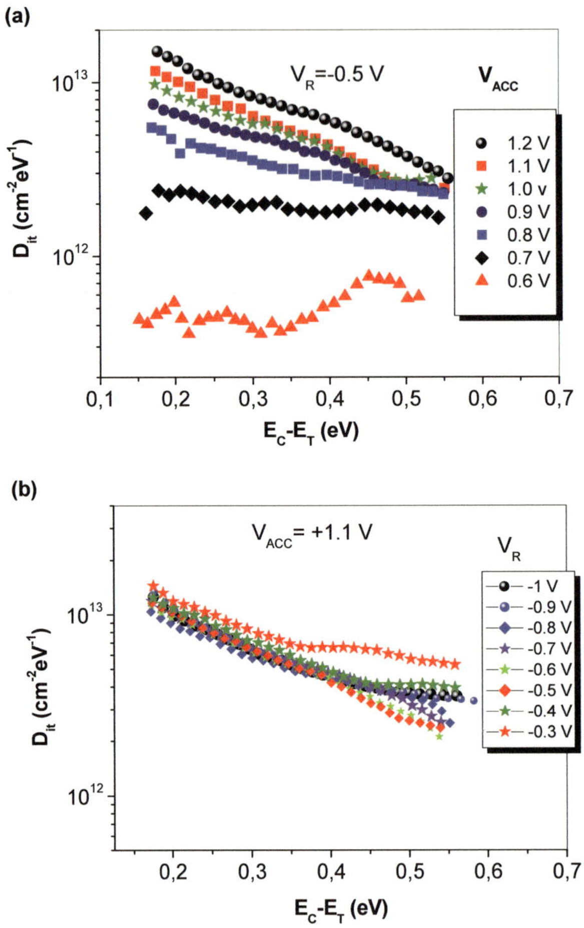

Figure 18. DLTS profiles obtained keeping constant the voltage of the reverse-emptying-pulse (a) and the accumulation-filling pulse (b).

In Figure 19 we show the DLTS values obtained for different energies as a function of gate voltage and the electric field at the Silicon Nitride film. The electric field has been evaluated according the expression:

$$F_{SiNx} = \frac{V_G - V_{FB}}{\frac{\varepsilon_{SiN_x}}{\varepsilon_{HfO_2}} t_{HfO_2} + t_{SiN_x}} \tag{13}$$

We clearly observed that for all the energies the relationship between D_{it} and electric field is linear:

$$\frac{dD_{it}}{dF_{SiNx}} = \eta(E_c - E_T) \tag{14}$$

The slope of Equation (14) is a function of energy. This dependency is plotted in Figure 20 and we have observed that the experimental points fit very well the following dependency.

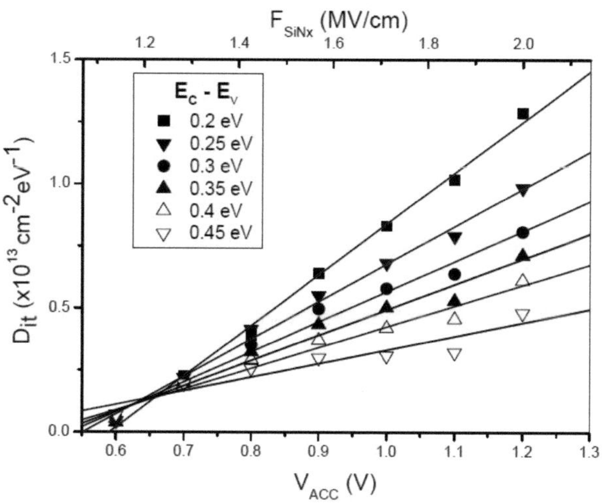

Figure 19. Experimental DLTS signal as a function of accumulation voltage and SiN_x electric field for different energies.

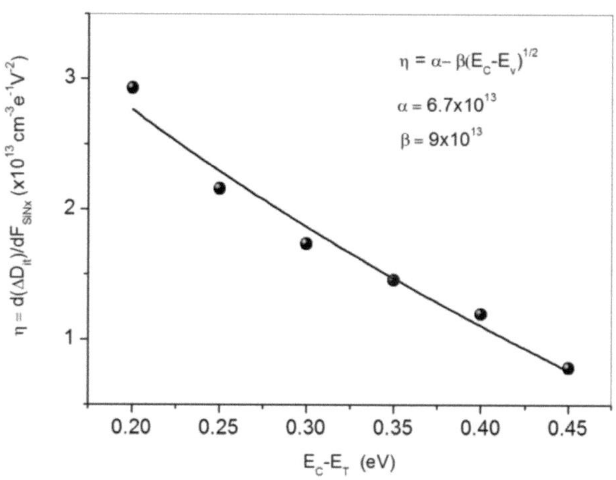

Figure 20. Variation with energy of the electric field barrier lowering parameter, η.

$$\eta(E_c - E_T) = \alpha - \beta\sqrt{E_C - E_T} \tag{15}$$

In summary, we can state that experimental DLTS profiles obey the following expression:

$$D_{it}^{*} = D_{it} + \eta F_{SiNx} = D_{it} + \left(\alpha - \beta\sqrt{E_c - E_T}\right) F_{SiNx} \tag{16}$$

where *Dit** is the as-measured apparent interface state profile. Dit is the true trap interface state density profile that is the obtained at low electric filed values. η is a parameter associated to the electric field lowering of the energy barrier between the silicon conduction band and traps located at the inner layer interface. This barrier is lower as higher the energy of the traps at the inner interface layer and this fact is included at the second term of parameter η.

The true interface state density, *Dit,* is plotted at Figure 21 as obtained for the lowest accumulation voltage values. These values do agree with those obtained when using HLCV technique. Moreover, this distribution show a profile consisting on broad gaussian peaks, as is usually reported for silicon nitride films [49-53].

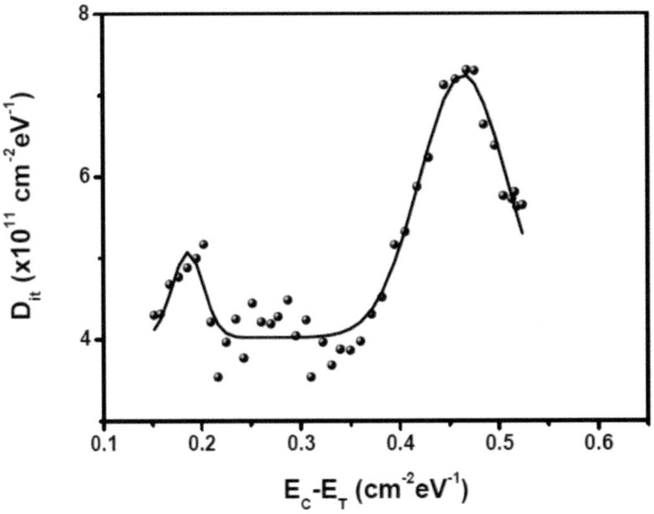

Figure 21. True interface state density profile as obtained at low electric fields (<1 MV.cm-1)

4.1.1. Band energy model

The energy diagrams of the MIS structures under accumulation and inversion are displayed in Figure 22. To construct them, we have included the published values of the bandgap and the conduction and valence band offsets of hafnium oxide and silicon nitride relative to silicon [54]. We also assume that defects exist at the HfO2/SiN_x:H inner layer interface (IL). DLTS measurements consist of applying accumulation pulses to fill the interface states in the upper half of the semiconductor bandgap followed by reverse pulses in which the interface states emit electrons to the conduction band yielding capacitance transients that are conveniently recorded and processed to obtain the D_{it} distribution. If the SiN_x:H film is thin enough, tunnelling between the semiconductor and the inner layer interface (IL) may occur. At accumulation, capturing electrons coming from the semiconductor band by direct tunnelling fills IL states. Then, when the reverse pulse is applied these defects emit the captured electrons to the semiconductor band. The emission process may occur in two different ways: IL states with energies above the silicon conduction band (light grey area) emit electrons by direct tunnelling (A). On the other hand, for energies ranging from the Fermi level to the semiconductor conduction band (dark grey area) tunnelling between the

IL states and the interface states (B). These interface states can emit electrons to the conduction band in a similar way as occurs in conventional DLTS (C). Electrons emitted according the (B)+(C) sequence increase the capacitance transient, obtaining an apparent increase in the measured interfacial state densities. Since all these mechanisms are tunnelling assisted, as thinner the silicon nitride films as higher their probability. In our experiment, the SiN_x:H layer thickness has been varied from around 3 to 6.6 nm. To roughly estimate the relationship between the tunnelling charging/discharging probabilities for two samples with different silicon nitride thickness (t_1 and t_2), we can use the following quantum mechanics expression:

$$\frac{p_1}{p_2} = \exp\left[\frac{2\pi\sqrt{2m_h\overline{\phi_V}}}{h}(t_1 - t_2)\right] \tag{17}$$

where m_h is the hole effective mass inside the barrier, $\overline{\phi_v}$ is the mean barrier height, t_1 and t_2 are the barrier thickness and h is the Plank's constant. For the h-well triangular barrier, $\overline{\phi_v} = \Delta E_V/2$, where ΔE_V is the valence band offset of silicon nitride relative to silicon. Gritsenko et al. [55] reported values of $\Delta E_V \approx 1.5$ eV and $m_h/m_0 = (0.3 \pm 0.1)$. Here m_0 is the free electron mass. These values yield a relation of $p_1/p_2 = 10^{-4}$ for two layers of 6 and 3 nm, respectively, so indicating that the IL trapping/detrapping mechanisms effect is negligible for thicker samples in comparison with the 3 nm-thick blocking layer samples where the very thin silicon nitride layer allows electron tunnelling from IL traps to the channel interface, so increasing the total charge emitted during the DLTS reverse pulses.

Moreover, as higher the electric field In Figure 22(b) higher filling-pulse (higher bias in the accumulation regime). In this case, a larger number IL traps has been filled. When biasing the sample in the inversion regime, a higher number of IL traps can contribute to the capacitance transient by direct tunnelling. This result agrees with results shown in Figure 18(a): the higher the filling pulse the higher the DLTS *Dit* results.

On the contrary, variations of the inversion bias do not change the total filled traps, and the emitted charge from the IL traps does not change significantly. The results shown in Figure 18(b) confirm this hypothesis: the measured D_{it} values hardly change when varying the reverse bias.

In samples with thicker SiN_x:H layer, IL traps cannot contribute to the DLTS capacitance transients, which take place in a relatively short time. However, the IL traps in these samples do exchange charge with the substrate in longer times, giving rise to the hysteresis phenomena not observed in the two thinnest samples. In fact, we can measure slow states inside the MIS insulator by the conductance transient technique (GTT) [56]. We measured the slow states inside the insulator and we observed only slow states in the two thickest samples: if these slow states were due to traps in the bulk SiN_x:H, they would appear in all the samples.

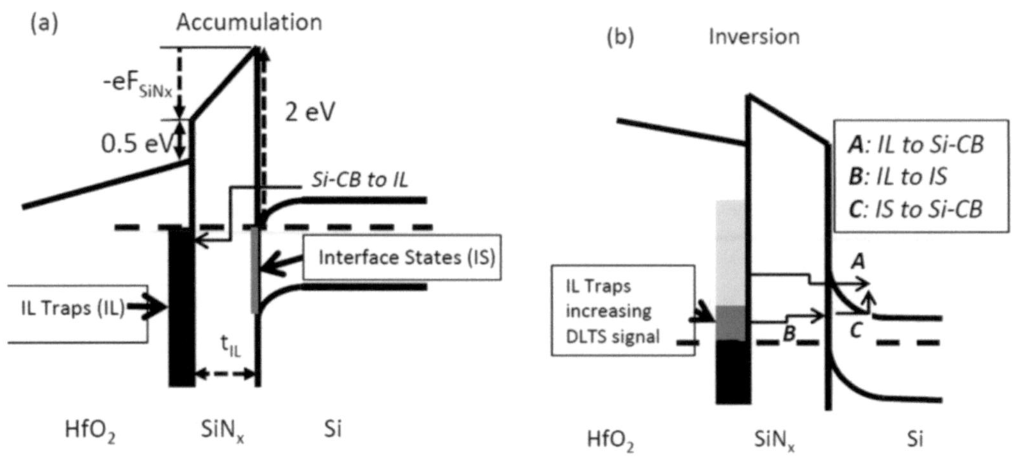

Figure 22. Energy band diagram of the HfO_2/SiN_x:H/n-Si MIS structures under accumulation (a) and inversion (b)

4.2. Flat-band voltage transients: Main fingerprints

In this section we summarize the main finger-prints of the flat-band voltage transients. We have obtained V_{FB} transients for many high-k dielectrics (HfO_2, hafnium silicate, Al_2O_3, TiO_2 and Gd_2O_3). In all cases, there is a direct relationship between the C-V curve hysteresis and the transient amplitude. Here we present a selection of our experimental work to show the information that can be extracted from the transients as well as the parameters affecting to their amplitude, shape, and time constant. We have observed that the main parameters affecting the transients are the experimental temperature, the dielectric film thickness, the dielectric material itself and, finally, the bias voltage and the setup time just before the flat band voltage condition is established in the sample.

4.2.1. Temperature and thickness dependencies

Figure 23 shows capacitance-voltage curves obtained at room temperature for as-deposited Al/Gd_2O_3/HF-etched-Si (a) and $Al/Gd_2O_3/SiO_2/Si$ (b) MIS structures with different Gd_2O_3 thickness. V_{FB} is negative in all cases indicating the existence of positive charge in the dielectric. In Figure 23(a) we see that V_{FB} moves to less negative values with thickness indicating that the charge centroid is closer to the interface for thicker films. That means that traps are preferentially created in the very first dielectric layers. Moreover, in Figure 23(b) we see that when a SiO_2 film is present, V_{FB} shows more negative values and weaker thickness dependence than when Gd_2O_3 films are directly deposited on HF-etched silicon. That must be due to the existence of non-mobile charge trapped at the interface between the high-k and SiO_2 films. V_{FB} transients for different thicknesses (Figure 24(a)) reveal time constants increasing with thickness below 5.7 nm. That indicates the existence of charge displacement mechanisms: the thinner the films the lower the distances to be covered for the mobile charges to reach the gate and/or the insulator-semiconductor interface.

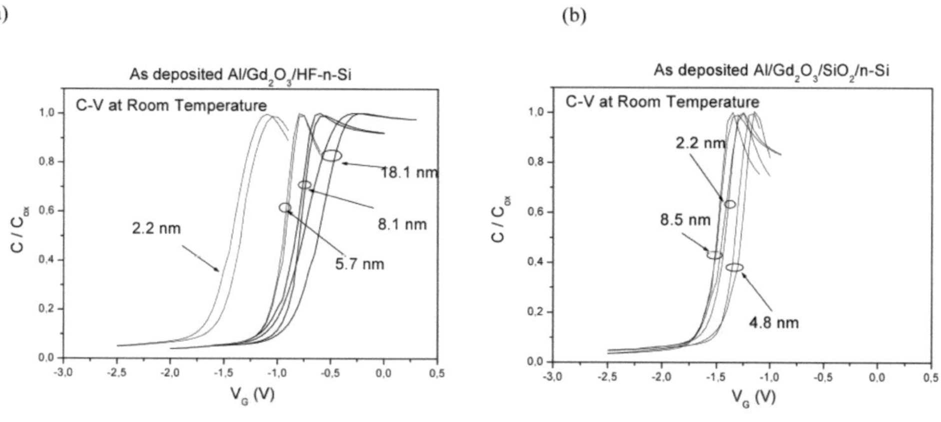

Figure 23. Normalized C-V curves of Al/Gd_2O_3/HF-etched-Si (a) and $Al/Gd_2O_3/SiO_2/Si$ (b) with different Gd_2O_3 thicknesses, measured at room temperature.

To characterize the time dependence of the transients, we have normalized them (Figure 24(b)) by dividing the experimental values by their value at 600 seconds. It is clear that the time constant is independent of the temperature, indicating that tunnelling mechanisms are involved in the conduction. As for temperature dependency of V_{FB} transients, we recorded transients at several temperatures (Figure 25(a)) and we observed that their magnitude follows an Arrhenius plot (Figure 25(b)) with activation energy in the range of the soft-optical phonon energies (W_{PH}) usually reported for high-k dielectrics. From our fits for different samples we have obtained that for Gd_2O_3 these energies are of about 55±10 meV. These values were obtained for both annealed and as-deposited samples and for Gd_2O_3 film thicknesses from about 2 to 20 nm.

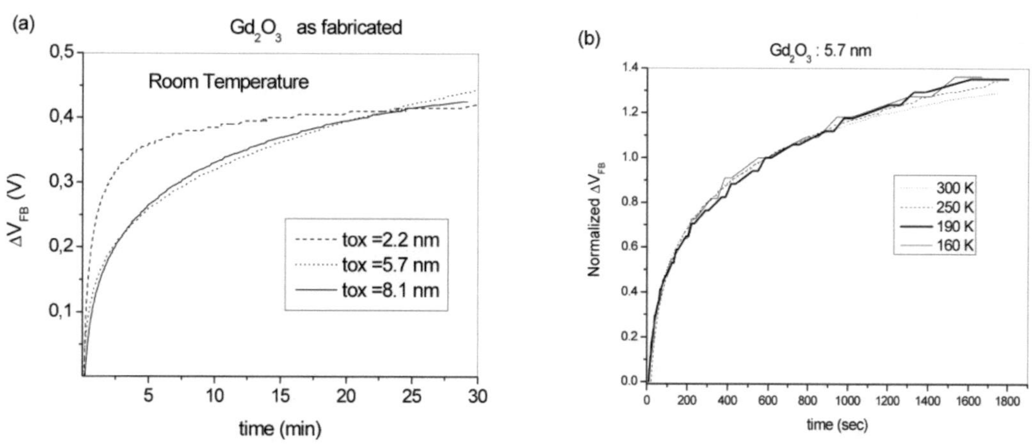

Figure 24. Flat-band voltage transients at different Gd_2O_3 thickness, t_{ox}, (a) and temperature (b) of Gd_2O_3-based MIS structures.

From all these observations we concluded that the flat band voltage transients under conditions without external stress are originated by phonon-assisted tunnelling between localized states: Phonons produce the ionization of traps existing in the bandgap of the insulator. Electrons and/or holes generated in this way move by hopping from trap to trap until they reach a defect location and neutralize the charge state of this defect. It is important to point out that the electrons (or holes) do not enter the conduction (or valence) band of the dielectric and the conduction takes place within the band gap.

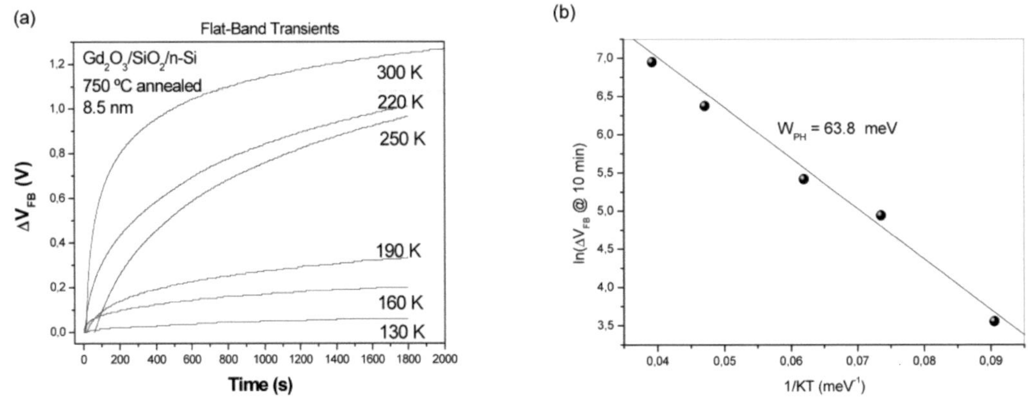

Figure 25. Flat-band voltage transients at different temperatures (a) and Arrhenius plot of the transient amplitude at 10 minutes (b) for an $Al/Gd_2O_3/SiO_2/Si$ sample.

4.2.2. Influence of the setup conditions

In this section we show how the V_{FB} transients are different depending on the bias regime of the sample just before the transient were recording. Transients are different depending if the samples are biased in accumulation or in inversion regimes. Setup time under these previous conditions also affect to the transient amplitude. To illustrate these influences we analyze here two cases based in HfO_2 films. The first one is an HfO_2 film directly grown on n-type silicon, and the second is a HfO_2/SiO_2 stack deposited on n-type silicon. Figure 26(a) shows the C-V curves at room and low temperature for an $Al/HfO_2/Si$ sample with a 250 Å HfO_2 layer grown by atomic layer deposition (ALD). We observe that blat band voltage is positive in all curves indicating the existence of negative charge in the dielectric. Moreover, C-V shows hysteresis at both temperatures. Both flat band voltage and hysteresis are bigger for low temperature. The amount of negative charge and hysteresis are higher at low temperature.

To explain that we suggest that there are positive and negative charges inside the dielectric having different activation energies. At low temperatures positively charged traps (PCT) are not ionized, whereas this temperature is high enough to ionize traps yielding negative charge (NCT). When temperature increases positive traps are ionized by emitting electrons that moves by hopping to the gate or to substrate, so partially compensating the total negative charge. Another point is that hysteresis is clockwise at all temperatures, that is, accumulation bias give places to an increase of the total negative charge. When sample is in accumulation, detrapping mechanisms occurs and traps remain ionized. At inversion, PCT trap electrons coming from the gate and NCTs trap holes coming from the inversion layer at the substrate. Since NCTs predominates the whole effect is that negative charge increases during accumulation and decreases at inversion. These arguments are also observed in the flat-band voltage transients (Figures 26(b) and (c)).

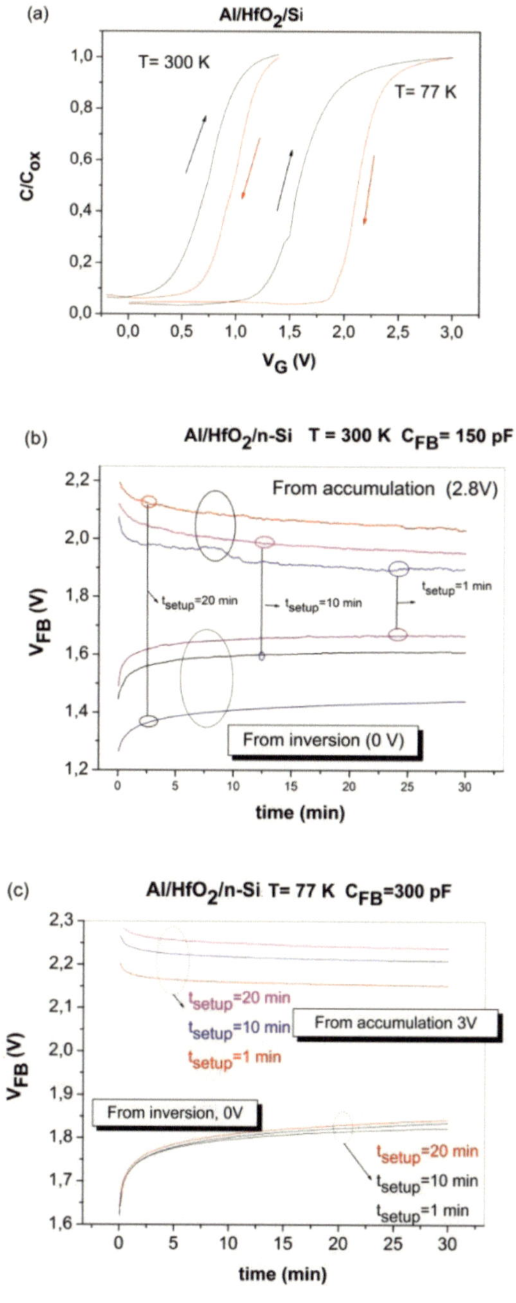

Figure 26. Normalized C-V curves (a) and Flat-band voltage transients at room temperature (b) and 77 K (c) of an Al/HfO_2 /Si sample grown by ALD

We see that transients are decreasing when coming from accumulation and increasing when the sample is previously biased in inversion. At flat-band conditions traps previously charged (PCTs in accumulation and NCTs at inversion) can emit the trapped charge giving place to the corresponding flat-band voltage variation. We see also that these effects are more important as the setup time is higher indicating that trapping and detrapping are not instantaneous because the time needed by free carriers to reach the trap locations. Another important point is that decreasing and increasing transients seem to reach the same final values but after very long times (very much longer than those used in our experimental records).

The second case presented here is a sample in which the dielectric is a stack of an 21 nm HfO_2 film grown by High-Pressure Reactive Sputtering (HPRS) and a SiO_2 buffer layer (3.4 nm-thick). In this case (Figure 27), C-V curves indicate that at room temperature there is positive charge at the dielectric, that is PCTs predominates over NCTs. Consequently, in accumulation the positive charge increases and decreases in inversion regime, giving place to the counter clock-wise hysteresis cycle observed at room temperature. At 77 K the PCTs are not ionized and the hysteresis cycles are due only to NCTs and, then, a clock-wise hysteresis cycle is obtained. This model is confirmed by the opposite trends shown by the flat-band voltage transients obtained at room and low temperature (Figures 27(b) and (c)). Low temperature curves are similar to those obtained in the previous case (Figure 26).

4.3. Conduction transient profiles of high-k dielectrics

In this section we review results obtained for several high-k dielectrics grown by atomic layer deposition (ALD) under different processing conditions. The most noticeable results provided by the experimental contour maps are outlined.

4.3.1. Hafnium-based dielectrics

HfO_2 is a promising gate dielectric material due to its high dielectric constant and excellent thermal stability. Figure 28 shows three-dimensional DIGS plots for HfO_2 atomic layer deposited on n-Si and over p-Si using chloride as metal (Hf) precursor. DIGS states are located at energies close to the majority band edge of the semiconductor. This can be explained in terms of the very nature of the conductance transient technique: majority band edges have the maximum majority carrier concentration, so states located at energies close to this position have the maximum probability to capture majority carriers. On the other hand, no conductance transients were observed for ultrathin samples (less than 40 Å). Kerber et al. [57] proposed the existence of a defect band in the HfO_2 layer. We find spatially distributed defect bands for films on both types of silicon substrates. These defect bands could be due to oxygen vacancies: when the capacitor structure is terminated by the oxide-Si interface, the electric field existing in the dielectric film makes oxygen vacancies (positively charged) to move towards locations farther away from the interface. That occurs in samples deposited on n-type silicon f the difference in semiconductor band bending at the interface [58]. Forming gas annealings (FGA) are usually employed in integrated circuit technology for passivation of defects (dangling bonds) on Si surface. Figure 29(a) shows DIGS density corresponding to post-metallization annealed (400ºC, 30 min) $Al/HfO_2/p\text{-}Si$ sample. Lower DIGS density is achieved, but D_{it} density is increased in this sample [59], indicating that thermal treatment partially moves the insulator defects to the interface. Ioannou-Sougleridis et al. [60] attributed instabilities observed in as-grown Y_2O_3 samples to slow traps, which were mostly removed after FGA. The same behaviour can affect our results.

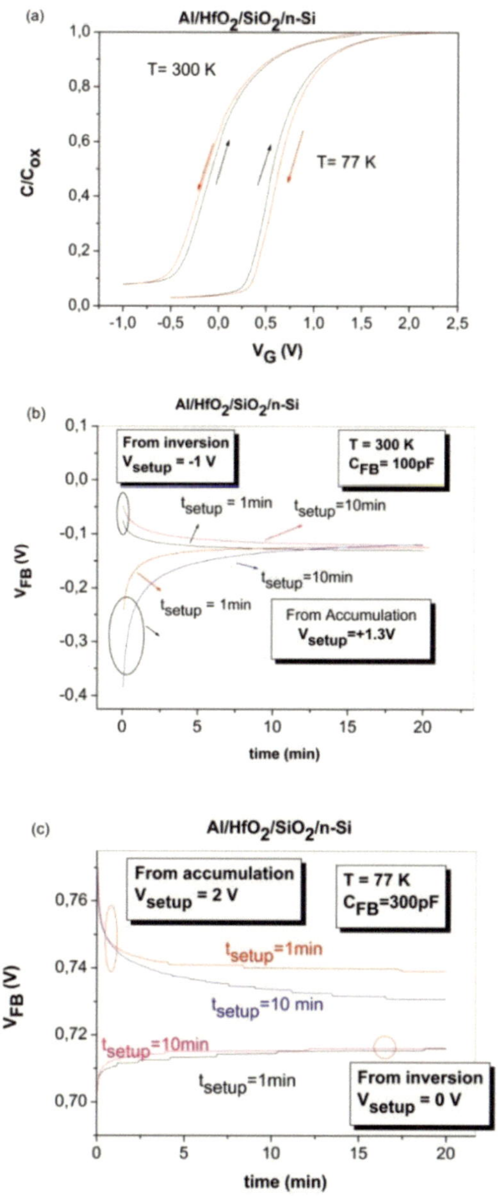

Figure 27. Normalized C-V curves and Flat-band voltage transients at room temperature (b) and 77 K (c) for an $Al/HfO_2/SiO_2/n\text{-}Si$ sample grown by HPRS

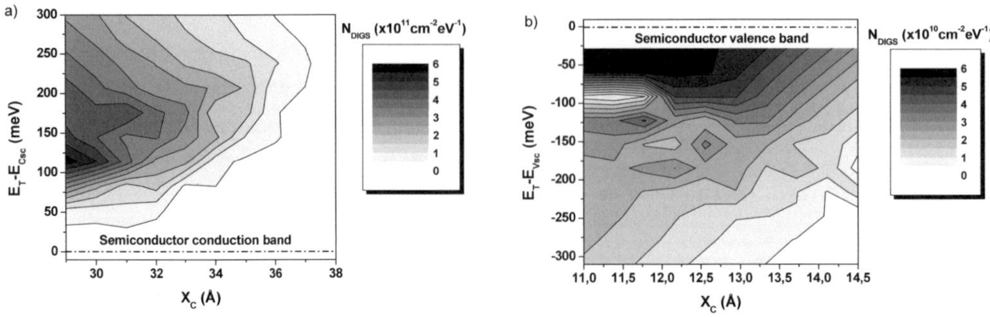

Figure 28. Three-dimensional DIGS plots for unannealed HfO2 atomic layer deposited on n-Si (a) and over p-Si (b)

Transition metal silicates, such as hafnium silicate, have also been the object of a considerable number of studies to replace SiO_2 because of their higher crystallization temperature. Figure 29(b) shows DIGS states obtained from as-deposited Al/$HfSi_xO_y$/n-Si structures grown using HfI_4 and $Si(OC_2H_5)_4$ as precursors. In this case contour lines have a more anisotropic shape than those for HfO_2 indicating less homogeneous distribution of DIGS defects. In fact, we can see two different local ordering at zones A and B. The boundary between these zones approximately follows the line $E_{Csc} - E_T = 588.22 - 15.42x_C$. Contour lines are parallel in zone A and perpendicular to this boundary, indicating some regularity in the defect distribution. On the other hand, DIGS density rapidly decreases to lower values in zone B, where uniformity is higher. When this sample is submitted to a post-deposition annealing at temperatures ranging from 700 to 800ºC, this two-region structure does not change [61].

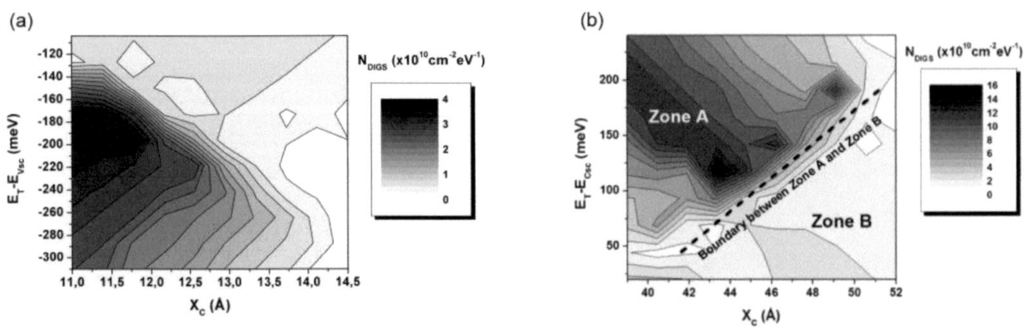

Figure 29. Contour plots of DIGS density obtained to 400 ºC-30 min. annealed Al/HfO_2/p-Si (oxide grown at 450 ºC) and Al/$HfSi_xO_y$/n-Si (silicate grown at 400 ºC)

4.4. Al_2O_3

The importance of Al_2O_3 as an insulating dielectrics is due to its large band gap (8.8 eV), excellent stability when deposited over silicon and its amorphousness (Al2O3 is a good glass former). We have studied Al/Al_2O_3/n-Si structures grown by atomic layer deposition at temperatures ranging from 300 ºC to 800 ºC. $AlCl_3$ and H_2O were used as precursors. DIGS states densities are listed in Table 2. The measured value is similar in all samples, but non measurable at 500 ºC. It is possible that Al_2O_3 grown at this temperature is free of residual defects and moreover, the amorphousness, high purity and structural homogeneity achieved cause low defect densities, making the conductivity signal difficult to measure. In Figure 30 one can see the contour plot corresponding to the sample grown at 300 ºC. The shape is similar to HfO_2 sample deposited on n-Si, but in the case of Al_2O_3 the maximum density appears near the interface which might cause faster defect detrapping. The highest quality sample in terms of DIGS states is that grown at 500 ºC, but if we consider also interface states densities obtained for these samples [62] the best sample would be that grown at 300 ºC. It is important to consider both D_{it} and DIGS densities before concluding the quality of the samples.

Growth temperature	**Maximum DIGS (× 10^{10} cm^{-2} eV^{-1})**
300	12
400	19
500	Undetectable
600	15
800	25

Table 2. DIGS densities obtained to $Al/Al_2O_3/n\text{-}Si$ structures grown at different temperatures.

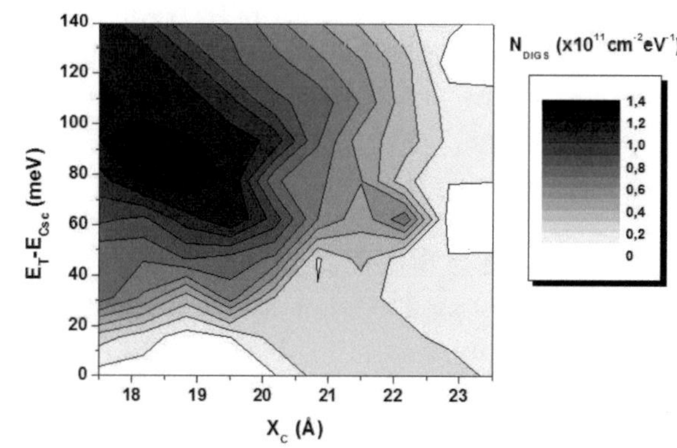

Figure 30. Contour plot of DIGS density obtained to $Al/Al_2O_3/n\text{-}Si$ (oxide grown at 300 ºC).)

4.5. TiO_2

TiO_2 is being extensively studied for memory and logic applications, because of its high dielectric constant, ranging from 40 to 86. We have studied TiO_2 atomic layer deposited on etched n-silicon and high-pressure reactive sputtered over SiO_2–covered Si. DIGS state densities and other growth parameters are listed in Table 3. All ALD samples have been annealed at 750 ºC, so the only differences are growth temperature and chemical precursors. H_2O seems to be more adequate as a precursor than H_2O_2 for the two grown temperatures. On the other hand, when titanium precursor is $Ti(OC_2H_5)$, carbon remains uniformly distributed in the film bulk [63]. In contrast, when $TiCl_4$ is used, chlorine remains in the film and accumulates near the interface [64]. Because of that, higher D_{it} and lower DIGS values are seen in the films grown with $TiCl_4$. To compare with the previous results, we grew TiO_2/SiO_2 dielectric thin films stacks on n-type silicon substrates. A 7 nm layer of SiO_2 was deposited by an Electron Cyclotron Resonance (ECR) oxygen plasma oxidation. Afterwards, 77.5 nm TiO_2 films were grown in an HPRS system at a pressure of 1 mbar during 3 hours and at a temperature of 200ºC. Finally, some samples were *in situ* annealed in oxygen atmosphere at temperatures ranging from 600 to 900ºC. Sputtered films exhibit lower DIGS densities, but the large band gap buffer layer (SiO_2) interposed between substrate and TiO_2 inhibits trap displacements from the interface to the dielectric bulk.

Figure 31 shows two contour maps corresponding to ALD sample grown from TiO_2 (Figure 31(a)) and to sputtered (600 ºC annealed) sample (Figure 31(b)). Defects are located closer to the interface in ALD films because the wider band gap SiO_2 interface layer is not present in this case.

4.6. Other materials: Mixtures

Mixtures, ternary or quaternary oxides are also studied in order to find replacement for SiO_2. Aluminum is a good glass former, so it can induce other dielectric layers to be amorphous, but at the expense of reducing the dielectric film permittivity. To avoid this fact, niobium is also mixed with dielectrics, due to its high permittivity. We have studied Hf-Al-O, Zr-Al-O, Hf-Al-Nb-O and Zr-Al-Nb-O mixtures.Ta_2O_5 layers have also been compared to Ta-Nb-O mixture. All these materials can be grown by ALD on p-silicon, using chlorides as precursors of hafnium and zirconium, $Al(CH_3)_3$ as aluminium precursor, and ethoxides for niobium and tantalum. Table 4 shows DIGS densities of these dielectric layers. In all cases niobium possibly acts as a barrier which inhibits trap displacement from the interface: in fact interface state densities are larger when Nb is incorporated and at the same time, DIGS state densities are reduced [65, 66]. Hf-Al-O behaves like Zr-Al-O due to the similarity between hafnium and zirconium. DIGS density for Ta_2O_5 has an intermediate value (~ 10^{11} cm^{-2} eV^{-1}), as seen in the contour plot in Figure 32. By comparing this plot with Al/HfO_2/p-Si plot, we realize that maximum DIGS reach deeper locations and lower energies for Ta_2O_5. This can be explained in terms of the larger valence band offset for HfO_2 or ZrO_2 with respect to Ta_2O_5.

TiO_2 atomic layer deposited over n-Si		
Precursors	T_G (ºC)	Maximum DIGS ($\times 10^{11}$ cm^{-2} eV^{-1})
$Ti(OC_2H_5)$, H_2O	275	0,1
$Ti(OC_2H_5)$, H_2O_2	225	3,5
$Ti(OC_2H_5)$, H_2O_2	275	1
$TiCl_4$, H_2O	225	2

TiO_2 sputtered over SiO_2	
Annealing	Maximum DIGS ($\times 10^{11}$ cm^{-2} eV^{-1})
No	Not detected
600 ºC	0,5
700 ºC	2,6
800 ºC	1,2
900ºC	Not detected

Table 3. DIGS densities obtained to TiO_2 deposited over n-silicon and over SiO_2

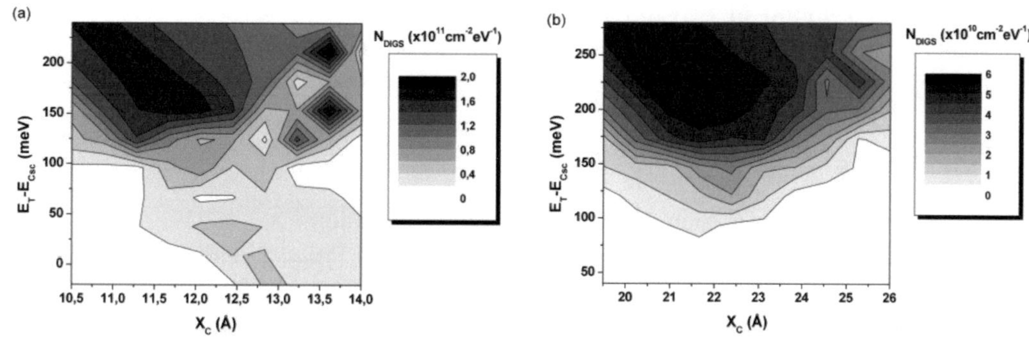

Figure 31. Contour plots of DIGS density obtained to ALD TiO_2 sample grown at 225 ºC from $TiCl_4$ on etched silicon (a) and TiO_2 sputtered on SiO_2–covered silicon (600 ºC annealed) (b).

	Maximum DIGS (x10^9 cm^{-2} eV^{-1})
Hf-Al-O	1200
Hf-Nb-Al-O	2
Zr-Al-O	2000
Zr-Nb-Al-O	Not detected
Ta_2O_5	120
Ta-Nb-O	Not detected

Table 4. DIGS densities obtained for different high-k dielectric mixtures

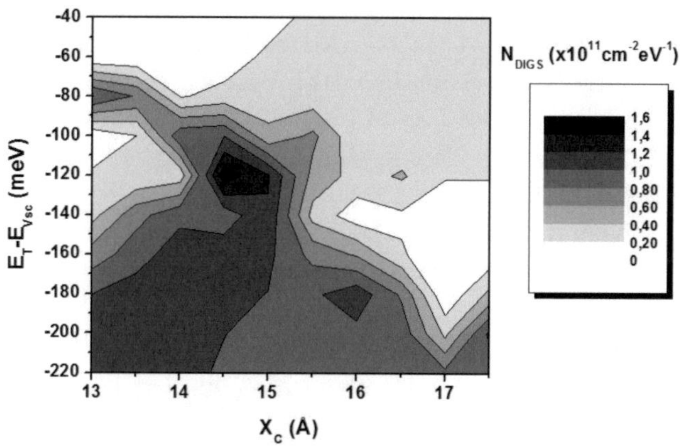

Figure 32. Contour plot of DIGS density obtained to Al/Ta_2O_5/p-Si (oxide grown at 300 ºC).

5. Conclusions and future trends

In this chapter we review several experimental techniques which allow detecting, measuring and identifying traps and defects in metal insulator interface, and at the bulk of the dielectric. The correlation between conduction mechanisms, defect location and preferential energy values provides very relevant information about the very nature of defects and, eventually, how these defects could be removed or diminished. Our techniques provide high resolution in two dimensions: defect energy (E) and depth relative to the interface (z). In the future, we want to combine these techniques with scanning probe microscopy in order to obtain high resolution in lateral dimensions (x,y) as well.

Author details

Salvador Dueñas*, Helena Castán, Héctor García and Luis Bailón
Dept. Electricidad y Electrónica, ETSI Telecomunicación, Campus "Miguel Delibes", Valladolid, Spain

Acknowledgement

The study here presented has been supported by the Spanish Ministry of Economy and Competitiveness through Grants TEC2008-06698-C02 and TEC2011-27292-C02.

6. References

[1] The International Technology Roadmap for Semiconductors, edition 2009, http://public.itrs.net.
[2] Wong H and Iwai H (2006) Microelectron. Eng., 83: 1867
[3] Wilk G D, Wallace R M, and Anthony J M (2001) J. Appl. Phys., 89: 5243
[4] Robertson J (2006) Rep. Prog. Phys, 69: 327
[5] Houssa M, Pantisano L, Ragnarsson L-A, Degraeve R, Schram T, Pourtois G, De Gendt S, Groeseneken G, and Heyns M M (2006) Mater. Sci. Eng. R., 51: 37
[6] Johnson R S, Lucovsky G, and Baum I (2001) J. Vac. Sci. Technol. A, 19: 1353
[7] Kim Y-B , Kang M-S, Lee T, Ahn J, and Choi D-K (2003), J. Vac. Sci.Technol. B, 21:2029.
[8] Rhee S J, Kang C Y, Kang C S, Choi R, Choi C H, Akbar M S, and Lee J C (2004) Appl. Phys. Lett.. 85: 1286.
[9] Lee B H, Kang L, Nieh R, Qi W-J, and Lee J C (2000) Appl. Phys. Lett., 76: 1926.
[10] Houssa M, Afanas'ev V V, Stesmans A, and Heyns M M (2000) Appl. Phys.Lett., 77: 1885.
[11] Wilk G D et al. (2002) Dig. Tech. Pap. - Symp. VLSI Technol. 2002: 88.
[12] Choi R, Kang C S, Cho H-J, Kim Y-H, Akbar M S, and Lee J C (2004) Appl. Phys. Lett., 84: 4839.
[13] Zhao C Z, Zhang J F, Chang M H, Peaker A R, Hall S, Groeseneken G, Pantisano L, De Gent S, and Heyns M (2008) J. Appl. Phys., 103: 014507.
[14] Kukli K, Ihanus J, Ritala M, and Leskela M (1996) Appl. Phys. Lett., 68: 3737.
[15] Johnson R S, Hong J G, Hinkle C, and Lucovsky G (2002) J. Vac. Sci.Technol. B, 20:1126.
[16] Lee J-H et al. (2002), Dig. Tech. Pap. - Symp. VLSI Technol. 2002: 84.
[17] Castgné R, Vapaille A (1971) Surface Science, 28: 157-193.
[18] Nicollian E H and A. Goetzberger A (1967) *Bell Syst. Tech. J*, 46: 1055–1133.

* Corresponding Author

[19] De Dios A, Castán H, Bailón L, Barbolla J, Lozano M, and Lora-Tamayo E (1990) Solid-State Electron., 33: 987–992.

[20] Duval E and Lheurette E (2003) Microelectron. Eng.,65: 103–112..

[21] Haddara H and Ghibaudo G (1988) Solid-State Electron, 31:1077–1082.

[22] Ma T P (2008) Applied Surface Science, 255: 672–675

[23] Chen W, Balasinski A, Ma T P (1993) IEEE Trans. Electron Devices, ED-40: 187.

[24] Tsuchiaki M, Hara H, Morimoto T, and Iwai H (1993) IEEE Trans. Electron Devices, ED-40: 1768.

[25] Chun C, and Ma T P (1998) IEEE Trans. Electron Devices,ED-45 (2): 512.

[26] Groeseneken G, H.E. Maes H E, Beltran N, and de Keersmaecker R F (1984) IEEE Trans. Electron Devices, ED-31: 42.

[27] Liu Z, and Ma T P(1999) Intl. Symp. VLSI-TSA: 195.

[28] Song L Y, Wang X W, Ma T P, Tseng H-H, and Tobin P J (2006) 2006 IEEE/SISC:

[29] Zafar S (2003), J. Appl. Phys., 93: 9298–9303.

[30] Yamasaki K, Yoshida M, and Sugano T (1979) Japanese Journal of Applied Physics, 18: 113-122.

[31] Johnson N M (1982) Journal of Vacuum Science and Technology, 21: 303-314.

[32] Dueñas S, Peláez R, Castán H, Pinacho R, Quintanilla L, Barbolla J, Mártil I, and González-Díaz G (1997) Appl. Phys. Lett., 71: 826.

[33] He L, Hasegawa H, Sawada T, and Ohno H (1998) J. Appl. Phys., 63: 2120.

[34] Barbolla J, Dueñas S, and Bailón L (1992) Solid-State Electron., 35: 285.

[35] Castán H, Dueñas S, Barbolla J, Redondo E, Blanco N, Mártil I, and González-Díaz G (2000) Microelectron. Reliab., 40: 845.

[36] Nicollian E H, and Goetzberg A (1967) Bell Syst. Technol. J.,46: 1055.

[37] Lucovsky G, Hong J G, Fulton C C, Zou Y, Nemanich R J, Ade H, Scholm D G, and Freeouf J L (2004) Phys. Status Solidi B, 241: 2221.

[38] Zafar S, Kumar A, Gusev E, and Cartier E (2005) IEEE Tans Dev Mater Rel , 5(1):45.

[39] Dueñas S, Castán H, García H, Bailón L, Kukli K, Hatanpää T, Ritala M, and Leskela M (2007) Microelectron. Reliab., 47: 653–656.

[40] O'Connor R, McDonnell S, Hughes G, Degraeve R, and Kauerauf T (2005) Semicond. Sci. Technol., 20: 668.

[41] http://www.intel.com/technology/45nm/index.htm

[42] http://spectrum.ieee.org/semiconductors/design/the-highk-solution

[43] http://www-03.ibm.com/press/us/en/pressrelease/23901.wss#release

[44] Houssa M, Gendt S D, Autran J L, Groeseneken G, and Heyns M M (2004) Appl. Phys. Lett., 85: 2101.

[45] Hakala M H, Foster A S, Gavartin J L, Havu P, Puska M J, and Nieminen R M (2006) J. Appl. Phys. 100, 043708.

[46] Dueñas S, Castán H, García H, Gómez A, Bailón L, Toledano-Luque M, Mártil I, González-Díaz G (2007) Semicond.. Sci. Technol. 22, 1344.

[47] Castán H, Dueñas S, García H, Gómez, Bailón L, Toledano-Luque M, del Prado A, Mártil I, and González-Díaz G (2010) J. Appl. Phys., 107: 114104; http://dx.doi.org/10.1063/1.3391181.

[48] Vuillaume D, Bourgoin J C, and Lannoo M (1986) Phys. Rev. B, 34: 1171.

[49] Aberle A G, Glunz S, and Warta W (1992) J. Appl. Phys., 71: 4422.

[50] Hezel R, Blumenstock K, Schiirner R (1984) J. Electrochem. Soc., 131: 1679.

[51] Schmidt J, Schuurmans F M, Sinke W C, Glunz S W, Aberle A G (1997) Appl. Phys. Lett., 71: 252.

[52] García S, Mártil I, González Díaz G, Castán H, Dueñas S, and Fernández M (1998) J. Appl. Phys., 83: 332.

[53] Schmidt J and Aberle A G (1999) J. Appl. Phys.,85: 3626.

[54] Robertson J (2000) J. Vac. Sci. Technol. B, 18: 1785.

[55] Gritsenko V A and Meerson E E (1998) Phys. Rev. B, 57: R2081.

[56] García H, Dueñas S, Castán H, Bailón L, Kukli K, Aarik J, Ritala M, and Leskelä M (2008) J. Non-Cryst. Solids, 354: 393.

[57] Kerber A, Cartier E, Pantisano L, Degraeve R, Kauerauf T, Kim Y, Hou A, Groeseneken G, and Schwalke U (2003) IEEE Electron Device Lett., 24: 87.

[58] Dueñas S, Castán H, García H, Barbolla J, Kukli K, Aarik J, and Aidla A (2004) Semicond. Sci. Technol., 19: 1141.

[59] Dueñas S, Castán H, García H, Barbolla J, Kukli K, and Aarik J (2004) J. Appl. Phys., 96: 1365.

[60] Ioannou-Sougleridis V, Vellianitis G, Dimoulas D (2003) J. Appl.Phys., 93: 3982.

[61] Dueñas S, Castán H, García H, Bailón L, Kukli K, Ritala M, Leskelä M, Rooth M, Wilhelmsson O, and Hårsta A (2006) J. Appl. Phys., 100: 094107.

[62] Dueñas S, Castán H, García H, de Castro A, Bailón L, Kukli K, Aidla A, Aarik J, Mändar H, Uustare T, Lu J, Hårsta A (2006) J. Appl. Phys., 99: 054902.

[63] Dueñas S, Castán H, García H, Barbolla J, San Andrés E, Toledano-Luque M, Mártil I, González-Díaz G, Kukli K, Uustare T, Aarik J (2005) Semicond.. Sci. Technol., 20: 1044.

[64] Ferrari S, Scarel G, Wiermer C, and Fanciulli M (2002) J. Appl. Phys., 92: 7675.

[65] Dueñas S, Castán H, Barbolla J, Kukli K, Ritala M, and Leskelä M (2003) Solid-State Electron., 47: 1623.

[66] Dueñas S, Castán H, García H, Barbolla J, Kukli K, Ritala M, and Leskelä M (2005) Thin Solid Films,474: 222.

Chapter 11

Empiric Approach for Criteria Determination of Remaining Lifetime Estimation of MV PILC Cables

I. Mladenovic and Ch. Weindl

Additional information is available at the end of the chapter

http://dx.doi.org/10.5772/51490

1. Introduction

Underground cables represent one of the biggest assets and investment demands of power utilities. In the same time they are the major source of faults and outages in medium voltage (MV) power networks. The oldest cable type, still present in a high percentage in today's MV power networks, is the paper insulated lead covered (PILC) cable. It was mainly laid in the period from 1920 to 1980, (Tellier, 1983), hereafter it has been systematically replaced by most distribution companies with thermoplastic polyethylene (PE) and finally cross-linked polyethylene (XLPE) cable types. Nevertheless, almost 95% of the MV power cable networks of "NUON Infra Noord-Holland" are made up of PILC cables, (E. F. Steennis, R. Ross, N. van Schaik, W. Boone & D.M. van Aartrijk, 2001), 65% of the network of one of the biggest energy supplier in Belgium, 56% in the urban areas in Bavaria (Germany), and ca. 50% of entire MV cable network of Germany. At the end of 20th century in Germany, as reported in (FGH - Forschungsgemeinschaft für Elektrische Anlagen und Stromwirtschaft e. V., 2006), there were more than 30% of cables over 30 years in service, and more than 15% over 45 years in field operation – almost all are PILC cables. Furthermore, this all corresponds to an cable network length of 110.000 km that consist only of cables which already have or soon will exceed the expected cable service life time of 40 years, and nearly 3,2 billion Euro of investments.

Within the years of service operation and especially when the predicted service lifetime is exceeded, the failure rate is expected to increase significantly. Sudden and unexpected cable failures mostly cause many incidental issues, additional costs and penalty payments. In order to optimize costs and to keep up or improve the reliability of the power system, more

The presented chapter is further discussed in Ph.D. thesis "Determination of the Remaining Lifetime of PILC Cables based on PD and tanδ Diagnosticis," Mladenovic, 2012 and more utilities and distribution companies decide for condition based asset management and maintenance strategies. A sophisticated knowledge of the components actual condition and an early detection and prediction of service failures are therefore the bases for an efficient planning of the maintenance strategy and the resulting investments. For this purpose various diagnostic systems are used, which are mostly based on the measurement of the partial discharge (PD) activity and/or other dielectric key values like e.g. the value of the dissipation factor tan(δ) at different test-voltage levels.

Unfortunately, there are still no well-established and physically - founded substantial criteria which define e.g. the probability of the next failure versus the PD or tanδ-levels for defined test voltages and test conditions. Hence,for further improvements of diagnostic systems and the prediction of failure-times a correlation of the field measurement data and parameters acquired under well-known laboratory conditions is necessary together with a following reference setting and interpretation. This could further lead to the development of physically oriented ageing models correlating the cable's level of lifetime consumption and several measured diagnostic parameters, their dependencies and development. The complex mathematical models can only be derived on the basis of a fundamental databank including cable specification data, service operation profiles and numerous electrical and diagnostic parameters monitored during and representing the complete cable life cycle. In this way, the assumption of the remaining life time will be based on numerous diagnostic measurements and parameters. A restriction to regular measurements and failure-time data out of the field would last in a monitoring process over several decades and more or less undefined, unknown or less reliable measurement conditions caused e.g. by the various influences of the equipment's temperature and its gradients on diagnostic parameters.

On this background, a system for artificial and accelerated ageing of MV PILC cables has been developed and realized, (Mladenovic & Weindl, Determination of the Characteristic Life Time of Paper-insulated MV-Cables based on a Partial Discharge and tan(δ) Diagnosis, 2008) (Dr.-Ing. Weindl & Dipl.-Ing. Mladenovic, 2009) (Mladenovic & Weindl, ICAAS – Integrated System for lasting Accelerated Aging of MV Cables, Data Monitoring and Acquisition, 2009) (Mladenovic I., 2009). The ICAAS (Integrated Cable Accelerated Ageing System) facilitates a realistic (50Hz) but accelerated ageing by applying pre-defined and concurrent thermal and electrical stressconditions with a highly sensitive and selective PD-detection and tanδ measurement. By controlling the technical and environmental conditions of the artificial ageing processes the ageing rapidity can be modified and increased. During the ageing experiment, a daily monitoring of the cable samples was realized by measuring the diagnostic parameters under pre-defined conditions and selective for each individual cable sample, (Freitag, Weindl, & Mladenovic, On-Line Cable Diagnostic Possibilities in an Artificial Aging Environment, 2011) (Freitag, Mladenovic, & Weindl, Fully Automated MV Cable Monitoring and Measurement System for Multi-Sample Acquisition of Artificial Aging Parameters, 2010). Moreover, the entire accelerated ageing process, all system-parameters and internal signals are monitored in close-meshed time intervals. Using a suitable set of pre-aged cabled samples, an ageing database of over 800GB was formed up that enables statistical approaches to determine the actual and integral ageing factor, the characteristics of the ageing process, the key ageing parameters, as well as their limits.

Once the databank is formed-up and before the statistical analysis is applied, there is a necessity to discuss the physical dependencies of the relevant electrical properties on the environmental and test conditions and the ageing process of the paper-mass insulation system. Herby, the structure and the chemical background of the insulation components of the PILC cables will be shortly presented, and the way it could influence the ageing rapidity, the development of partial discharges and the thermal breakdown. The discussion of the behavior of the relative permittivity (ε_r) and conductivity (κ) with varying temperature and humidity, voltage and frequency as well as the influence of the impurities, cavities and bubbles presence on the electrical properties and ageing of the insulation material will follow. Hence, the main dependencies of the diagnostic parameters and their development can be explained, (Mladenovic & Weindl, Determination of the Characteristic Life Time of Paper-insulated MV-Cables based on a Partial Discharge and tan(δ) Diagnosis, 2008) (Mladenovic & Weindl, Dependencies of the PD- and tan(δ)-Characteristics on the Temperature and Ageing Status of MV PILC Cables, 2011) (Mladenovic & Weindl, Dependency of the Dissipation Factor on the Test-Voltage and the Ageing Status of MV PILC Cables, 2011) (Mladenovic & Weindl, Development of the Partial Discharges Inception Voltage for Different Sets of Pre-Aged PILC Cable Samples, 2010) (Mladenovic & Weindl, Influence of the thermal stress on the diagnostic parameters of PILC cables, 2010).

The approach presented here, the entire ageing experiment, the data analyses and conclusions are systematically elaborated in (Weindl, Verfahren zur Bestimmung des Alterungsverhaltens und zur Diagnose von Betriebsmitteln der elektrischen Energieversorgung, 2012) Mladenovic, Ph.D., Determination of the Remaining Lifetime of PILC cables based on PD and tan(δ) diagnosis, 2012, to be published) (Freitag, Ph.D., to be published).

2. Insulation system of PILC cables

The insulation system of PILC cables is a complex and inhomogeneous structure of mass impregnated paper layers. During the operation, the electrical field is distributed so that the thin mass layers overtake a bigger part of the electrical field strength. The paper will keep the separation distance and will be a barrier to the impurities from layer to layer. An insulating paper (e.g. kraft paper consists of about 90% of long-chained macromolecules, Figure 1) i.e. cellulose fibrils, is formed by the polymerization of the glucose molecules. Cellulose molecules arranged in fibrils have an immense tensile strength.

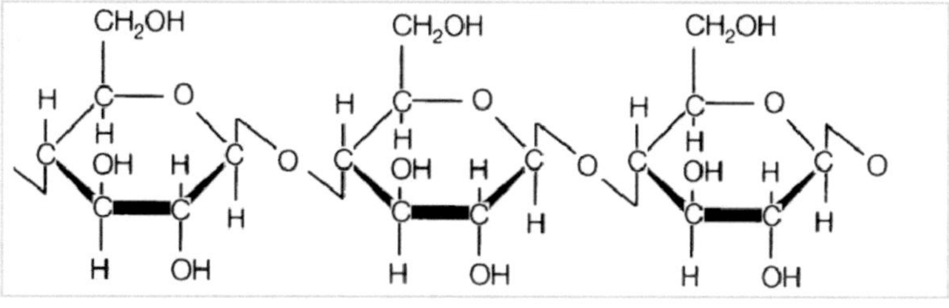

Figure 1. Cellulose molecule $(C6H10O5)_n$, (Colebrook)

The length of the cellulose chains defines the degree of polymerization (DP), the number of glucose units that make up one polymer molecule:

$$DP = \frac{molecular\ weight}{molecular\ weight\ of\ base\ unit} \quad (1)$$

Properties of the cellulose and therefore paper are strongly dependent on DP.

For example, a new insulation paper has a DP of 1100-1300, which decrease with the operation time or ageing, reducing as a consequence the mechanical strength of the paper. A DP of less than 500 indicates significant thermal degradation, and finally, a DP of 200 is concerned as the limit of the mechanical strength and end of paper life, (Küchler, 2009), since weakening of mechanical properties could lead to e.g. cable failure during short circuits.

The structure of the impregnating compound was changed over the decades of manufacturing PILC-cables. At the beginning of 20th century, mineral oils have been used, followed by oil-rosin up to lastly non-draining (non-migratory) compounds in the time before World War II, (Bennett, October 1957). This poly-isobutylene compound – MIND (mass-impregnation with non-draining compound) held up to today. It presents the differently proportional mixtures of natural or synthetic resin, paraffin, bitumen and oil, (www.wikieduc.ch, 2010). Beside it is much more practical for handling; process of oxidation in MIND compound is much more slowly than in insulating oils. Also its dielectric loss angle δ is less dependent on the temperature. One of its disadvantages, as given in (Bennett, October 1957), is that brand new cables could contain numerous voids, due to the high expansion coefficient of the material. Also, the compound retains its non-draining properties up to 70°C, nowadays improved to 90°C, or even 100°C, (Kock & Strauss, 2004).

3. Ageing mechanisms of mass-impregnated paper

During the operation, the insulation of the PILC cables is exposed to numerous effects like temperature, electrical field, moisture presence (invasion), mechanical stress, which causes over the time steady insulation degradation. Herby, the paper presents the weaker component, and it's therefore the main cause of the insulation ageing. As presented in (Bennett, October 1957) the degradation of the mass is an oxidative and slow process. Moreover, oxygen presence in the cable insulation is not worth of consideration differing it from the oil insulation in transformer. There are several mechanisms of the impregnated paper degradation like hydrolytic, oxidation, thermal and electrical degradation (Emsley & Stevens, 1994). In Figure 2 the degradation of cellulose is shown.

During field operation, there is a regular thermal degradation of the components insulation. The maximal nominal operating temperature of the 12/20 kV three core PILC cables is 65°C, (Glaubitz, Postler, Rittinghaus, Seel, Sengewald, & Winkler, 1989). Temperatures in the range lower than 200°C, in the absence of oxygen and moisture tends to open glucose rings and break the glycosidic bonds. Finally, it results in free glucose molecules (decrease of DP) and release of moisture, carbon monoxide and carbon dioxide, as shown in Figure 2. Additional presence of moisture again accelerates the process of cellulose decomposition. If the temperature would exceed 200°C, other reactions would occur, including even destruction of the solid components.

Moreover, gasses formed through the thermal degradation would fill the cavities (if there), presenting in this way an electrical weakness within the insulation and a potential PD source. An increase of the electrical stress would intensify PD activity producing additional gases in the locally heated area, leading further to the expanding of weak region. However, this chain process can also be interrupted by voids migration or by the voids refilling with the mass, as soon as the mass viscosity reaches a necessary value and it moves under the pressure caused by a temperature increase. The process of "moving" or "disappearing" PD sources makes the diagnostic of the PILC cables based on PD very complex, unreliable and incomplete.

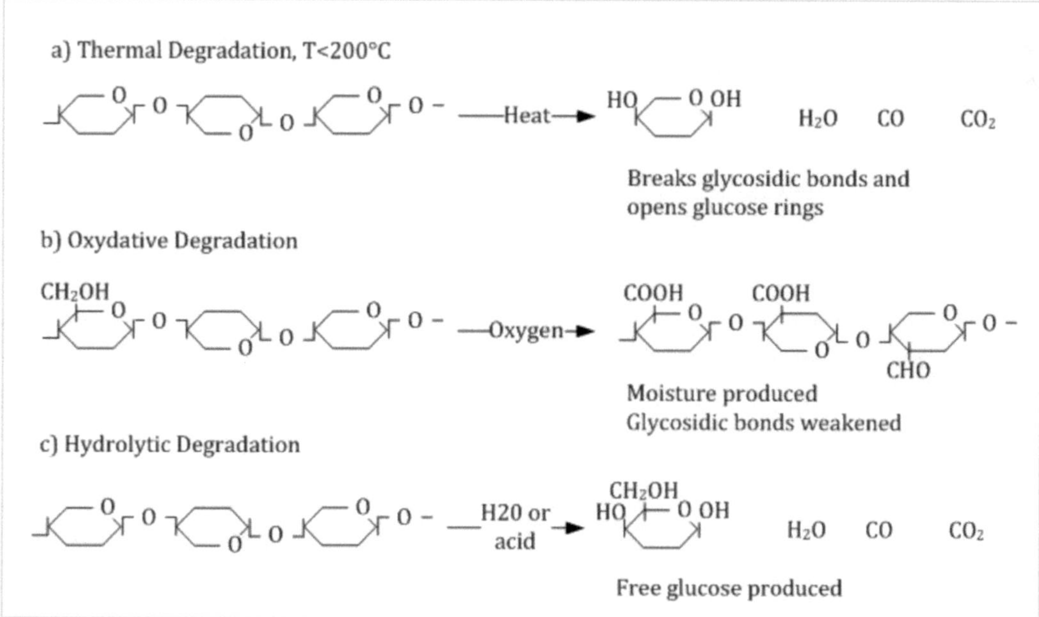

Figure 2. Degradation of cellulose (Unsworth & Mitchell, 1990)

The presence of moisture leads to hydrolysis, the most dominant degradation process, where chemical connections are divided through the influence of water. Moisture could penetrate the insulation due to mechanical damages in the lead sheath, or it is a self-product of cellulose degradation through the other degradation processes, initiating in this way a chain reaction. This process could be additionally accelerated by higher temperatures. According to (Glaubitz, Postler, Rittinghaus, Seel, Sengewald, & Winkler, 1989) the maximal allowed short circuit temperature of three core PILC cables is 155°C. Therefore, the temperatures that could lead to the pyrolysis of the cellulose are not common in the field. In

(Emsley & Stevens, 1994) the chemical mechanisms of low-temperature (<200°C) degradation of cellulose is thoroughly described.

Moreover, through the transients, short circuits or load variations in field operation, the cable temperature could vary and change very suddenly. As it was shown in (Soares, Caminot, & Levchik, 1995) there is no influence of the temperature increase (heating rate) on activation energy; it was opined that thermal decomposition of the kraft paper mostly proceeds through steady depolymerisation. Anyway, hydrolytic degradation is the most powerful degradation in cellulose, and due to the moisture outcome it could be initiated by both, thermal or oxidative degradation.

Obviously, ageing is a complex chain cause-reaction-cause process. Summarized, it results in the decrease of DP, increase of moisture content, and appearance of different gasses, resulting in the changing of electrical and mechanical properties of the insulation system.

4. PILC cables – Diagnostic methods

Diagnostic methods are nondestructive measurements of the dielectric properties in the purpose of a condition determination. Among electrical diagnostic methods, which refer to the behavior of the insulation in an electrical field, there are also chemical diagnostic methods, optical, acoustical, magnetic, mechanical, etc. defined in several standards (e.g. (High-voltage test techniques – Partial discharge measurements, 2000) (High voltage test techniques - Measurement of partial discharges by electromagnetic and acoustic methods "Proposed Horizontal Standard", 2011)). Electrical diagnostic methods of PILC cables could be divided in two main groups dependent on the type of applied voltage, AC or DC as shown in Figure 3. Measurements of cable performance with an AC test-voltage are Partial Discharge (PD), Oscillating Wave Test System (OWTS) and tanδ measurements, where measurements of PDs and tanδ are usually performed by 0,1Hz (Very Low Frequency (VLF) (IEEE Guide for Field Testing of Shielded Power Cable Systems Using Very Low Frequency (VLF), 2004)), and in some cases at 50Hz. Besides, some diagnostic measurements of tanδ like e.g. Frequency Response Analyses (FRA) operate in a wide range of frequencies, (Neimanis & Eriksson, 2004). The DC methods are based on the analyses of the diagnostic properties in time domain, (Mladenovic, Determination of the Remaining Lifetime of PILC cables based on PD and tan(δ) diagnosis, 2012, to be published), like in e.g. return voltage measurement (RVM), polarization/depolarization current measurement (PDC) and the less used Decay Voltage Method (DVM) and Isothermal Relaxation Current (IRC).

The only measurement with a local character are measurements of partial discharges, means it can localize the fault in progress, but it does not give the information about the general insulation condition. In (Densley, 2001) diagnostic tests for PILC cable system with their advantages and limitations are listed.

It has to be also mentioned, that a direct comparison of the values measured under different test-conditions e.g. ambient temperature, or even with different diagnostic systems is not always plausible. In any case, the results of the different diagnostic measurements are more or less complementary to one other and summarized could deliver more complete and more reliable information about insulation condition.

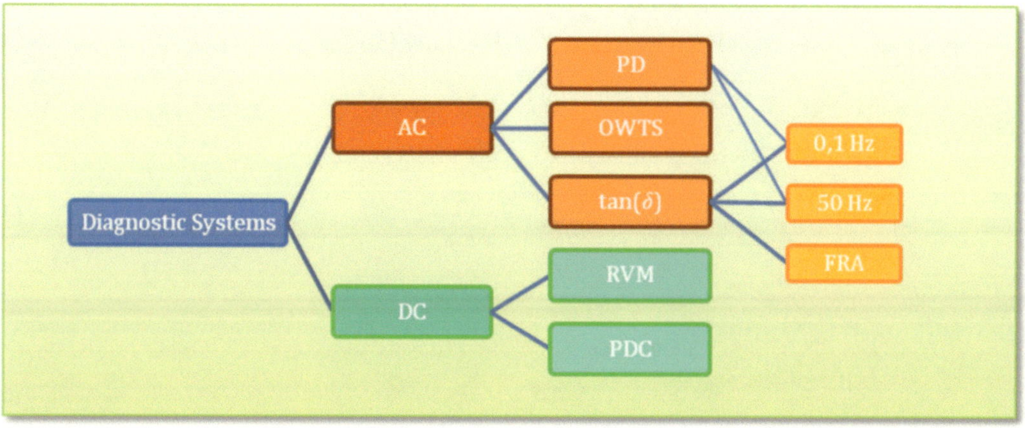

Figure 3. Principal overview of diagnostic systems for power cable

4.1. Dielectric losses and dissipation factor

The dissipation factor is the tangent of the loss angle defined as the phase shift between leakage current and the applied test voltage to 90°.

Losses in the dielectric are caused by the moving of the charges under the influence of applied electrical field. The real losses (P_δ) are determined through the current component I_δ and the reactive power (Q_c) through the component $\underline{I}_c$, Figure 4. Thus, dissipation factor $\tan\delta$ could be determined as:

$$\tan\delta = \frac{P_\delta}{Q_c}. \quad (2)$$

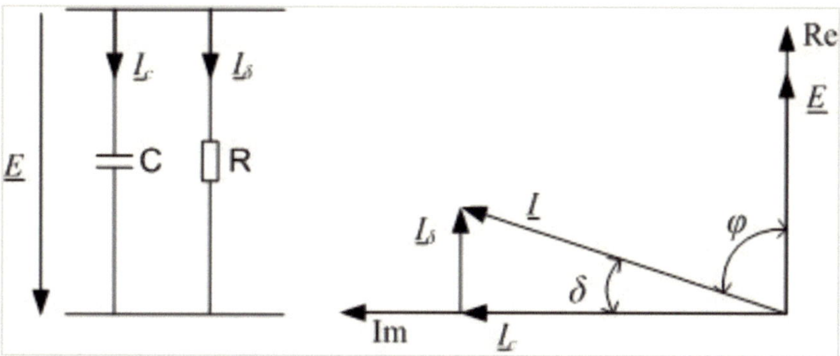

Figure 4. Simplified equivalent circuit and vector diagram of a real dielectric

Hereby two kinds of losses could be defined: polarization losses and losses due to the conductivity:

$$\tan\delta = \frac{\kappa + \omega\varepsilon_0\varepsilon_r''}{\omega\varepsilon_0\varepsilon_r'} = \tan\delta_{con} + \tan\delta_{pol}, \tag{3}$$

where:

$$\tan\delta_{con} = \frac{\kappa}{\omega\varepsilon_0\varepsilon_r'} \tag{4}$$

and

$$\tan\delta_{pol} = \frac{\varepsilon_r''}{\varepsilon_r'}. \tag{5}$$

Assuming a very low value of the dissipation loss angle δ and regarding to (3) it can be approximated that:

$$\delta = \delta_{con} + \delta_{pol}. \tag{6}$$

It is important to mention that if there is some PD activity within the dielectric, it will cause additional losses:

$$\tan\delta = \tan\delta_{con} + \tan\delta_{pol} + \tan\delta_{PD}. \tag{7}$$

Analog to (6), it can be approximated:

$$\delta = \delta_{con} + \delta_{pol} + \delta_{PD}. \tag{8}$$

4.2. Partial discharges

Partial discharges are gas-discharges of Townsend type and streamers which occur under specific conditions in gas filled cavities within the insulation. The presence of the cavities leads to a local change of the electrical permittivity and conductivity, disturbing the homogeneity of the electrical field and therefore to a decrease of electrical stability.

Exposed to an electrical AC field a random ionization process within the cavity will start, which produces a free 'start' electron needed for the development of PD. The electrical field forces the electron moving to the (local) anode within the void, ionizing on the way more gas atoms and producing therefore positive ions and more electrons. In this way the ionization events multiply causing the start of an electron avalanche. These are called

Townsend discharges. In the same time a cloud of positive ions, which are heavier and therefore slowly, moves towards the cathode, tending to reduce the cavity voltage under its breakdown value. If the voltage increases more Townsend discharges will occur again, means if the sinusoidal voltage continues to increase exceeding again the breakdown voltage of the gas, new discharge will occur to stabilize the cavity voltage. The process of charge transferring will begin tapering off near the breakdown voltage. Townsend discharges predominantly occur between paper layers and within the butt gaps, (Mladenovic, 2012, to be published), (Robinson, 1990).

On the other hand, if the applied voltage is high enough and the cavity is of enough depth so that the mean free path of the electrons is shorter than the distance between the electrodes, Townsend avalanches can develop into stream-discharges. If one start-electron creates the avalanche of 10^6 up to 10^8 electrons, it will come to a change of the electrical field in the surrounding area, since the heavier positive ions remain at the avalanche-tail while the electrons form the tip of the streamer. Through the enhanced electrical field and space charges on the avalanche-tip it will come more rapidly to new avalanches and more photons, accelerating the charge transfer regardless to the reduction of the cavity voltage. If the producing of the free electrons trough the impact ionization is faster than the electron summation, a conducting channel within the cavity will be formed and the transfer will stop as all the charges stored on the faces of the cavity are transferred, i.e. cavity voltage drops to zero, (Küchler, 2009) (Robinson, 1990).

Even fewer in number, single streamers could cause the bigger damage to the insulation (paper) due to the higher released energy. "The tip of the streamer has a radius of the order of tens of micrometers and the concentration of energy within the dimension results in penetration of the paper layers by pinholes", (Robinson, 1990). Moreover, in (Lemke, 2008) PDs are classified as: "the pulse charge created by glow discharges, oft referred to as Townsend discharges, is usually in the order of few pC. Streamer discharges create pulse charges between about 10pC and some 100 pC. A transition from streamer to leader discharges may occur if the pulse charge exceeds few 1000pC". Dependent on the intensity and the repetition rate of the local breakdowns, its surrounding within the dielectric could be "infected" and thus the weak area could expand, finally leading to the breakdown of the complete dielectric.

Beside the strength of applied voltage, relevant for the occurrence and sustain of these discharges, i.e. partial discharges, is the voltage frequency and insulation temperature. The nature of the discharges, its detection, PDs damaging power and insulation degradation exposed to PD activity, detection and measurement of PD, their intensity, repetition rate, etc. are systematically presented e.g. in (Niemeyer, 1995) (Bartnikas, 2002).

Partial discharges are defined in IEC 60270 as: "localized electrical discharges that only partially bridge the insulation between conductors and which can or cannot occur adjacent to a conductor. Partial discharges are in general a consequence of local electrical stress concentrations in the insulation or on the surface of the insulation. Generally, such discharges appear as pulses having duration of much less than 1 μs."

Gas filled cavities, locally increased moisture content; impurities within the dielectric, rifts, etc. are all classified as local weakness. A consequence of their presence is a local decrease of the relative permittivity which results in the decrease of a local breakdown voltage. In Figure 5 a gas filled cavity within a dielectric and its equivalent circuit are shown. C_1 is the capacity of the cavity with relative permittivity ε_{r1}, C_3 and C_2 are inner dielectric capacities with relative permittivity ε_r.

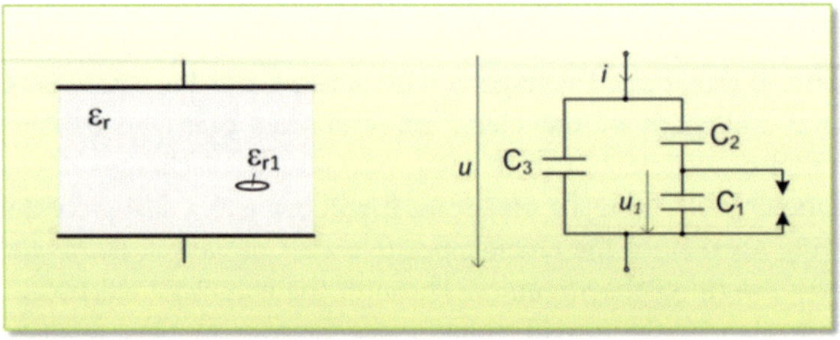

Figure 5. Equivalent circuit of the cavity within the dielectric and PD activity

Here u_1 is indicated as the breakdown voltage of the cavity. If, by applying an ACvoltage u, the voltage over capacitance C_1 exceeds ignition voltage u_1, the capacity will discharge as symbolically shown in Figure 5 over spark gap. This process will repeat with the frequency 40 kHz-1MHz (up to 10MHz, (Kuhnert, Wieznerowicz, & Wanser, 1997)), as long as the absolute value of the voltage over C_1 is higher than the ignition voltage of the void, and it can be notified as discharge pulses in the measured current i. Theoretically, PDs are essentially Townsend discharges, although their form and some features could vary as shown in (Bartnikas, 2002). However, in the praxis, measurement systems concern pulse-type discharges, their inception voltage, intensity, repetition rate i.e. frequency, etc. Generally, for an insulation material it can be stated that PD activity could cause progressive deterioration and cause irreversible mechanical weakening through chemical reactions and physical changes, (Mladenovic, Determination of the Remaining Lifetime of PILC cables based on PD and tan(δ) diagnostics, 2012, to be published).

However, in the complex insulation systems like PILC cables, or oil cables, the effect of self-healing is well known and refers to the e.g. oil/mass refilling of the cavities, or cavities displacing. Still, if PD occurs between the paper layers, they could punctures it, (Robinson, 1990), causing irreparable chemical changes which will influence the surrounded material by exposing it to an increased electrical stress, heating, and so on, until the further carbonization and break down, (Stanka, 2011). In (Bartnikas, 2002), the mechanism and nature of PD is well elaborated, and its detection and measurement systems are chronologically shown and discussed, since first reported measurement methods in 1933.

The measurement or rather detection of these pulses can be done using electrical, mechanical, optical and chemical methods, (High voltage test techniques - Measurement of partial discharges by electromagnetic and acoustic methods "Proposed Horizontal

Standard", 2011). The PD measurements applicable on the MV and HV cables are based on the electrical detection methods, which are shown in (High-voltage test techniques – Partial discharge measurements, 2000). For diagnostic purposes, the most appropriate are the measurements on the nominal frequency of 50/60Hz. In this way measured inception voltage or PD intensity etc. corresponds to those during normal network operation. Hereby measurements are commonly run on different voltage levels, for example 1, 1,3, 1,7, 2, and 2,5 times the line to ground voltage. Beside 50/60 Hz measurement systems, there are diagnostic systems operating on 0,1Hz, and so-called Oscillating Wave Test System (OWTS) presented in (Petzold & Zakharov, 2005), (Petzold & Gulski, 2006). Newly, many researches and developments are directed to the improving of on-line PD measurements, (A. N. Cuppen, E. F. Steennis, & P. C. J. M. van der Wielen, 2010) (P. A. A. F. Wouters, P. C. J. M. van der Wielen, J. Veen, P. Wagenaars, and E. F. Steennis, 2005) (Boltze, Markalous, Bolliger, Ciprietti, & Chiu, 2009) (Tian, Lewin, Wilkinson, Sutton, & Swingler) (Ambikairajah, Phung, Ravishankar, Blackburn, & Liu, 2010).

Finally, presence of PD could accelerate the ageing rapidity hardly and shorten the remaining lifetime of the insulation material, (Robinson, 1990). However, also for the purpose of the interpretation of the PD measurements, there are still no classifications or criteria, and hence the correspondence between PD activity levels and recommended actions. Nevertheless, beside PD magnitudes numerous other related and derived PD quantities are considered for the data interpretation, (Lemke, 2008): PD Inception Voltage (PDIV), PD Extinction Voltage (PDEV), PD magnitudes on different voltage levels, PD repetition rate, PD repetition frequency, phase angle, average discharge current, discharge power, etc. Due to the numerous still unknown influencing factors on the PD activity, the physical interpretation of the measured data is still a very complex process. For example:

- the number of pulses and their amplitudes vary from half-cycle to half-cycle.
- The PD repetition rate per second should increase with a rising of the test-voltage. It could also indicate the size of the source cavities, since small voids yield to very few PD pulses per half-cycle until the number increase with cavity size.
- The temperature, i.e. pressure increase in the cavity could yield to extinguishing of PD activity.

Finally, it is given in (High Voltage - VLF Hipot Instruments): "There is no reliable consensus on what are good versus bad PD levels. Splices can exhibit very high levels of partial discharge yet last for years, while those showing lower PD levels might fail sooner. One must keep in mind what the purpose of the test is. Whether using partial discharge or $\tan\delta$ techniques, the point of the test is to grade all cables tested on a scale from high quality to low".

5. Dependencies of the diagnostic parameters

The structural dependencies of the $\tan\delta$, the relative permittivity ε_r and the conductivity κ are shown in Figure 6. A temperature increase results in an increase of the mobility of the (polarized) molecules, the charge carriers and the number of electrons having at least the necessary energy to overcome the potential barrier or in other words the activation energy E_a, (Küchler, 2009). Therefore, there is a constant rise of conductivity, (Bayer, Boeck, Möller, & Zaengl, 1986):

$$\tan\delta_{con} = Ae^{-E_a/kT} \quad (9)$$

where are: A - the pre-exponential factor, T is the absolute temperature and k Boltzmann constant.

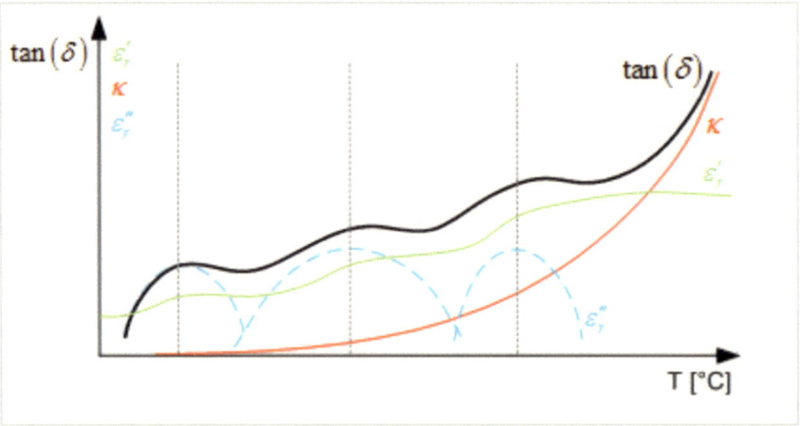

Figure 6. Structural dependencies of the dissipation factor tanδ, relative permittivity ε_r, and conductivity κ on temperature with three different polarization mechanisms (dashed blue line)

On the other hand, the polarization processes show different resonant phenomena over a wider temperature range. Appropriate temperatures drive different polarization processes. Even one polarization process like the orientational polarization can have e.g. two resonant temperatures due to the presence of different molecular structures, like e.g. moisture in the paper-mass insulation system. Therefore, with an increasing temperature the relative permittivity rises stepwise at discrete temperatures as a consequence of increased dipole mobility. Finally, with higher temperatures ε_r decreases again due to the thermal agitation which results in a partial disorganization of the dipole arrangements caused by the field. The activity of PD, if any, will change according to ideal gas and Paschen's law, (Mladenovic, Determination of the Remaining Lifetime of PILC cables based on PD and tan(δ) diagnostics, 2012, to be published). Also, conductive channels within insulation, if present, will develop faster leading to the complete breakdown of the insulation.

Since the temperature cannot be adjusted in field measurements, the temperature dependency of tanδ is inappropriate to be used as a diagnostic criteria directly. Although, it is very important to know its characteristic behavior for different ageing situations and to have reference tanδ-temperature-condition profiles, so that the measured parameters can be evaluated correctly. According to (Bayer, Boeck, Möller, & Zaengl, 1986) the tanδ of MV

PILC cables (viscosity of 50 mm²/s by 100°C) reaches its maximum in the vicinity of -8°C and the minimum in the region of 30°C. It was shown and discussed in (Mladenovic & Weindl, Influence of the thermal stress on the diagnostic parameters of PILC cables, 2010) that temperatures around 30°C are also not optimal for diagnostic measurements of MV PILC cables based on the dissipation factor.

Unlike to the temperature, the test voltage can be adjusted and is therefore used as a parameter for some diagnostic methods. According to equation (1), the differential tan(δ) values-$\Delta\tan\delta$ measured at two times the nominal voltage U_2 and the nominal voltage U_1 are partially used for this purpose.

Theoretically, it is to be expected that the dissipation factor raises with an increase of the voltage since these means an injection of more energy, enhancing the energy of the charge carriers and a multiplication of the ions. Therefore, the conductivity and in this way the $\tan\delta$ is principally rising with increasing test voltages. It was shown in (Bayer, Boeck, Möller, & Zaengl, 1986) on the example of epoxy resin, that ε_r increases with rising test voltages, and that the gradient is steeper for higher temperatures.

The dominating ageing process in paper-mass insulation systems is a degradation of the cellulose which results in a higher moisture content. Since the conductivity and permanent resistivity of the water is much higher than of cellulose-mass it can be expected that cables with higher moisture content show stronger dependency of the $\tan\delta$ on voltage.

The behavior of PD activity with varied test-voltage is already well known and widely used as a diagnostic criterion for the detection of numerous failures that can occur in cables and cable garnitures.

Moreover, a stepwise increase of the dissipation factor with the test-voltage is often interpreted as a result of increased PD activities within the insulation. Anyway, the rate of the losses caused by PD activity in the total dissipation losses is very indistinct, since it is defined by the cable length, the number of weaknesses, the PD intensity, etc., (Mladenovic & Weindl, Comparison of the parametric Partial Discharges and Dissipation factor Characteristics of MV PILC Cables, 2012).

6. Artificial ageing experiment

For a successful development of reliable ageing models, it is of prime importance to have different but constant ageing conditions and access to the regularly measured and monitored parameters up to the failure events. Therefore the characteristic key-values of the PD and $\tan\delta$ were acquired selectively for each cable at least daily over the complete artificial ageing period of two years. Beside the main field of thermo-electrically aged cable samples, selected cables were set under thermal stress only, while another group cables was electrically aged. In this way it should be possible to determine the parameters in the ageing models and the influence of each stress type on the ageing rapidity. The thermal ageing can principally be modeled by Arrhenius law, the electrical ageing by e.g. the inverse power law. When concurrent stress conditions are applied combined and complex ageing models have to be developed and analyzed. In addition, the Weibull distribution function is fitted to the fault behavior or measurement data characteristics. Therefore, the most probable lifetime, i.e. the most probable time to the next failure and its dependency on the values of the diagnostic parameters, load conditions or cable temperatures can be evaluated.

6.1. Functional principle of ICAAS

As part of the project, thoroughly described in (Weindl, Verfahren zur Bestimmung des Alterungsverhaltens und zur Diagnose von Betriebsmitteln der elektrischen Energieversorgung, 2012), the fully automated and Integrated Cable Accelerated Ageing System (ICAAS), incorporating a realistic accelerated ageing model, was developed, Figure 7. It involves voltage and current generation, PD detection and measurement, tanδ measurement, as well as some other diagnostic parameters (Dr.-Ing. Weindl & Dipl.-Ing. Mladenovic, 2009) (Mladenovic & Weindl, IEEE Electrical insulation Magazine, 2012). Artificial ageing and diagnostic measurements are performed on each cable sample with nominal line-to-line voltage $U_n = 20\ \text{kV}$ and nominal current of $I_n = 239\ \text{A}$. In order to ensure realistic accelerated ageing processes, the ageing and measurements were made under conditions comparable to normal service conditions at 50 or 60 Hz (IEEE Trial-Use Guide for Accelerated Aging Tests for Medium-Voltage Extruded Electric Power Cables Using Water-Filled Tanks, 1998), ("IEEE Guide for Field Testing and Evaluation of the Insulation of Shielded Power Cable Systems, 2001).

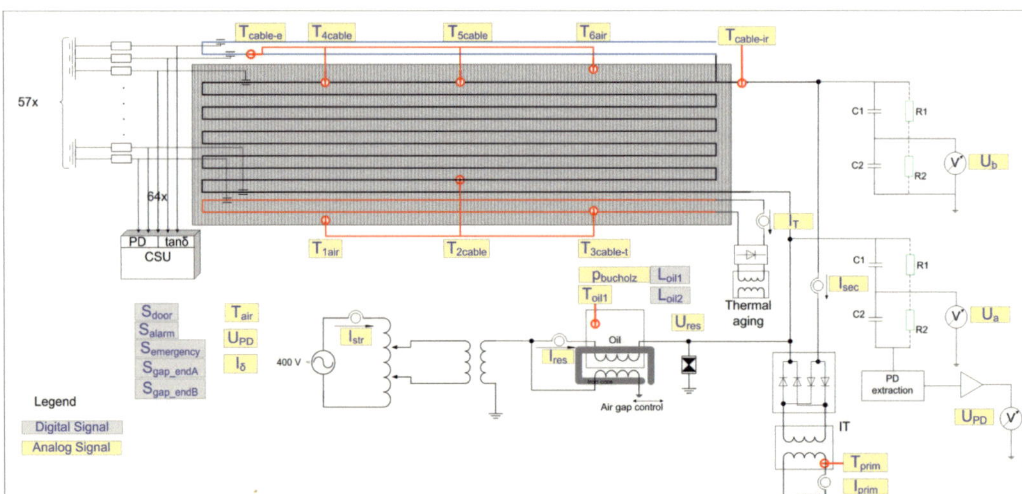

T_{1air}, T_{6air}, T_{2cable}, T_{4cable} and T_{5cable} are temperatures within the thermal tank. $T_{cable\text{-}e}$, $T_{3cable\text{-}t}$ are the temperatures of electrically and thermally aged cables respectively, and $T_{cable\text{-}ir}$ is the temperature of the termination. T_{prim}, T_{oil} and T_{air} are respectively the temperatures of the primary windings of the current generating transformer, the oil and the air. I_{str}, I_{res}, I_{prim} and I_{sec} are respectively the controlling currents in the autotransformer, the resonant circuit, and the primary and secondary sides of the current generating transformer. U_{PD} and I_{δ} are respectively the diagnostic currents through the insulation. U_a and U_b are (redundant) ageing voltages measured by the voltage divider.

Figure 7. Simplified structure of the ICAAS ageing, measurement and control system

Since most of the system components are highly specialized, and therefore rather expensive or not available on the market, almost all of the ICAAS system was designed and built in the Institute's laboratories. Voltages of up to four times the nominal operating voltage, and currents large enough (up to 500 A rms) to heat the cable conductors to the desired temperatures above 100 °C, have to be generated by the developed ageing system. The voltage generation is bases on a resonant system (Figure 8). The series resonant circuit consists of the cable capacitance and a purpose-developed variable inductance coil with more than 3000 windings in more than 20 layers, and its inductance range up to 580 H, Figure 9.

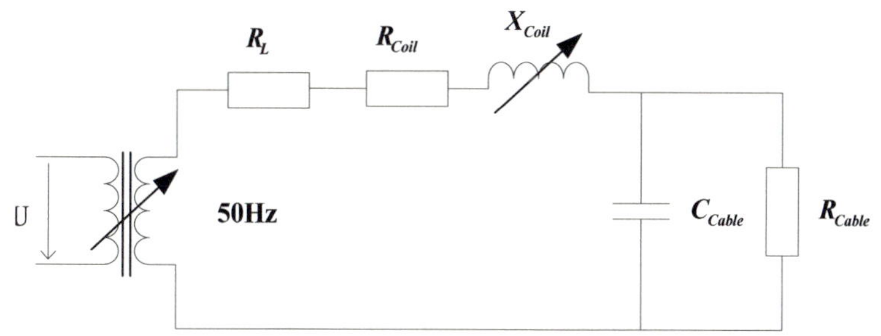

Figure 8. Structure of the voltage generating resonant circuit

Figure 9. The resonant coil - a key ICAAS component

Other requirements of the resonance circuit are a duty cycle of 100%, exceeding the specifications of many of the standard test systems on the market, and absence of partial discharge up to approximately 50 kV.

The conductor current is generated using a custom-made pulse-width-modulated high current transformer, with an adequate electrical strength between the primary and the secondary side connected to the cable samples. In this way the ageing voltages up to 50 kV can be applied in parallel. Preselected voltage profiles and defined load patterns, as well as other technical parameters, are controlled by a custom-built measurement and control system. In Figure 7 some of the main analog values which are measured, controlled and saved in pre-defined time intervals are shown in yellow, while the digital signals are shown in grey. In order to record all the necessary data and to control the ageing conditions, a Supervisory Control and Data Acquisition (SCADA) system was designed and developed (Freitag C., Entwicklung und Implementierung eines Steuerungs-, Regelungs- und Messsystems zur Realisierung einer automatisierten Versuchsanlage für die beschleunigte Alterung von Mittelspannungskabeln, 2008). It handles more than 100 analog/digital input/output values.

6.2. Selection of cable samples and ageing parameters

After the realization of ICAAS the first 24 installed cable samples were used within an pre-test lasting for 6-9 months, in order to check and adjust the operation of the ageing and measurement systems, and to point out the most appropriate ageing conditions.

The test field was made up of nearly 100 PILC cable samples, with a nominal voltage U_n of 20 kV and a length of 13,5m, which were arranged in selective and representative age-groups. Most of the cables that were investigated had been in service for times between 20 and 60 years. Others were brand new or had been stored for 10 years. In Figure 10 cables stored in the ICAAS thermal tank, installed and prepared for the artificial ageing are shown. Up to 64 samples can be artificially aged and monitored simultaneously.

Figure 10. Cable samples in the thermal tank of the ICCAS system

On the other hand, an inappropriate choice of the ageing conditions could result in too rapid ageing, caused perhaps by physical and chemical processes within the insulation system which do not occur under normal field operation. Thus, too high ageing currents and too high conductor temperatures could cause abnormal degradation of the paper insulation (cellulose) of the cables. Also, too low ageing temperatures would result in a low fault rate within the intended ageing period (two years). The determination of the electrical and environmental ageing conditions during the preliminary work is described in detail in (Mladenovic & Weindl, Determination of the Environmental Conditions for the Accelerated Ageing of MV-PILC Cables, 2009).

6.3. Measurement techniques

Of the greatest importance for the ageing experiments are, beside the selection of ageing parameters, the measurements of the ageing voltage, the cable conductor temperature, the leakage currents (mainly capacitive) through the cable insulation, the environmental temperature and measurements of the diagnostic parameters. In Figure 11 a partial overview over some ICAAS components is shown.

Figure 11. Partial overview over central components of the ICAAS system (transformers for voltage and current generation, overvoltage protection, rectifiers, etc.) and cable samples

Within the project period, several diagnostic measurement techniques have been developed, optimized and performed on each cable sample. Measurements of the dissipation factor on 50Hz, (Freitag C., to be published), with an accuracy of better then 10^{-5}, as well as partial discharges are performed regularly in pre-defined time intervals (at least daily). Measurements of return voltage and polarization /depolarization currents have been carried out several times during complete ageing period.

The initial condition of all cable samples was documented using PD analysis, tan(δ) measurements and return voltage measurements (RVM), (Mladenovic & Weindl, IEEE Electrical insulation Magazine, 2012). The resulting data are to be used as reference indicators for the later measurements. Anyway, the measurement of the PD is unique due to its local character, but should be correlated with measurements indicating the general cable condition like $\tan\delta$. However, a direct comparison of the measurement values is not possible due to their complex and not negligible dependency on environmental and test parameters, (Weindl, Mladenovic, Scharrer, & Patsch, 2010) (Mladenovic & Weindl, Dependencies of the PD- and tan(δ)-Characteristics on the Temperature and Ageing Status of MV PILC Cables, 2011) (Mladenovic & Weindl, Dependency of the Dissipation Factor on the Test-Voltage and the Ageing Status of MV PILC Cables, 2011) (Mladenovic & Weindl, Comparison of the parametric Partial Discharges and Dissipation factor Characteristics of MV PILC Cables, 2012). For this reason series of parametric studies under controlled test conditions were made, with the aim of documenting the dependence of $\tan\delta$, PD and return voltage on the cable condition, test voltage and cable temperature. In the following, some characteristic profiles of the dissipation factor and PD for different ageing groups will be presented and discussed.

7. Parametric studies

The measurements, which enable the analyses of the diagnostic parameters dependency on the test and the cable conditions, were accomplished under defined and monitored environmental conditions, at the network frequency of 50 Hz, at selective test voltages in the range from 0.4 U_n to 2.2 U_n and within a temperature range from 10°C to over 90°C. All measured values of the dissipation factor are presented in a normalized form of $\tan(\delta)/\tan(\delta)_n$, where $\tan(\delta)_n$ is the measured value on brand new (reference) cable at room temperature.

7.1. Parametric studies of the dissipation factor

Initial parametric study has pointed out three characteristic profiles of the dissipation factor, as shown in Figure 12. The occurrence of the resonant temperatures were dominant for unused but stored cables, since strong temperature dependencies were more related for cables with a long operation history.

According to Figure 6, a slight increase of the $\tan\delta$ in the region of 30-40°C in Figure 12 a) and b) could be interpreted as an optimal temperature for particular polarization processes, and temperature dependent profiles as a dominancy of $\tan\delta_{con}$. Differing to a temperature dependency that cannot be used as a diagnostic criterion (but must be considered), it is often assumed, that avoltage dependency is related to ionization processes and PD activity.

After approximately 6 weeks of artificial ageing the measurement procedure was repeated. The results are shown in Figure 13, respectively to Figure 12. The effects of the ageing on the absolute value of the dissipation factor are readily identifiable. Its increase is more intensive for cables with longer operation history, Figure 13 c). Moreover, the resonances are not

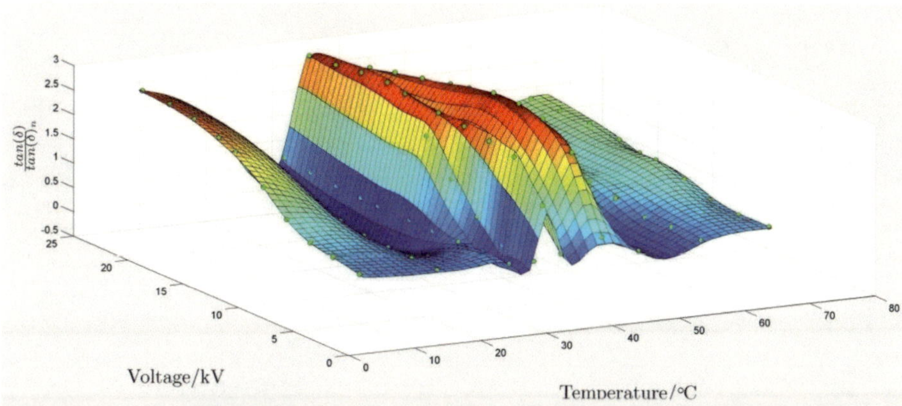

(a)

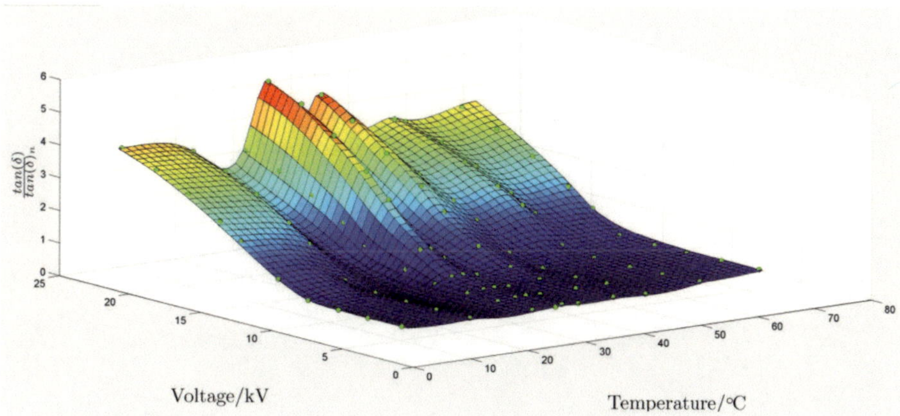

(b)

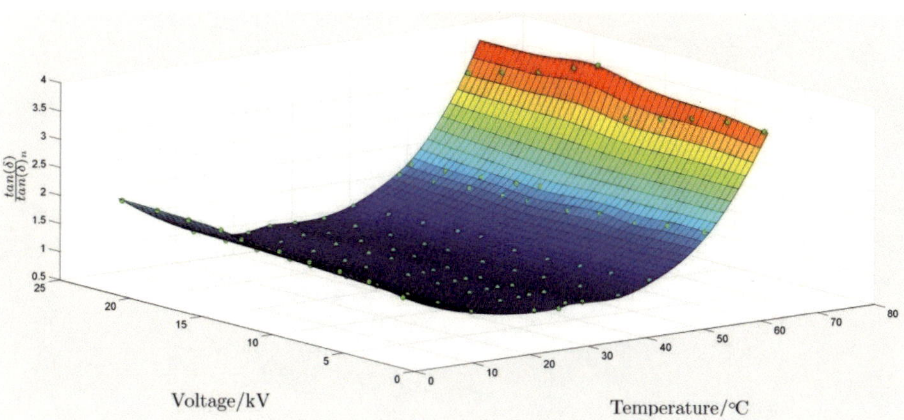

(c)

Figure 12. Selected initial profiles of $\tan\delta$ showing a resonant temperature region (a) and the voltage and temperature dependency for the different cables

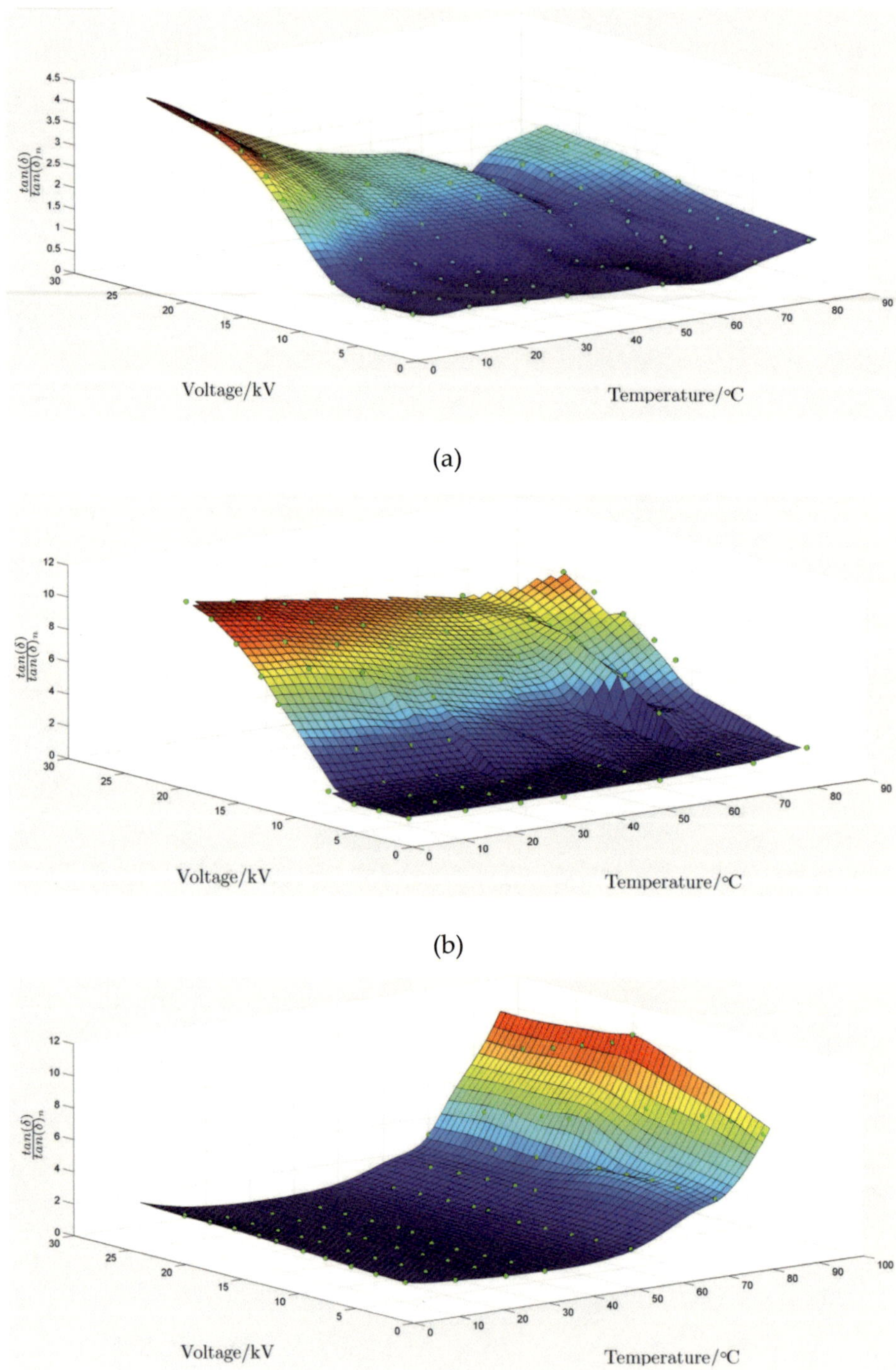

(a)

(b)

(c)

Figure 13. $\tan(\delta)$ profiles from Figure 12 after six weeks of ageing present in the same way as in Figure 12 a) and b). Since these cables have no service history (brand new and 10 years stored cable, respectively), presence of the voids at the beginning of the experiment can be assumed. During the ageing process, mass allocation through the temperature variation proceeds and could lead to the grouping of the voids and fast failure or voids releasing in terminations, as it was shown in (Mladenovic & Weindl, Determination of the Environmental Conditions for the Accelerated Ageing of MV-PILC Cables, 2009). In further studies the PD activity, simultaneously measured to the tanδ, and the corresponding portion of losses will be presented.

7.2. Parametric studies of PD

Before PD profiles for specific cables are discussed, it is of the significant importance to affirm the evaluation and weighting of the single PD values out of the numerous impulses within one single measurement. For example, in Figure 14 two very different PD profiles from the same measurement cycle are presented. PD values are hereby evaluated from the impulses with highest repetition rate Figure 14 a) or from the average of the 50 maximal values Figure 14 b. In order to determine the profile which represents the "real" PD activity in a better way, each single measurement has to be analyzed.

In Figure 15 single measurements for voltages u_0 up to $2{,}2 \cdot U_0$ with the step $0{,}2 \cdot U_0$ and variable temperatures are presented. Concerning the region of high temperatures, where is the deviation between figures a) and b) (Figure 14) the most significant, and comparing them to the single measurements shown in Figure 15, it can be concluded that better correlation of the profiles is reached by a calculation as shown in Figure 14 b). Moreover, the analyses of other PD measurements were used to confirm this experimental thesis.

Finally, in the Figure 16 and Figure 17 PD-Temperature-Voltage profiles are shown respectively to the dissipation factor profiles in Figure 12 and Figure 13. It is obvious that a temperature region with amplified PD activity can be detected. Besides, only very intensive PD activity with a high density could cause losses within the material that would be recognizable in total dissipation losses. Due to the low absolute value of the initial dissipation factor this is visible by brand new cables, Figure 12 a) and Figure 16 a).

On the other hand, less intensive PDs in case Figure 16 b) and c) do not leave some obvious marks on the total dielectric losses in Figure 12 b) and c). Moreover, an amplified PD activity is present by brand new and stored-unused cable sample, what could confirm the assumption of the void presence, as mentioned before.

After six weeks of artificial ageing, there was a significant development of the PD activity, especially in the region of high temperatures. Anyway, it cannot be stated that areas with maximal PD activity correspond to those of highest dissipation losses. For example, very strong PD activity is present in the area of higher temperatures and voltages in Figure 17 b). On the other hand, the dissipation losses of this cable show a dominant voltage dependency and almost no temperature dependency except on high voltages, whereby the maximal value lay in the region of low temperatures - differing therefore to PDs.

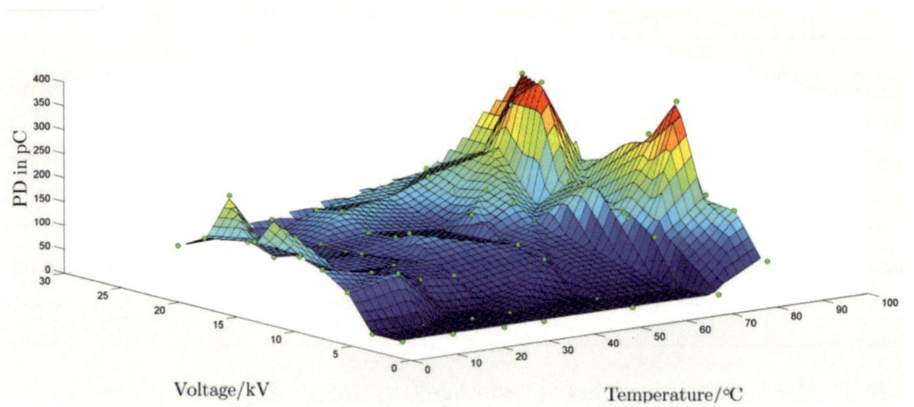

(a)

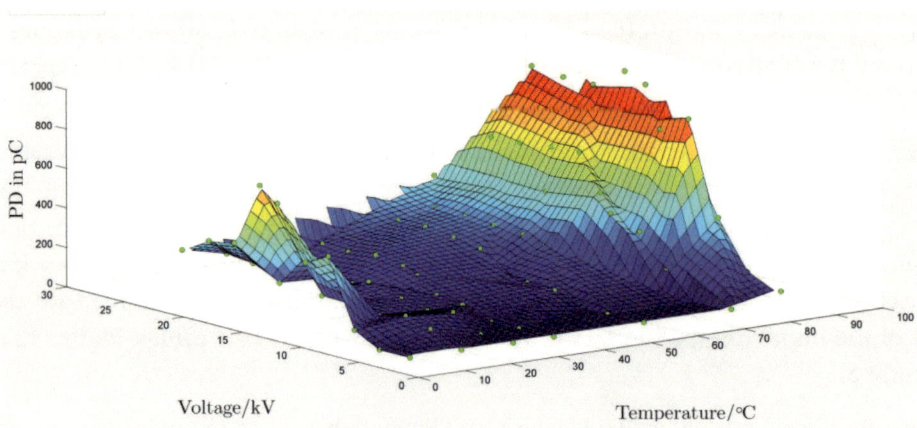

(b)

Figure 14. Two different PD profiles calculated from the same measurement cycle a) evaluated from the values with the highest repetition rate and b) evaluated from the average of the 50 maximal PD impulses

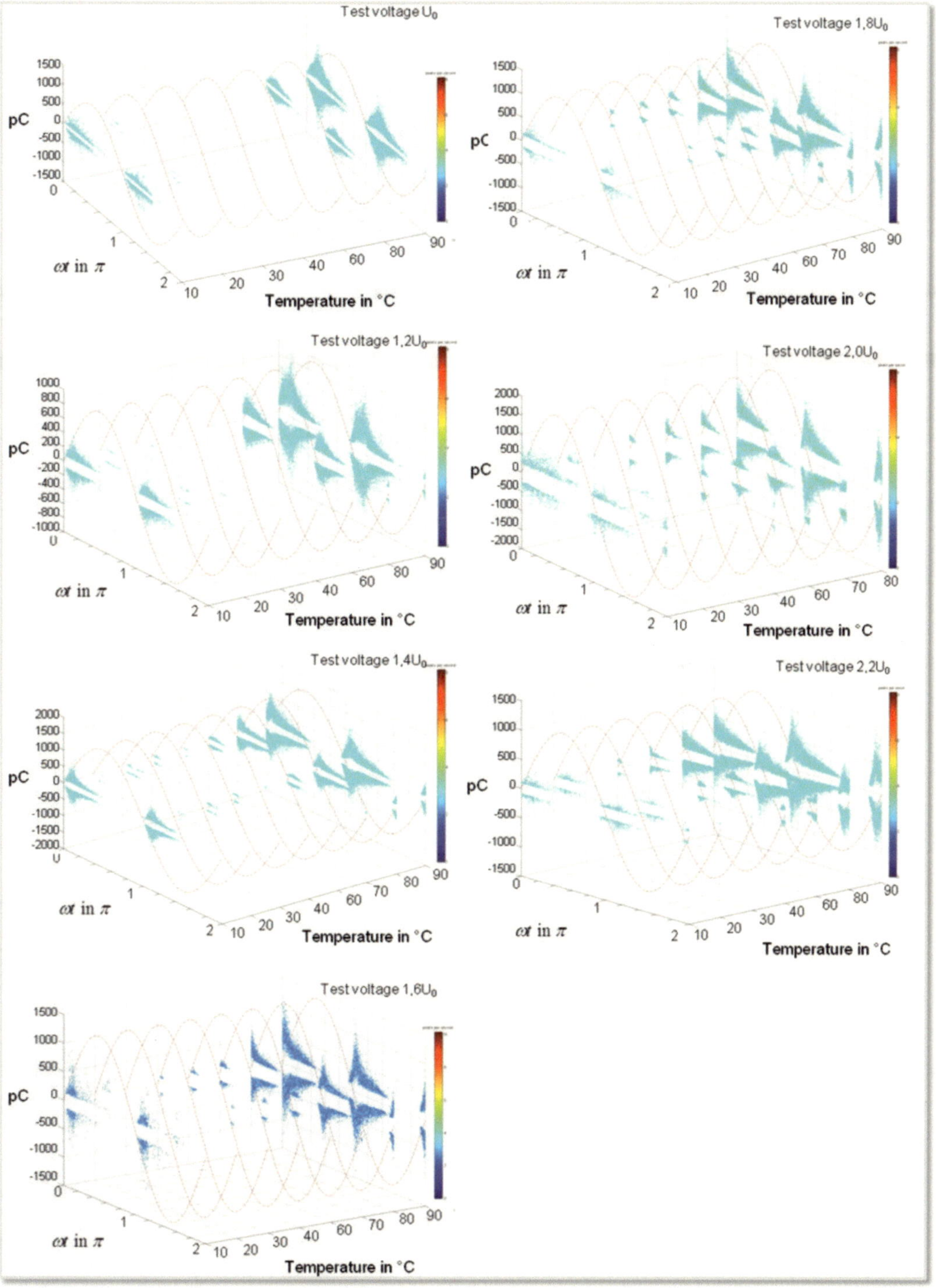

Figure 15. Background of PD profiles in Figure 14 - single PD measurements for variable test-voltages and temperatures

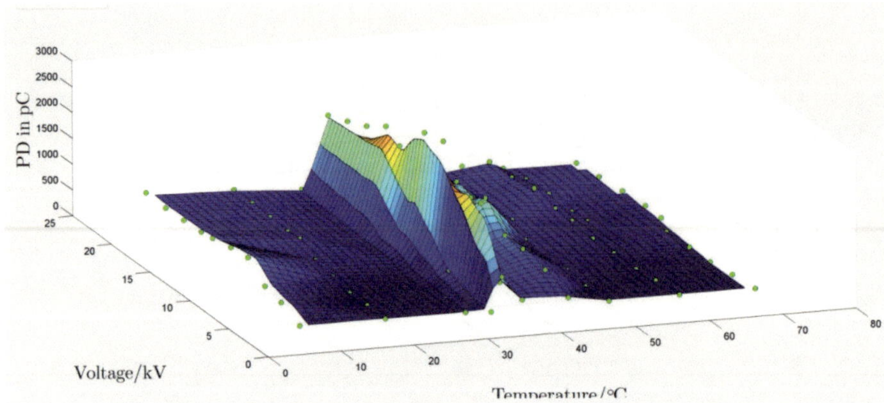

(a)

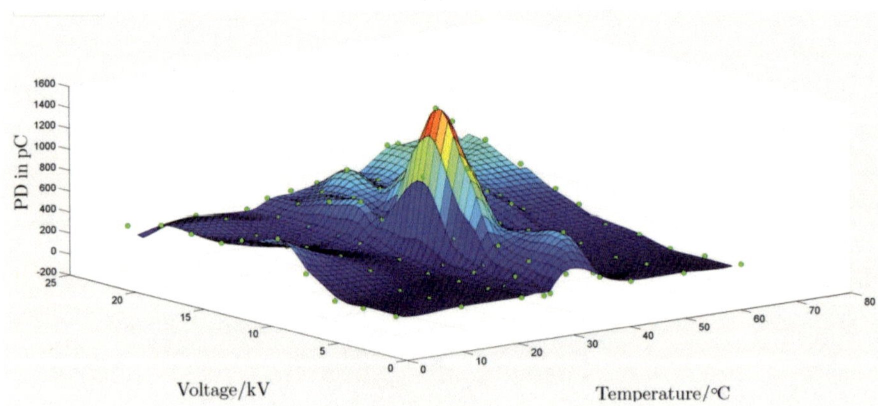

(b)

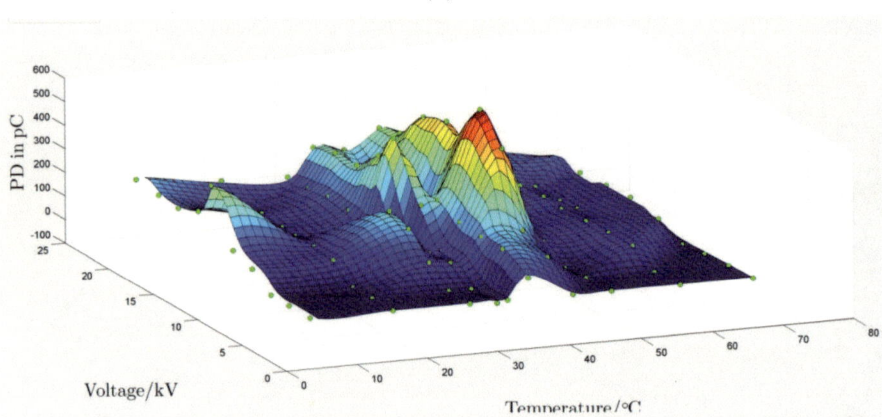

(c)

Figure 16. PD profiles simultaneously measured to the tanδ acquisition shown in Figure 12

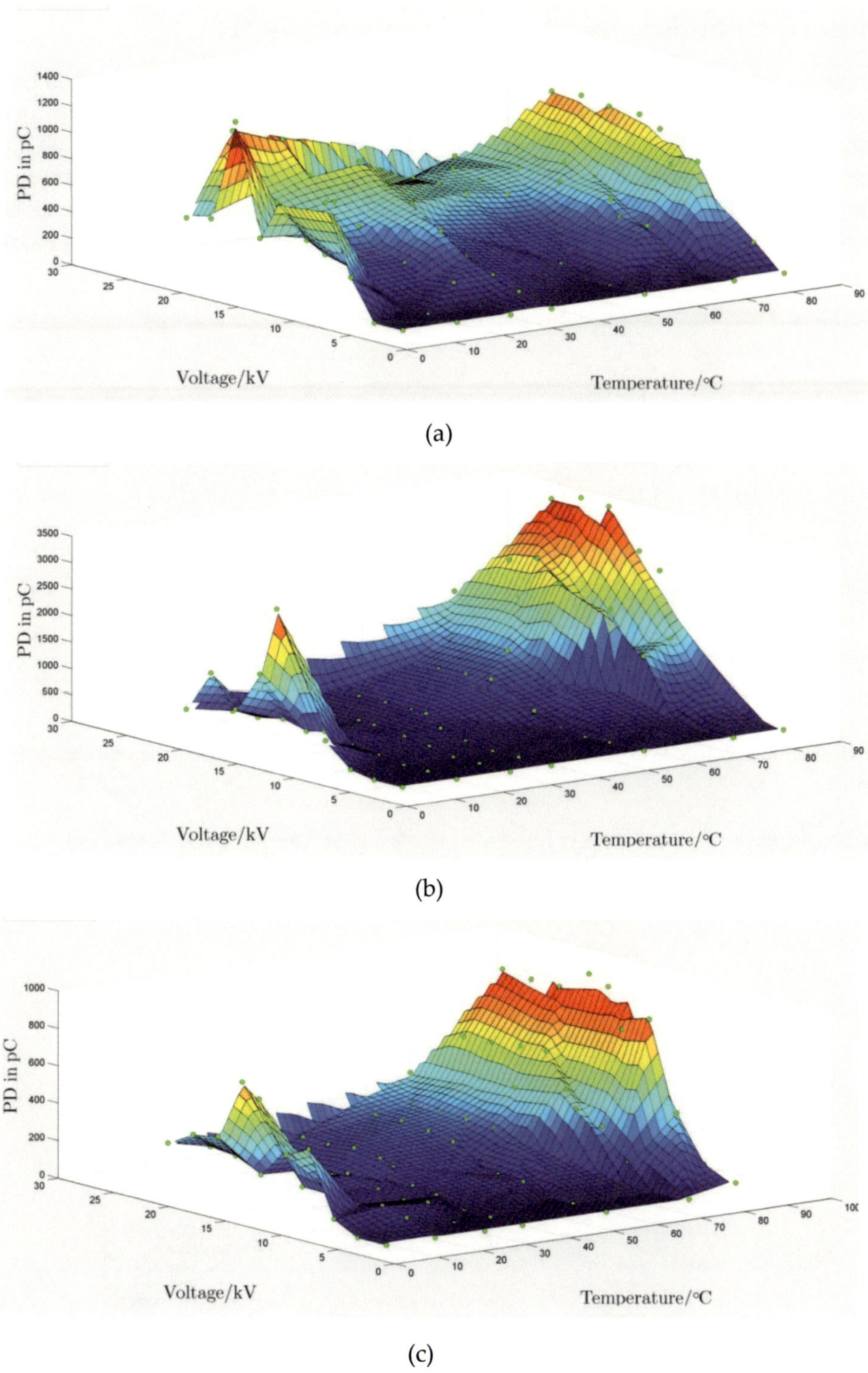

Figure 17. PD profiles simultaneously measured to the tanδ shown in Figure 13

8. Failure distribution over diagnostic parameters

Concerning a failed cable group from the same generation the Weibull distribution has been calculated. The Weibull plots in figure 4 show the distributions of cable failures over the normalized dissipation factor for variable test conditions. It can be noticed that the test voltage does not influence the Weibull curves significantly. Also, there is no strong curve deviation comparing the Weibull plot of one cable generation to all failed cables. The dominating influence factor is the temperature, showing more suitable test (diagnostic) conditions in higher temperature regions due to a better allocation of the values.

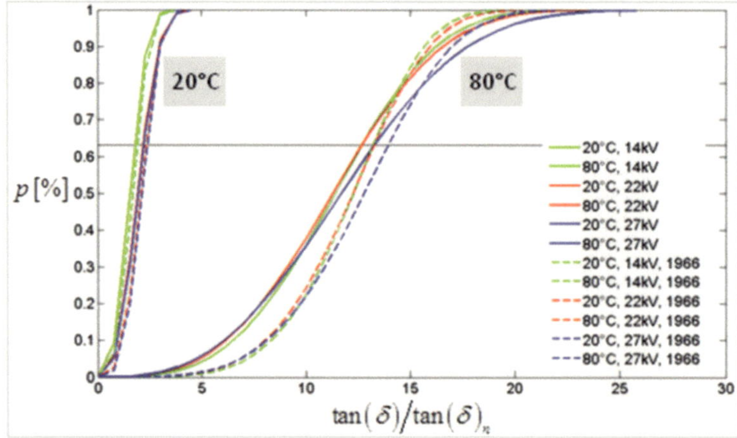

Figure 18. Weibull plots for variable tests conditions shown for all failed cables and one cable generation

Nevertheless, experimental results have also shown that temperatures lower than 20 °C are suitable for diagnostic measurements, too. Anyway, regarding to Figure 18 and Figure 19, the critical $\tan(\delta)/\tan(\delta)_n$ -interval measured at 20°C is in the region of 1,8-2,4, and at 80°C it is within 13-14.

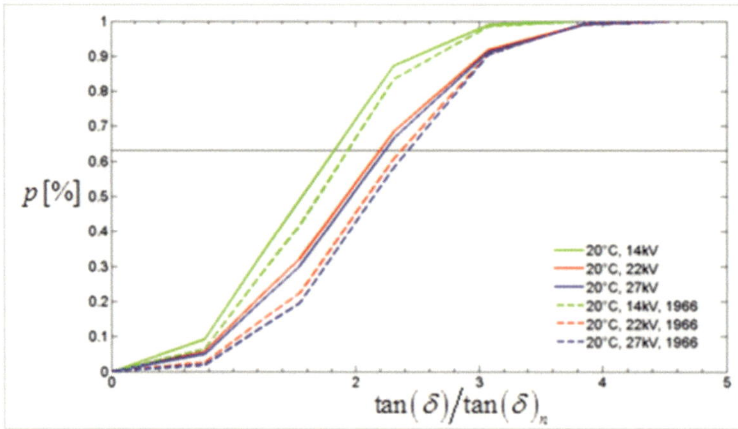

Figure 19. Weibull plot for 20°C, variable test-voltage and for all failed cables or one cable generation

Finally, the difficulty of the data analyzing of such a complex data collection is only indicated. In principal, in order to fit the most suitable distribution function and to determine the most probable remaining lifetime of the sample, the data must be carefully and precisely selected. Thereafter, it is possible to evaluate parameters of the ageing models, ageing factor and as a final point to apply these results in field measurements. In this way the cables condition can be estimated and life-consumption prognoses can be developed.

9. Conclusion

In a long-lasting ageing experiment, the ageing characteristics and diagnostic possibilities of the insulation system of MV PILC cables have been investigated. The study could be divided in several phases: developing, realization and verification of the entire ageing system (ICAAS) and its hardware and software components; selection of the cable samples and installation in ICAAS; selection of the ageing parameters; ageing experiment with regular measurements of the diagnostic parameters; regular parametric studies on the tanδ, PD, RVM, PDC in different stages of the ageing process; correlation between electrical, environmental and condition parameters; chemical analyses of the mass and paper samples; developing of the ageing/life models; reliability analyses, etc.

In the presented chapter dependencies of diagnostic parameters, like the PD and the tanδ characteristics are presented and discussed. The temperature region 20°C to 40°C is indicated as mainly inappropriate for measurements of cable's electrical properties and therefore diagnostic studies. PD diagnostics deliver a unique information with a mostly local character and should be executed and interpreted independently to other diagnostic measurements since considerable correlations between PD activity and tanδ profiles cannot be stated.

Additionally, Weibull plots for identical ageing-groupsof cables on different temperatures (20°C and 80°C) are shown, and the characteristic tanδ values are pointed out.

Finally, one of the major future research objectives is and will be the further interpretation and development of theoretical models describing the correlations between the data from the determined sophisticated knowledge databank. Moreover, the developed models and correlations will be verified through already started diagnostic field studies.

Author details

I. Mladenovic and Ch. Weindl
University of Erlangen-Nuremberg, Institute of Electrical Power Systems, Germany

Acknowledgement

The authors would like to thank the following cooperating companies for the financial and organizational support of the entire project: N-ERGIE AG (Germany), N-ERGIE Netz GmbH (Germany), N-ERGIE Service GmbH (Germany), Bayerische Kabelwerke AG (Germany).

10. References

1407, I. S. (1998). IEEE Trial-Use Guide for Accelerated Aging Tests for Medium-Voltage Extruded Electric Power Cables Using Water-Filled Tanks.

(2010). Retrieved from www.wikieduc.ch: www.wikieduc.ch/images/7/71/Isolierstoffe.doc

400, I. S. (2001). "IEEE Guide for Field Testing and Evaluation of the Insulation of Shielded Power Cable Systems.

400.2, I. S. (2004). IEEE Guide for Field Testing of Shielded Power Cable Systems Using Very Low Frequency (VLF).

60270, I. (2000). High-voltage test techniques – Partial discharge measurements.

62478, I. (2011). High voltage test techniques - Measurement of partial discharges by electromagnetic and acoustic methods "Proposed Horizontal Standard".

Ambikairajah, R., Phung, B. T., Ravishankar, J., Blackburn, T. R., & Liu, Z. (2010). Smart Sensors and Online Condition Monitoring of High Voltage Cables for the Smart Grid. *roceedings of the 14 th International Middle East Power Systems Conference (MEPCON'10)*, (p. ID 289). Cairo University, Egypt.

Bartnikas, R. (2002, October). Partial Discharges - Their Mechanism, Detection and Measurement. *IEEE Transactions on Dielectrics and Electrical Insulation, 9* (5), pp. 763-808.

Bayer, M., Boeck, W., Möller, K., & Zaengl, W. (1986). *Hochspannungstechnik - Theoretische und praktische Grundlagen für die Anwendung.* Soringer.

Bennett, G. E. (October 1957). Paper Cable Saturants: European Preferences and Selection. *AIEE Summer General Meeting*, (pp. 687-696). Montreal, Canada.

Boltze, M., Markalous, S., Bolliger, A., Ciprietti, O., & Chiu, J. (2009). On-line partial discharge monitoring and diagnosis at power cables. *76th Annual International Doble Client Conference.*

Colebrook, M. (n.d.). Retrieved from Green spirit: www.greenspirit.org.uk/resources/cellulose.gif

Cuppen, A. N., Steennis, E. F., & van der Wielen, P. C. (2010). Partial Discharge Trends in Medium Voltage Cables measured while in-service with PDOL. *Transmission and Distribution Conference and Exposition, 2010 IEEE PES.* New Orleans, LA, USA.

Densley, J. (2001, January/February). Ageing mechanisms and Diagnostics for Power Cables - An Overview. *IEEE Electrical Insulation Magazine, 17* (1), pp. 14-22.

Dr.-Ing. Weindl, C., & Dipl.-Ing. Mladenovic, I. (2009). Bestimmung von Restlebensdauer von Massekabeln anhand einer Teilentladungs- und tan(δ)-Diagnose. *Diagnostik elektrischer Betriebsmittel - ETG Kongress.* Dusseldorf, Germany.

Emsley, A. M., & Stevens, G. C. (1994). Kinetics and mechanisms of the low-temperature degradation of cellulose. *Cellulose* (1), 26-56.

FGH - Forschungsgemeinschaft für Elektrische Anlagen und Stromwirtschaft e. V.. (2006). *Zustandsdiagnose von Papiermasse-Kabelanlagen in Verteilungsnetzen.* Technischer Bericht 300, Mannheim.

Freitag, C. (to be published). *Entwicklung und Anwendung optimierter Verfahren der Verlustfaktorbestimmung zur Diagnose von Mittelspannungskabeln (working title).* Doktoral Thesis, University of Erlangen-Nuremberg, Erlangen.

Freitag, C. (2008). *Entwicklung und Implementierung eines Steuerungs-, Regelungs- und Messsystems zur Realisierung einer automatisierten Versuchsanlage für die beschleunigte Alterung von Mittelspannungskabeln.* Diploma Thesis, University of Erlangen-Nürnberg, Erlangen, Germany.

Freitag, C., Mladenovic, I., & Weindl, C. (2010). Fully Automated MV Cable Monitoring and Measurement System for Multi-Sample Acquisition of Artificial Aging Parameters. *ICREPQ.* Spain.

Freitag, C., Weindl, C., & Mladenovic, I. (2011). On-Line Cable Diagnostic Possibilities in an Artificial Aging Environment. *ICREPQ 2011.* Spain.

Glaubitz, W., Postler, H., Rittinghaus, D., Seel, G., Sengewald, F., & Winkler, F. (1989). *Kabel und Leitungen für Starkstrom - Teil 2* (4. ed.). (L. H. Stubbe, Ed.) Germany: Siemens Aktiengesellschaft.

High Voltage - VLF Hipot Instruments. (n.d.). Tan δ (Delta) Cable Testing.

Kock, J. d., & Strauss, C. (2004). *Practical Power Distribution for Industry* (1 ed.). (Newnes, Ed.) Oxford, UK: Elsevier.

Küchler, A. (2009). *Hochspannungstechnik.* Berlin: Springer.

Kuhnert, Wieznerowicz, & Wanser. (1997). *Eigenschaften von Energiekabeln und deren Messung* (2. Edition ed.). Frankfurt am Main, Germany: VWEW.

Lemke, E. (2008). *Guide for partial discharge measurements in compliance to IEC 60270.* Germany.

Mladenovic I., W. C. (2009). New System for MV-Cable ageing under controlled conditions. *Congrès international des Réseaux électriques de Distribution 2009, CIRED.* Tschech Republic.

Mladenovic, I. (2012, to be published). *Determination of the Remaining Lifetime of PILC cables based on PD and tan(δ) diagnostics.* Doctoral Thesis, University of Erlangen-Nuremberg, Erlangen.

Mladenovic, I., & Weindl, C. (2012, January/February). Artificial Aging and Diagnostic Measurements on Medium-Voltage, Paper-Insulated, Lead-Covered Cables. *IEEE Electrical insulation Magazine, 28* (1), pp. 20-26.

Mladenovic, I., & Weindl, C. (2012). Comparison of the parametric Partial Discharges and Dissipation factor Characteristics of MV PILC Cables. *IEEE International Symposium on Electrical Insulation - ISEI 2012.* Puerto Rico.

Mladenovic, I., & Weindl, C. (2011). Dependencies of the PD- and tan(δ)-Characteristics on the Temperature and Ageing Status of MV PILC Cables. *Electrical Insulation Conference - EIC.* Annapolis, USA.

Mladenovic, I., & Weindl, C. (2011). Dependency of the Dissipation Factor on the Test-Voltage and the Ageing Status of MV PILC Cables. *Jicable11.* Versailles, France.

Mladenovic, I., & Weindl, C. (2008). Determination of the Characteristic Life Time of Paper-insulated MV-Cables based on a Partial Discharge and tan(δ) Diagnosis. *14th International Power Electronics and Motion Control Conference EPE-PEMC.* Poland.

Mladenovic, I., & Weindl, C. (2009). Determination of the Environmental Conditions for the Accelerated Ageing of MV-PILC Cables. *9th International Conference on Properties and Applications of Dielectric Materials, ICPADM.* China.

Mladenovic, I., & Weindl, C. (2010). Development of the Partial Discharges Inception Voltage for Different Sets of Pre-Aged PILC Cable Samples. *International Symposium on Electrical Insulation, ISEI.* San Diego, USA.

Mladenovic, I., & Weindl, C. (2009). ICAAS – Integrated System for lasting Accelerated Aging of MV Cables, Data Monitoring and Acquisition. *IEEE Conference on Electrical Insulation and Dielectric Phenomena 2009, CEIDP.* Virginia Beach, USA.

Mladenovic, I., & Weindl, C. (2010). Influence of the thermal stress on the diagnostic parameters of PILC cables. *International Conference on Condition Monitoring and Diagnosis, CMD.* Japan.

Neimanis, R., & Eriksson, R. (2004, January). Diagnosis of moisture in oil/paper distribution cables - Part I: Estimation of moisture content using frequency-domain spectroscopy. *IEEE Transactions on Power Delivery, 1*, pp. 9-14.

Niemeyer, L. (1995). A Generalized Approach to Partial Discharge Modeling. *IEEE Transactions on Dielectrics and Electrical Insulation, 2* (4), 510-528.

Petzold, F., & Gulski, E. (2006). Experiences with PD offline Diagnosis on MV cables – Knowledge Rules for Asset Decisions. *MNC-CIRED, Asia Pacific Conference on T&D Asset Management.*

Petzold, F., & Zakharov, M. (2005). PD Diagnosis on Medium Voltage Cables with Oscillating Voltage (OWTS). *Power Tech.* St. Petersburg, Russia.

Robinson, G. (1990, March). Ageing characteristics of paper-insulated power cables. *Power Engineering Jurnal*, pp. 95-100.

Soares, S., Caminot, G., & Levchik, S. (1995). Comparative study of the thermal decomposition of pure cellulose and pulp paper. *Polymer Degradation and Stability, 49*, pp. 275-283.

Stanka, M. (2011). *Alterungsmechanismen ausgesuchter Isoilierstoffe für Betriebsmittel der Hoch- und Mittelspannungstechnik.* Diploma Thesis, University of Erlangen-Nuremberg, Erlangen.

Steennis, E. F., Ross, R., van Schaik, N., Boone, W., & van Aartrijk, D. (2001). Partial discharge diagnostics of long and branched medium-voltage cables. *IEEE 7th International Conference of Solid Dielectrics - ICSD.* Eindhoven, the Netherlands.

Tellier, R. (1983). Hundert Jahre Energiekabel - Rückschau und Ausblick. *Elektrizitätswirschaft, 3.*

Tian, Y., Lewin, P. L., Wilkinson, J. S., Sutton, S. J., & Swingler, S. G. Continuous Online Monitoring of Partial Discharges in High Voltage Cables. http://eprints.soton.ac.uk/38246/1/2797.pdf.

Unsworth, J., & Mitchell, F. (1990, August). Degradation of Electrical Insulating Paper Monitored with High Performance Liquid Chromatography. *IEEE Transactions on Electrical Insulation, 25* (4), pp. 737-746.

Weindl, C. (2012). *Verfahren zur Bestimmung des Alterungsverhaltens und zur Diagnose von Betriebsmitteln der elektrischen Energieversorgung.* University of Erlangen-Nuremberg, Erlangen.

Weindl, C., Mladenovic, I., Scharrer, T., & Patsch, R. (2010). Development of the p-factor in an Accelerated Ageing Experiment of the MV PILC Cables. *10th International Conference on Solid Dielectrics, ICSD.* Germany.

Wouters, P. A., van der Wielen, P. C., Veen, J., Wagenaars, P., & Steennis, E. F. (2005). Effect of Cable Load Impedance on Coupling Schemes for MV Power Line Communication. *IEEE Transactions on Power Delivery, Vol. 20, No. 2, April 200, 20* (2), 638-645.

Chapter 12

Electromechanical Control over Effective Permittivity Used for Microwave Devices

Yuriy Prokopenko, Yuriy Poplavko, Victor Kazmirenko and Irina Golubeva

Additional information is available at the end of the chapter

http://dx.doi.org/10.5772/50584

1. Introduction

Ferroelectrics are well known tunable dielectric materials. Permittivity of these materials can be controlled by an applied electrical bias field. Controllable permittivity leads to alteration of characteristics of tunable microwave components such as propagation constants, resonant frequency etc. However, dielectric losses in the ferroelectric-type tunable components are comparatively high and show substantial increase approaching to the millimetre waves due to fundamental physical reason.

Alternative way to achieve controllability of characteristics in tunable microwave devices is mechanical reconfiguration. In this case the alteration of microwave characteristics can be attained by displacement of dielectric or metallic parts of devices. Mechanical tuning is very promising to produce low insertion loss combined with good tunability in microwave subsystems. In the case of ferroelectric technique of tuning, microwaves interact with the ferroelectric material which is a part of microwave line. That is why transmitted energy is partially absorbed by this material. On the contrary, mechanical system of control is located out of microwave propagation route so it does not contribute to the microwave loss. Moreover, it will be shown that dielectric losses have a trend to reduce in such devices. Mechanical control is valid at any frequency range, including millimetre wave range.

Transformation of microwave characteristics could be described in terms of medium's effective dielectric permittivity (ε_{eff}). Effective dielectric permittivity of inhomogeneous medium can be defined as dielectric permittivity of homogeneous medium, which brings numerically the same macro parameters to the system of the same geometrical configuration. Effective permittivity is convenient parameter to describe devices with TEM wave propagating, where propagation constant is proportional to $\sqrt{\varepsilon_{eff}}$, however it can be used to describe other devices as well.

Application of piezoelectric or electrostrictive actuators opens an opportunity for electromechanical control over effective dielectric permittivity in microwave devices [1]. However for such applications a tuning system should be highly sensitive to rather small displacement of device components. The key idea how to achieve such a high sensitivity of system characteristics to small displacement of device's parts is to provide a strong perturbation of the electromagnetic field in the domain influenced directly by the mechanical control. For that, a tunable dielectric discontinuity (the air gap) should be created perpendicularly to the pathway of the electric field lines. This air gap is placed between the dielectric parts or the dielectric plate and an electrode. An alteration of the air gap dimension leads to essential transformation in the electromagnetic field, and revising of components' characteristics such as resonant frequency, propagated wave phase, and so on.

The goal of this chapter is to describe electromagnetic field phenomena in structures suitable for electromechanical control of effective dielectric permittivity.

2. Dispersion properties of one-dimensional dielectric discontinuity

Simplest structure suitable for electromechanical alteration of microwave characteristics is presented in Figure 1. In this structure two dielectrics are placed between infinite metal plates. The thickness d of the dielectric in the domain 2 may be variable.

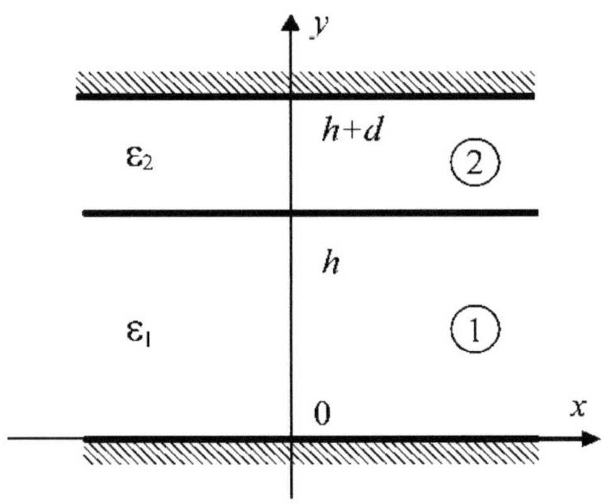

Figure 1. One-dimensional dielectric discontinuity

Electromagnetic field of this structure can be described in terms of LM and LE modes. Transverse wavenumber of the LM mode can be found from dispersion equations:

$$\begin{aligned} &\frac{\beta_{y1}^{e}}{\varepsilon_1}\tan\left(\beta_{y1}^{e}h\right)+\frac{\beta_{y2}^{e}}{\varepsilon_2}\tan\left(\beta_{y2}^{e}d\right)=0; \\ &\left(\varepsilon_1-\varepsilon_2\right)k^2={\beta_{y1}^{e}}^2-{\beta_{y2}^{e}}^2, \end{aligned} \tag{1}$$

where $\beta_{y1(2)}^{e}$ is the transverse wavenumber in the region 1 or 2 respectively, $\varepsilon_{1(2)}$ is the permittivity of the region 1 or 2 respectively, h and d are thicknesses of regions 1 and 2 respectively, $k = \frac{\omega}{c}$ is the wavenumber in free space, ω is the circular frequency, c is the light velocity in vacuum.

Using equations (1) calculations of the transverse wavenumbers are carried out in a wide range of the permittivities and thicknesses of dielectrics in domains 1 and 2. It is found that the transverse wavenumbers as the solutions of equations (1) depend on frequency, permittivities of dielectrics and sizes of domains 1 and 2. Computed results are shown in Figure 2. This figure illustrates a dependence of normalized transverse wavenumber of domain 1 for fundamental LM mode versus the normalized air gap while permittivity of domains 2 is $\varepsilon_2 = 1$.

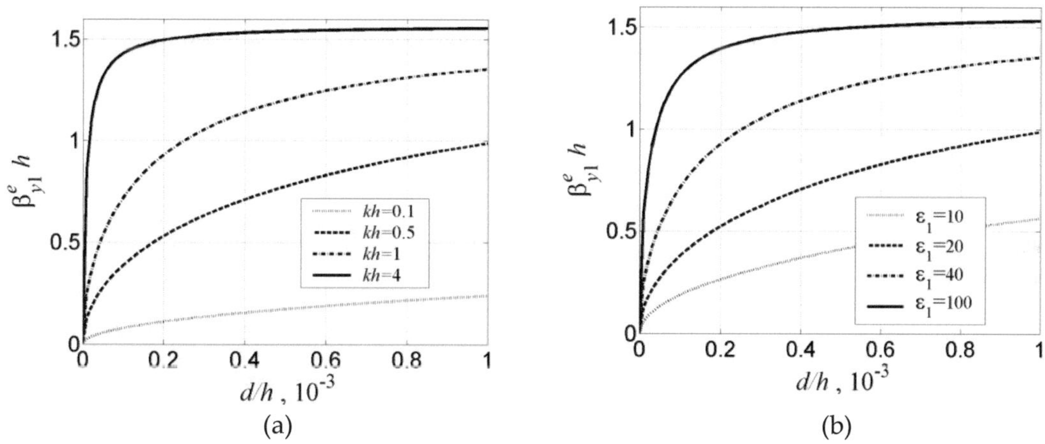

Figure 2. Normalized transverse wavenumber of fundamental LM mode versus normalized air gap size: (*a*) for certain normalized wavenumbers kh while $\varepsilon_1 = 80$; (*b*) for various permittivities of dielectric in domain 1 while normalized wavenumber is $kh = 2$.

As it is seen, transverse wavenumber of LM mode is very sensitive to variation of air gap between dielectric and metal plate. The change in only tenth or even hundredth part of percent from size of dielectric in domain 1 is sufficient for considerable alteration of transverse wavenumber. Required absolute change of air gap for significant alteration is not more than tens or hundreds micrometres depending on wavelength band and permittivity of dielectric in domain 1.

In contrast to LM mode distribution of electromagnetic field of LE mode is significantly less sensitive to variation of air gap. Transverse wavenumbers β^m_{y1} and β^m_{y2} of the LE mode are solutions of the dispersion equations:

$$\begin{aligned} &\beta^m_{y1} \cot\left(\beta^m_{y1} h\right) + \beta^m_{y2} \cot\left(\beta^m_{y2} d\right) = 0; \\ &\left(\varepsilon_1 - \varepsilon_2\right) k^2 = {\beta^m_{y1}}^2 - {\beta^m_{y2}}^2. \end{aligned} \tag{2}$$

Figure 3 illustrates a dependence of transverse wavenumber β^m_{y1} of the basic LE mode on normalized air gap thickness. As it is seen, for the LE mode required change of air gap is comparable with size of dielectric in domain 1 and quantitatively the alteration is appreciably less than for the LM mode.

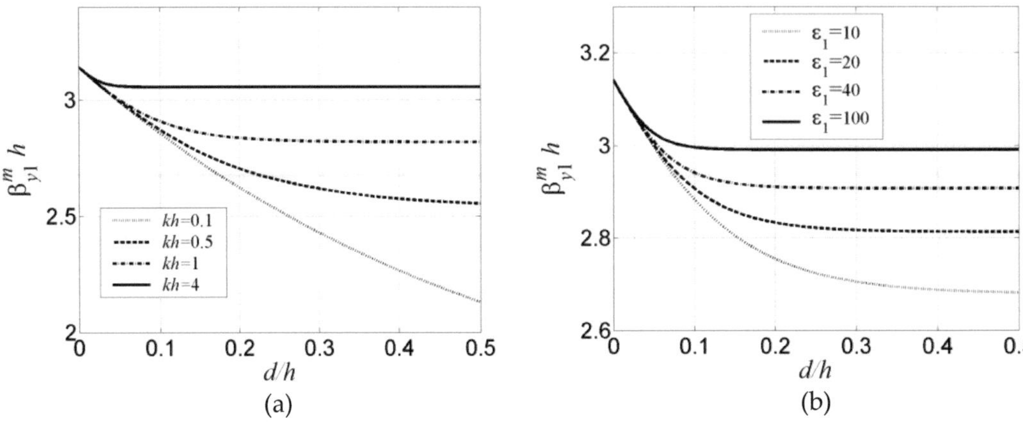

Figure 3. Normalized transverse wavenumber of basic LE mode versus normalized air gap size: (*a*) for certain normalized wavenumbers *kh* while $\varepsilon_1 = 80$; (*b*) for various permittivities of dielectric in domain 1 while normalized wavenumber is $kh = 2$.

Peculiarity of the LM mode is existence of E_y -component of electrical field which is directed normally to the border of dielectric discontinuity. For the LE mode the component E_y is equal to zero. Therefore to achieve considerable alteration of electromagnetic field a border should be located between dielectric and air to perturb normal component of the electric field. This principle should be applied to all of electromechanically controlled microwave devices.

If the domain 1 contains lossy dielectric characterized by the loss tangent tanδ, then transverse wavenumber is a complex value and its imaginary part defines dielectric loss. Figure 4 demonstrates dependences of imaginary part of normalized transverse wavenumber of domain 1 for fundamental LM mode versus the normalized air gap.

Negative values of the imaginary part say, that dielectric losses in the structure would be reduced in comparison with homogeneous structure. Moreover, for certain frequency and air gap size the dielectric loss reaches a minimum. This effect is fundamental and is observed in more complicated tunable structures.

Rigorous simulation of electromechanically controllable microwave devices requires solving of scattering problem on dielectric wedge placed between metal plates, Figure 5. Solution of the problem by the boundary element method (BEM) is discussed below.

3. Scattering on dielectric wedge placed between metal plates

Let's consider an incident wave impinging from domain 1 loaded by dielectric with permittivity ε_3 upon dielectric discontinuity in domain 2, Figure 5. Electromagnetic field of this structure can be described in terms of LM and LE modes represented by y-component of electrical Γ^e and magnetic Γ^m Hertz vectors.

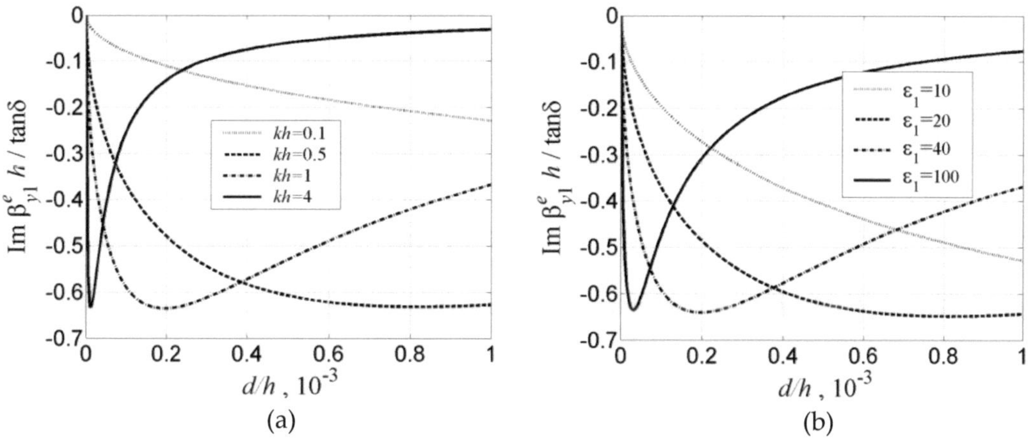

Figure 4. Imaginary part of normalized transverse wavenumber of fundamental LM mode versus normalized air gap size: (*a*) for certain normalized wavenumbers *kh* while ε_1=80; (*b*) for various permittivities of dielectric in domain 1 while normalized wavenumber is $kh = 2$.

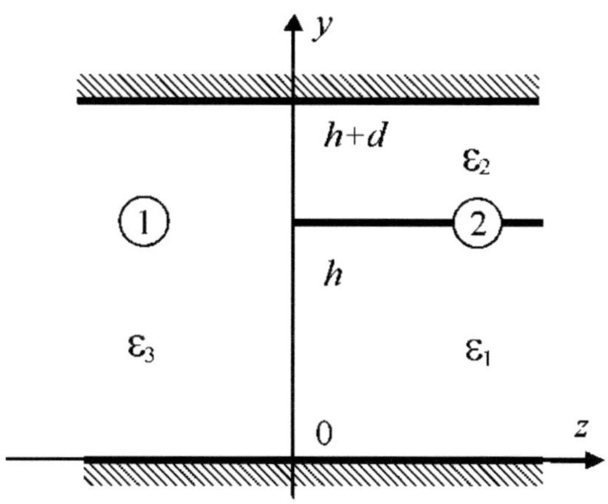

Figure 5. Structure that illustrated scattering problem

An incident wave in domain 1 is described by a sum of partial waves of LM and LE types:

$$\Gamma^{e+} = \sum_{i=0}^{n_e} c_i^e Y_{1i}^e(y) X^e(x) e^{-j\beta_{z_i} z},$$
$$\Gamma^{m+} = \sum_{i=1}^{n_m} c_i^m Y_{1i}^m(y) X^m(x) e^{-j\beta_{z_i} z}, \qquad (3)$$

where $c_i^{e(m)}$ are amplitudes of partial waves, $Y_{1i}^e(y)$ and $Y_{1i}^m(y)$ are eigen functions of the domain 1, $X^e(x)$ and $X^m(x)$ are solutions of the Helmholtz equation

$\frac{d^2X^{e(m)}(x)}{dx^2}+\beta_x^2X^{e(m)}(x)=0$, β_x is a constant, $\beta_{zi}=\sqrt{\varepsilon_3k^2-\beta_{y1i}^2-\beta_x^2}$ is the propagation constant in domain 1, $\beta_{y1i}=\frac{i\pi}{h+d}$ is an eigen value of the domain 1, n_e is the quantity of incident LM modes, n_m is the quantity of incident LE modes. Eigen functions of the domain 1 are equal to

$$Y_{1i}^e(y)=\begin{cases}\sqrt{\frac{1}{h+d}}, & i=0\\ \sqrt{\frac{2}{h+d}}\cos(\beta_{y1i}y), & i\neq 0\end{cases}, \tag{4}$$

$$Y_{1i}^m(y)=\sqrt{\frac{2}{h+d}}\sin(\beta_{y1i}y). \tag{5}$$

Reflected wave is represented as series of the 1st domain's eigen functions as

$$\begin{aligned}\Gamma^{e-}&=\sum_{i=0}^{\infty}a_{1i}^eY_{1i}^e(y)X^e(x)e^{j\beta_{zi}z},\\ \Gamma^{m-}&=\sum_{i=1}^{\infty}a_{1i}^mY_{1i}^m(y)X^m(x)e^{j\beta_{zi}z},\end{aligned} \tag{6}$$

where $a_{1i}^{e(m)}$ are amplitudes of eigen modes.

Total electromagnetic field in the 1st domain is expressed as a composition of incident (3) and reflected (6) waves.

Electromagnetic field in the 2nd domain is represented as series of domain's eigen functions as

$$\begin{aligned}\Gamma_2^e&=\sum_{i=0}^{\infty}a_{2i}^e\rho(y)Y_{2i}^e(y)X_i^e(x)e^{-j\beta_{zi}^ez},\\ \Gamma_2^m&=\sum_{k=0}^{\infty}a_{2i}^mY_{2i}^m(y)X_i^m(x)e^{-j\beta_{zi}^mz},\end{aligned} \tag{7}$$

where $a_{2i}^{e(m)}$ are amplitude of eigen modes in the domain 2, $\rho(y)=\begin{cases}\sqrt{\varepsilon_1}, & y\leq h\\ \sqrt{\varepsilon_2}, & h\leq y\leq h+d\end{cases}$ is the weight function, $\beta_{zi}^{e(m)}=\sqrt{\varepsilon_1k^2-\beta_{y1i}^{e(m)\,2}-\beta_x^2}=\sqrt{\varepsilon_2k^2-\beta_{y2i}^{e(m)\,2}-\beta_i^2}$ is the propagation constant of the domain 2, k is the propagation constant in free space, $\beta_{y1(2)i}^e$ are i-th solutions of the equations (1), $\beta_{y1(2)i}^m$ are eigen values of magnetic Hertz vector in the domain 2 computed from the equations (2), $Y_{2k}^{e(m)}(y)$ are eigen functions of the 2nd domain.

Eigen functions of the domain 2 are equal to

$$Y_{2i}^{e}(y)=\begin{cases}\dfrac{\cos\left(\beta_{y1i}^{e}y\right)}{N_i^e\varepsilon_1\cos\left(\beta_{y1i}^{e}h\right)}, & 0\le y\le h\\ \dfrac{\cos\left(\beta_{y2i}^{e}(y-h-d)\right)}{N_i^e\varepsilon_2\cos\left(\beta_{y2i}^{e}d\right)}, & h\le y\le h+d\end{cases},\tag{8}$$

$$Y_{2i}^{m}(y)=\begin{cases}\dfrac{\sin\left(\beta_{y1i}^{m}y\right)}{N_i^m\sin\left(\beta_{y1i}^{m}h\right)}, & 0\le y\le h\\ \dfrac{\sin\left(\beta_{y2i}^{m}(h+d-y)\right)}{N_i^m\sin\left(\beta_{y2i}^{m}d\right)}, & h\le y\le h+d\end{cases},\tag{9}$$

where $N_i^e=\sqrt{\dfrac{\dfrac{h}{2}+\dfrac{\sin\left(2\beta_{y1i}^{e}h\right)}{4\beta_{y1i}^{e}}}{\varepsilon_1\cos^2\left(\beta_{y1i}^{e}h\right)}+\dfrac{\dfrac{d}{2}+\dfrac{\sin\left(2\beta_{y2i}^{e}d\right)}{4\beta_{y2i}^{e}}}{\varepsilon_2\cos^2\left(\beta_{y2i}^{e}d\right)}}$, $N_i^m=\sqrt{\dfrac{\dfrac{h}{2}-\dfrac{\sin\left(2\beta_{y1i}^{m}h\right)}{4\beta_{y1i}^{m}}}{\sin^2\left(\beta_{y1i}^{m}h\right)}+\dfrac{\dfrac{d}{2}-\dfrac{\sin\left(2\beta_{y2i}^{m}d\right)}{4\beta_{y2i}^{m}}}{\sin^2\left(\beta_{y2i}^{m}d\right)}}$.

Using orthogonality property of eigen functions, amplitudes of eigen functions were expressed via indeterminate functions which are proportional to tangential components of electrical and magnetic field at the boundary of spatial domains:

$$\begin{aligned} f^e(y)&=\frac{E_y}{X^e(x)}=\frac{\dfrac{\partial^2\Gamma_i^e}{\partial y^2}+\varepsilon_i(y)k^2}{X^e(x)},\\ f^m(y)&=\frac{Z_0H_y}{X^m(x)}=Z_0\frac{\dfrac{\partial^2\Gamma_i^m}{\partial y^2}+\varepsilon_i(y)k^2}{X^m(x)},\end{aligned}\tag{10}$$

where i=1,2, $\varepsilon_i(y)=\begin{cases}\varepsilon_3, & i=1\\ \varepsilon_1, \quad 0\le y\le h\\ \varepsilon_2, h\le y\le h+d & ,i=2\end{cases}$, $Z_0=\sqrt{\dfrac{\varepsilon_0}{\mu_0}}$ is the characteristic impedance of free space.

Equality requirement for another tangential components of electromagnetic filed reduces the scattering problem to set of Fredholm integral equations of the first kind for functions $f^e(y)$ and $f^m(y)$:

$$\int_0^{h+d}\left(G_j^e(y,y')f^e(y)+G_j^m(y,y')f^m(y)\right)dy=\phi_j(y),\quad j=1,2,\tag{11}$$

where kernels of integral equations $G_j^e(y,y')$ and $G_j^m(y,y')$ are expressed via eigen functions of domains 1 and 2:

$$G_1^e(y,y') = \pm\beta_x \sum_{i=0}^{\infty} \left(\frac{Y_{1i}^e(y') \dfrac{dY_{1i}^e(y)}{dy}}{\varepsilon_3 k^2 - \beta_{y1i}^2} - \frac{\rho^2(y') Y_{2i}^e(y') \dfrac{dY_{2i}^e(y)}{dy}}{\left(\varepsilon_1 k^2 - \beta_{y1i}^{e\;2}\right)} \right), \tag{12}$$

$$G_1^m(y,y') = -k \sum_{i=0}^{\infty} \left(\frac{\beta_{zi} Y_{1i+1}^m(y') Y_{1i+1}^m(y)}{\varepsilon_3 k^2 - \beta_{y1i+1}^2} + \frac{\beta_{zi}^m Y_{2i}^m(y') Y_{2i}^m(y)}{\varepsilon_1 k^2 - \beta_{y1i}^{m\;2}} \right), \tag{13}$$

$$G_2^e(y,y') = \omega\varepsilon_0 \sum_{i=0}^{\infty} \left(\frac{\varepsilon_3 \beta_{zi} Y_{1i}^e(y') Y_{1i}^e(y)}{\varepsilon_3 k^2 - \beta_{y1i}^2} + \frac{\varepsilon_2(y) \rho^2(y') \beta_{zi}^e Y_{2i}^e(y') Y_{2i}^e(y)}{\varepsilon_1 k^2 - \beta_{y1i}^{e\;2}} \right), \tag{14}$$

$$G_2^m(y,y') = \mp \frac{\beta_x}{Z_0} \sum_{i=0}^{\infty} \left(\frac{Y_{1i+1}^m(y') \dfrac{dY_{1i+1}^m(y)}{dy}}{\varepsilon_3 k^2 - \beta_{y1i+1}^2} - \frac{Y_{2i}^m(y') \dfrac{dY_{2i}^m(y)}{dy}}{\varepsilon_1 k^2 - \beta_{y1i}^{m\;2}} \right), \tag{15}$$

where ω is the circular frequency, ε_0 is the dielectric constant in vacuum, μ_0 is the magnetic constant. Sign in (12) depends on relation between signs in $\dfrac{dX^e(x)}{dx}$ and $X^m(x)$. If $\dfrac{dX^e(x)}{dx} = \beta_x X^m(x)$ then the sign "+" shall be applied in (12). However if $\dfrac{dX^e(x)}{dx} = -\beta_x X^m(x)$ then the sign of (12) shall be "−".

Functions $\phi_j(y)$ are described by incident partial waves:

$$\phi_1(y) = -2\omega\mu_0 \sum_{i=1}^{n_{1m}} \beta_{zi} c_{1i}^m Y_{1i}^m(y), \tag{16}$$

$$\phi_2(y) = 2\omega\varepsilon_0\varepsilon_3 \sum_{i=0}^{n_{1e}} \beta_{zi} c_{1i}^e Y_{1i}^e(y). \tag{17}$$

The set of integral equations (11) was solved by Galerkin method. Functions $f^e(y)$ and $f^m(y)$ were expanded in respect to basis $\varphi_0^{e(m)}(y)$, $\varphi_1^{e(m)}(y)$, ... and set of integral equations was reduced to a system of linear algebraic equations by ordinary Galerkin procedure.

For small values of d/h eigen functions of domains 1 and 2 were selected as a basis of the Galerkin method. However, for large values of d/h to improve convergence for proper selection of basis it is necessary to take into account that in close proximity to dielectric edge electromagnetic field behaves according to the law:

$$E \sim r^{\nu - \frac{1}{2}}, \tag{18}$$

where r is the distance to dielectric edge, $\nu = \dfrac{1}{2} - \dfrac{\arctan\sqrt{\eta^2 - 1}}{\pi}$,

$\eta = \dfrac{s}{t - \sqrt{t^2 + 2s\left(\varepsilon_2\left(\varepsilon_1 + \varepsilon_3\right)^2 + \varepsilon_1\left(\varepsilon_2 + \varepsilon_3\right)^2\right)}}$, $s = 2\left(\varepsilon_1(\varepsilon_2^2 + \varepsilon_3^2) + \varepsilon_2(\varepsilon_1^2 + \varepsilon_3^2) + \varepsilon_3(\varepsilon_1 + \varepsilon_2)^2\right)$,

$t = \varepsilon_3(\varepsilon_1 - \varepsilon_2)^2$.

To satisfy (18) the Gegenbauer polynomials $C_n^{\nu}(y)$ shall be used as a basis of the Galerkin method. As a consequence, scattered electromagnetic field is calculated from computed solution for functions $f^e(y)$ and $f^m(y)$.

Multimode scattering matrix can be computed from the equations:

$$S_{11}^{j_{e(m)}k_{e(m)}} = \sqrt{\frac{\int_0^{h+d} \mathbf{E}_{\perp 1k_{e(m)}}^{-} \times \mathbf{H}_{\perp 1k_{e(m)}}^{*-} \cdot \mathbf{e}_z dy}{\int_0^{h+d} \mathbf{E}_{\perp 1j_{e(m)}}^{+} \times \mathbf{H}_{\perp 1j_{e(m)}}^{*+} \cdot \mathbf{e}_z dy}}, \tag{19}$$

$$S_{21}^{j_{e(m)}k_{e(m)}} = \sqrt{\frac{\int_0^{h+d} \mathbf{E}_{\perp 2k_{e(m)}}^{+} \times \mathbf{H}_{\perp 2k_{e(m)}}^{*+} \cdot \mathbf{e}_z dy}{\int_0^{h+d} \mathbf{E}_{\perp 1j_{e(m)}}^{+} \times \mathbf{H}_{\perp 1j_{e(m)}}^{*+} \cdot \mathbf{e}_z dy}}, \tag{20}$$

where indices j_e and j_m define numbers of incident LM and LE modes, but indexes k_e and k_m define numbers of scattered LM and LE modes, $\mathbf{e}_z$ is the unit vector of z-axis, $\mathbf{E}_{\perp 1}^{+}, \mathbf{H}_{\perp 1}^{+}$ are transverse components of electrical and magnetic field in the domain 1 forward propagated, $\mathbf{E}_{\perp 1}^{-}, \mathbf{H}_{\perp 1}^{-}$ are transverse components of electrical and magnetic field in the domain 1 back propagated, $\mathbf{E}_{\perp 2}^{+}, \mathbf{H}_{\perp 2}^{+}$ are forward propagated transverse components of electrical and magnetic field in the domain 2.

Figure 6 demonstrates a comparison of computed components of scattering matrix by the proposed (BEM) and finite-difference time-domain (FDTD) methods for the structure characterized by the parameters: ε1=10, ε2=1, ε3=1, d/h=0.01, $\beta_x h = \pi / 6$. As it is seen, there is good agreement between proposed and FDTD methods. However computing time for BEM is much less than for FDTD method due to lower order of resulting system of linear algebraic equation.

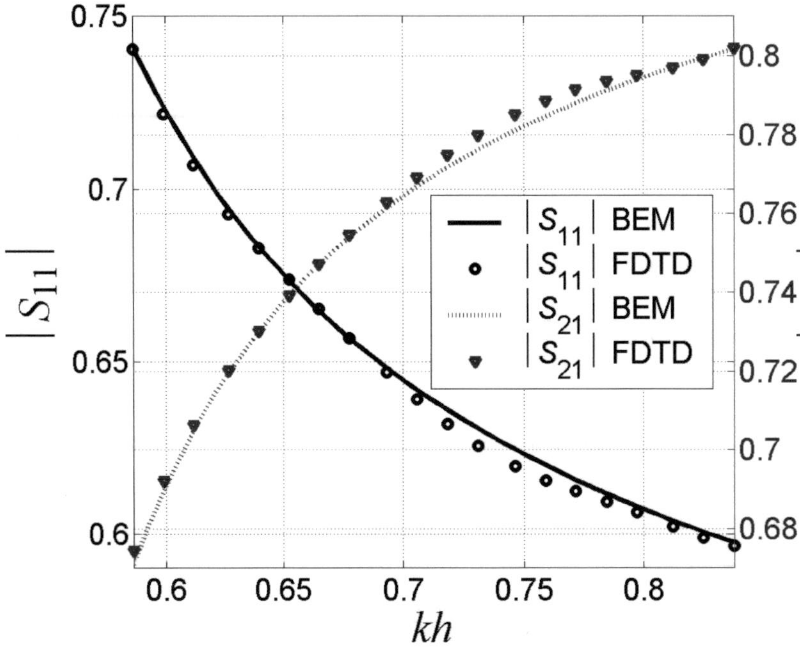

Figure 6. Reflection (S_{11}) and transmission (S_{21}) coefficients computed by boundary element (BEM) and finite-difference time-domain (FDTD) methods. ε1=10, ε2=1, ε3=1, d/h=0.01, $\beta_x h = \pi / 6$

4. Effective permittivity of one-dimensional dielectric discontinuity

Transverse wavenumber defines a propagation constant of the structure presented in Figure 1, which contains a discontinuity. Effective permittivity of the structure can be stated as such permittivity of homogeneous structure, which gives numerically the same propagation constant as in inhomogeneous structure. The effective permittivity of basic LM mode can be easily recomputed from transverse wavenumber by the equation:

$$\varepsilon_{eff} = \varepsilon_1 - \frac{{\beta_{y1}^e}^2}{k^2} . \tag{21}$$

As it follows from the equation (21), nature of the dependence of effective permittivity on distance between metal plate to dielectric is determined by the function $\beta_{y1}^e(d)$. Let's consider alteration limit of effective permittivity while displacement of metal plate under the dielectric. It follows from equation (21) that relative alteration of effective permittivity can be derived from the equation:

$$\delta\varepsilon_{eff} = \frac{\varepsilon_1 - \varepsilon_{eff}}{\varepsilon_1} = \frac{\tilde{\beta}_{y1}^{e\;2}}{\varepsilon_1 \tilde{k}^2}, \tag{22}$$

where $\tilde{\beta}_{y1}^e = \beta_{y1}^e h$ is normalized transverse wavenumber in domain filled by dielectric with permittivity ε_1 and $\tilde{k} = kh$ is normalized propagation constant in free space.

Utmost value of the normalized transverse wavenumber is equal to $\pi/2$ (Figure 2). Therefore for large values of normalized propagation constant $\tilde{k}$ maximal alteration of normalized effective permittivity is restricted by the value $\delta\varepsilon_{eff\max} < \frac{\pi^2}{4\varepsilon_1 \tilde{k}^2}$. Criterion of large and small values of $\tilde{k}$ will be determined below.

As it follows from (22) relative alteration of effective permittivity is increased while normalized propagation constant is reduced. Utmost range of the alteration can be found on the assumption of $\tilde{k} \to 0$. At this assumption as it is seen from (1) normalized transverse wavenumber $\tilde{\beta}_{y1}^e$ tends to zero as well. In this case the equations (1) can be solved analytically:

$$\lim_{\tilde{k} \to 0} \tilde{\beta}_{y1}^e = \sqrt{\frac{(\varepsilon_1 - \varepsilon_2)\frac{d}{h}}{\frac{\varepsilon_2}{\varepsilon_1} + \frac{d}{h}}}\,\tilde{k}. \tag{23}$$

Substitution of (23) into (22) gives utmost value of relative alteration of effective permittivity:

$$\lim_{\tilde{k} \to 0} \delta\varepsilon_{eff} = \frac{\varepsilon_1 - \varepsilon_2}{\varepsilon_2 \frac{h}{d} + \varepsilon_1}. \tag{24}$$

It is seen that on the assumption of $\tilde{k} \to 0$ effective permittivity can be controlled from ε_1 to ε_2. If medium in domain 2 (Figure 1) is air, then the range of effective permittivity alteration due to displacement of metal plate is from ε_1 to 1. Such high controllability is not available in other methods of control including electrical bias control of ferroelectrics.

Graphically dependence (24) is presented in Figure 7. This dependence demonstrates utmost controllability of effective permittivity of the dielectric discontinuity by alteration of distance between metal plate and dielectric if medium in the domain 2 is air. For certain permittivity ε_1 the dependence is an upper asymptote for other values of $\tilde{k}$. For $\tilde{k} \neq 0$ analogous dependences stand below those presented in Figure 7.

Due to limitation of effective permittivity alteration from ε_1 to ε_2 and utmost value of normalized transverse wavenumber equal to $\pi/2$, to achieve utmost controllability the normalized propagation constant shall satisfy to requirement:

$$\tilde{k} < \frac{\pi}{2\sqrt{\varepsilon_1 - \varepsilon_2}}. \quad (25)$$

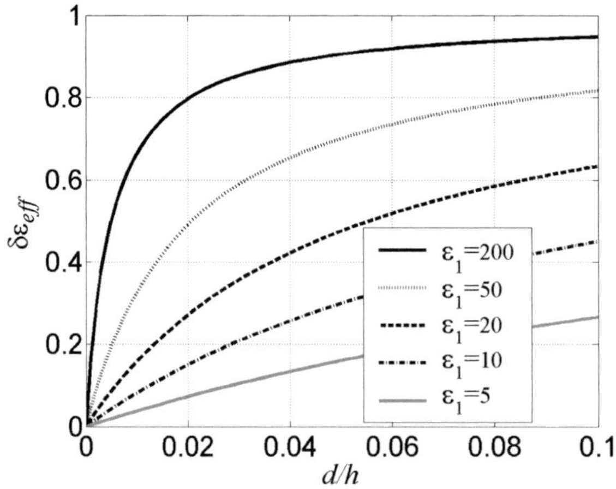

Figure 7. Dependence of relative alteration of effective permittivity on distance between metal plate and dielectric on the assumption of $\tilde{k} \to 0$.

Neglect of requirement (25) leads to decrease of controllability range and ε_{eff} is limited by the value $\varepsilon_1 - \frac{\pi^2}{4\tilde{k}^2} > \varepsilon_2$. The same conclusion can be derived from analysis of the formula (24) on the assumption of $\frac{d}{h} \to \infty$ and solution of equations (1).

Figure 8 demonstrates influence of normalized propagation constant on utmost range of effective permittivity alteration. This picture reflects dependence of relative alteration of effective permittivity on normalized distance between metal plate and dielectric with permittivity $\varepsilon_1 = 50$. For this permittivity of dielectric the requirement (25) is transformed to $\tilde{k} < \frac{\pi}{14} \approx 0.224$. As it is seen in Figure 8 if the last requirement is not satisfied then the range of effective permittivity alteration is considerably reduced and if $\tilde{k} = 0.6$ is only equal to 10% from ε_1, and if $\frac{d}{h} > 0.02$ then effective permittivity is almost independent on distance between metal plate and dielectric. Similar phenomenon in waveguides filled by multilayer dielectric if phase velocity of electromagnetic wave is almost independent on sizes of high-permittivity dielectrics was named as dielectric effect or effect of dielectric waveguide.

As it is seen in Figure 8 dependence of alteration of effective permittivity on normalized propagation constant has opposite trend than analogous dependence of alteration of transverse wavenumber (see Figure 2, *a*): controllability of transverse wavenumber is increased while propagation constant is risen up but controllability of effective permittivity is reduced at the same condition. The point is that the controllability of effective permittivity depends on ratio of transverse wavenumber to propagation constant rather than transverse wavenumber.

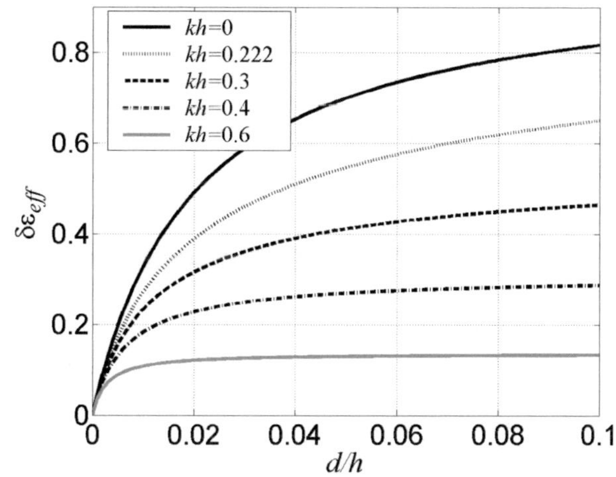

Figure 8. Dependence of relative alteration of effective permittivity on distance between metal plate and dielectric for certain normalized propagation constant *kh* while $\varepsilon_1 = 50$, $\varepsilon_2 = 1$.

Requirement (25) can be considered as criteria of smallness of normalized propagation constant. If this requirement is satisfied then the normalized propagation constant may be considered as small, otherwise as large.

Hereby to increase controllability of effective permittivity one should reduce normalized propagation constant. It can be done by two ways. The first method is decreasing of working frequency. However this way has limitation because for many implementations the frequency shall exceed cutoff frequency and cannot be reduced. The second way is to reduce thickness of dielectric h. It follows from (25) that efficient controllability of effective permittivity the thickness should to satisfy the requirement:

$$h < \frac{\pi}{2k\sqrt{\varepsilon_1 - \varepsilon_2}}. \tag{26}$$

If requirement (26) is not satisfied then the range of effective permittivity alteration is decreased according to the law close to $\sim h^{-2}$. Moreover required displacement of metal plates for effective permittivity control would be increased.

Effective permittivity model simplifies understanding and simulation of phenomena in controllable microwave devices. This model accurately describes wavelength of fundamental mode in controllable structure. However accuracy of scattering problem description should be investigated. Let's compare scattering matrix derived from effective permittivity model and rigorous solution of scattering problem by the BEM described above.

Scattering matrix for effective permittivity approach can be found from equations:

$$S_{11} = \frac{\beta_{z1} - \beta_{z2}}{\beta_{z1} + \beta_{z2}}, \quad S_{21} = \frac{2\sqrt{\beta_{z1}\beta_{z2}}}{\beta_{z1} + \beta_{z2}}, \tag{27}$$

where β_{z1} is the propagation constant in domain 1 (Figure 5) but β_{z2} is the propagation constant in domain 2 filed by uniform dielectric with permittivity ε_{eff}.

Comparison of two techniques is demonstrated in Figure 9. Good agreement for reasonable parameters set is observed.

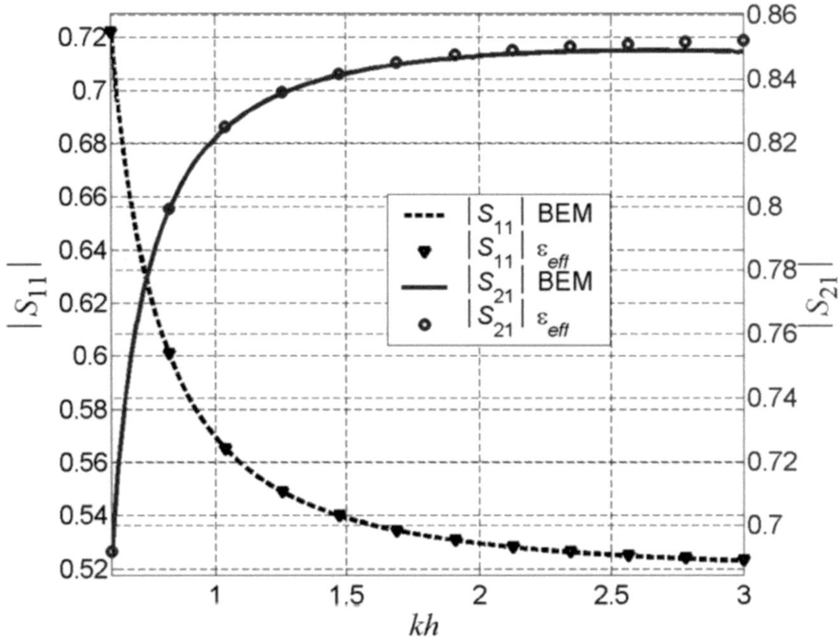

Figure 9. Comparison of *S*-parameters computation using boundary element method (BEM) and effective permittivity approach for the structure with parameters: $\varepsilon_1 = 10$, $\varepsilon_2 = 1$, $\varepsilon_3 = 1$, $d/h = 10^{-3}$

Hereby effective permittivity approach is efficient method for investigation of controllable microwave structures. Below this technique is extended for microstrip and coplanar lines.

5. Effective permittivity of microstrip and coplanar lines

Microstrip and coplanar lines are the most widely used waveguide types in modern microwave systems. They interconnect oscillators, amplifiers, antennas and so on. Sections of transmission lines also used as coupling element for resonators. Usually characteristics of the transmission line are defined at design time and remain constant in fabricated device. However, transmission lines can get some agility. For example, movement of dielectric body above microstrip or coplanar line surface results in propagation constant change [2]. We have shown that the more efficient control can be achieved, if one of the conductors is detached from substrate's surface, Figure 10. Because controllable discontinuity crosses electric field strength lines, it results in higher sensitivity.

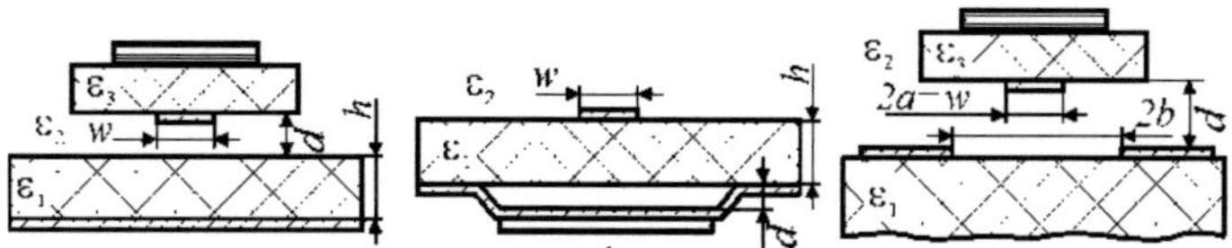

Figure 10. Mechanically controllable microstrip (*a, b*) and coplanar (*c*) lines

Conventional methods of microstrip lines analysis, such as Whiller equations [3], Hammerstad equations [4], and their extensions to coplanar lines [5] exploit symmetry of the line with aid of conformal mappings. These methods also introduce effective permittivity to relate quasi-TEM wave propagation characteristics to those of equivalent TEM wave. Transmission lines in Figure 10 still possess symmetry, but their rigorous analysis becomes cumbersome. Thus, numerical techniques could be applied to accurately calculate electromagnetic field distribution.

Electromagnetic problem can be solved using electric and magnetic scalar φ^e, φ^m and vector $\mathbf{A}^e$, $\mathbf{A}^m$ potentials:

$$\mathbf{E} = -i\omega\mathbf{A}^e - \nabla\varphi^e; \quad \mathbf{H} = \frac{\nabla\times\mathbf{A}^e}{\mu\mu_0}; \tag{28}$$

$$\mathbf{E} = \frac{\nabla\times\mathbf{A}^m}{\varepsilon\varepsilon_0}; \quad \mathbf{H} = -i\omega\mathbf{A}^m - \nabla\varphi^m. \tag{29}$$

Using these potentials one can introduce electromagnetic filed distribution types with one of components being zero. If electrical vector potential oriented along z axis ($\mathbf{A}^e = A^e\mathbf{e}_z$, where $\mathbf{e}_z$ is z-axis unit vector), then A^e and φ^e functions define E-type field, or TM-mode, for which $H_z = 0$. Similarly, if magnetic vector potential oriented along z axis ($\mathbf{A}^m = A^m\mathbf{e}_z$), then A^m and φ^m functions define H-type filed, or TE-mode, for which $E_z = 0$. Equation (28) is more convenient in the systems with dielectric only discontinuities, but with uniform permeability.

Equations (28), (29) allow ambiguity in relation of vector and scalar potentials with electromagnetic field components. For example, if $\mathbf{A}^e$ and φ^e define certain electromagnetic filed distribution, then $\mathbf{A}^e + \nabla\phi$ and $\varphi^e + \phi$, where ϕ is differentiable function, define the same distribution. This ambiguity is removed applying Lorentz's calibration:

$$\nabla\left(\varepsilon\nabla\varphi^e\right) + \varepsilon^2\mu\frac{\omega^2}{c^2}\varphi^e = 0. \tag{30}$$

In case of axial symmetry and absence of external currents solution of (30) may be presented in the form:

$$\varphi^e = \psi(x,y)Z(z),$$

where $\psi(x,y)$ is distribution of scalar potential in Oxy plane, $Z(z)$ is distribution along propagation direction Oz. Then (30) splits in two equations with two mentioned distribution functions. In most practical cases electric field component along direction of propagation is much smaller and could be neglected. This is so called quasi-TEM mode. Thus 3D electromagnetic problem reduces to 2D plane problem:

$$\nabla \cdot (\varepsilon \nabla \psi) + \beta^2 \psi = 0, \tag{31}$$

where $\varepsilon\mu\frac{\omega^2}{c^2} = \beta^2 + \beta_z^2$. Applying appropriate boundary conditions the problem is solved numerically using two dimensional finite element method (2D FEM). Then one may calculate electromagnetic field distribution as:

$$E_x = -\frac{\partial \psi}{\partial x}; \qquad E_y = -\frac{\partial \psi}{\partial y}; \qquad E_z = -i\frac{\beta^2}{\sqrt{\varepsilon\mu\frac{\omega^2}{c^2} - \beta^2}}\psi;$$

$$H_x = Z_0^{-1}\sqrt{\frac{\varepsilon}{\mu}}\frac{\sqrt{\varepsilon\mu}\frac{\omega}{c}}{\sqrt{\varepsilon\mu\frac{\omega^2}{c^2} - \beta^2}}\frac{\partial \psi}{\partial y}; \qquad H_y = -Z_0^{-1}\sqrt{\frac{\varepsilon}{\mu}}\frac{\sqrt{\varepsilon\mu}\frac{\omega}{c}}{\sqrt{\varepsilon\mu\frac{\omega^2}{c^2} - \beta^2}}\frac{\partial \psi}{\partial x},$$

where $Z_0 = \sqrt{\frac{\mu_0}{\varepsilon_0}} \approx 120\pi\Omega$ is free space characteristic impedance. In most practical cases $\beta_z << \sqrt{\varepsilon\mu\frac{\omega^2}{c^2} - \beta^2}$, thus $E_z \approx 0$ and wave is close to TEM.

Having solution of (31) we introduce effective permittivity ε_{eff} relating total power in the system under consideration to power in the system with uniform filling:

$$\varepsilon_{eff} = \frac{\sum_{i=1}^{N}\left(\varepsilon_i \iint_{S_i}\left(\left(\frac{\partial \psi}{\partial x}\right)^2 + \left(\frac{\partial \psi}{\partial y}\right)^2\right)dxdy\right)}{\iint_{S}\left(\left(\frac{\partial \psi_1}{\partial x}\right)^2 + \left(\frac{\partial \psi_1}{\partial y}\right)^2\right)dxdy},$$

where S_i is i-th domain area with permittivity ε_i, S is line's cross section total area, ψ_1 is distribution of scalar potential in regular line with $\varepsilon_1 = 1$. Quasi-static approximation gives results, which coincide well with rigorous solution by 3D FEM and finite difference in time domain (FDTD) method (Figure 11). However, at small displacements of conductor above substrate rigorous solutions faced convergence difficulties, especially FDTD method.

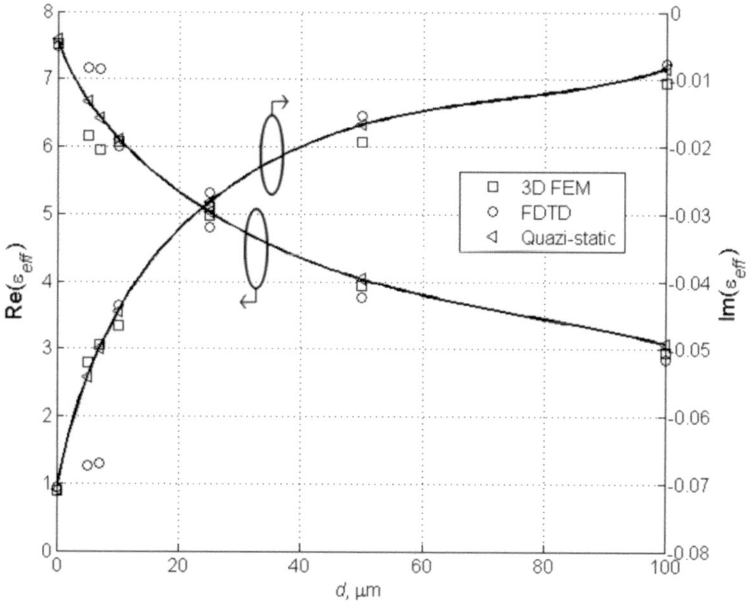

Figure 11. Comparison of effective permittivity calculation in microstrip line using 3D FEM, FDTD and quasi-static approximation ($\varepsilon_1 = 12$, $w = 0.5$ mm, $h = 1.5$ mm)

In conventional microstrip line most of electromagnetic field is confined in substrate between the strip and ground plane. When conductor is lifted above substrate as in Figure 10 *a*, *b*, certain part of electromagnetic filed redistributes from substrate to the air filled domains close to the strip. Because of lower permittivity energy stored in air filled domains is lower comparing to that one in substrate. This leads to decrease of the system's effective permittivity, as it is shown in Figure 12. Effective permittivity of the line defines wavelength in the system or, equivalently, propagation constant. Thus, mutual displacement of transmission line parts results in change of propagation constant. Described method of effective permittivity control has strong sensitivity. As seen in Figure 12 displacement by 10% of substrate's thickness may change effective permittivity more then by half.

Redistribution of electromagnetic energy to air filled domains also changes loss in the system. Because air is almost lossless medium, the portion of energy confined in air filled domains experience practically no dielectric loss. Consequently more energy reaches output terminal, resulting in lower effective loss, Figure 12.

Presented quasi-static approximation can be applied for analysis of coplanar line as well. Dependencies of effective permittivity and loss on coplanar line with lifted signal strip qualitatively similar to those of coplanar line, Figure 13.

Derived values of effective permittivity and loss then can be used to design device similarly to strict TEM-mode devices. Controllability of effective permittivity for more complicated microwave devices was presented in [1].

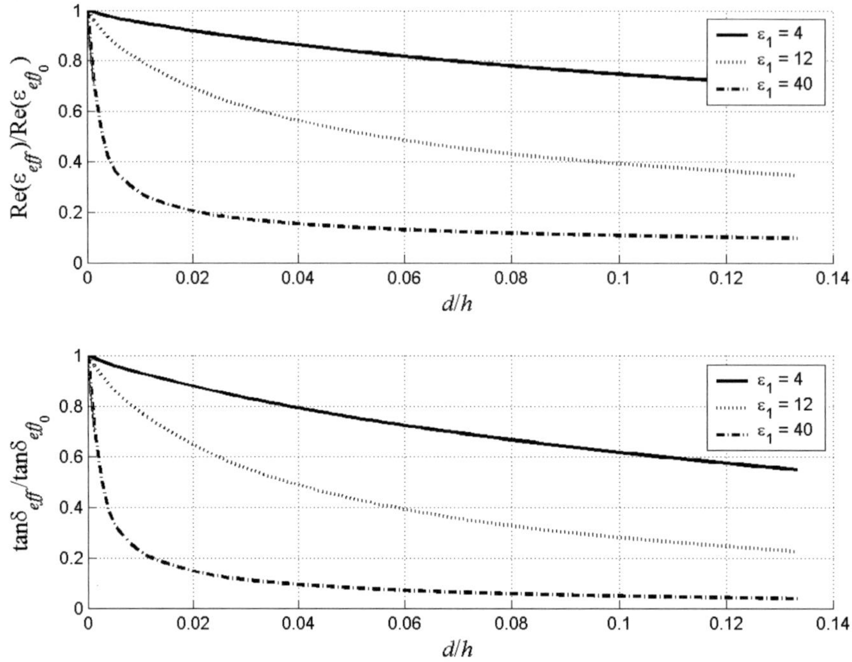

Figure 12. Effective permittivity in near 50Ω microstrip line with micromechanical control ($w/h = 2$). ε_{eff_0} and $\tan\delta_{eff_0}$ are effective permittivity and loss tangent at zero $d/h = 0$

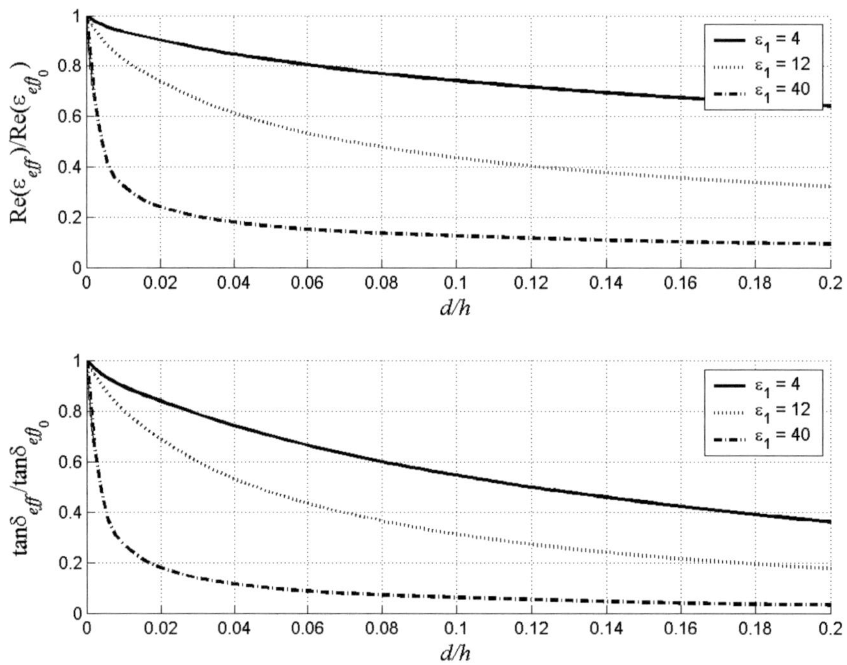

Figure 13. Effective loss in near 50Ω coplanar line with micromechanical control ($b/a = 0.72$). ε_{eff_0} and $\tan\delta_{eff_0}$ are effective permittivity and loss tangent at $d/h = 0$

Demonstrated high sensitivity of effective permittivity to microwave device parts displacement opens an opportunity to employ piezoelectric or electrostrictive actuators to control characteristics of the microwave devices by the electromechanical manner. Properties of materials for piezoelectric and electrostrictive actuators are discussed in the next section.

6. Piezoelectric and electrostrictive materials for actuators

Application of usual piezoelectric ceramics for the microwave device tuning was described previously [1, 2]. However, in a strong controlling field piezoelectric ceramics show electromechanical hysteresis that produces some inconveniences. Much more prospective are relaxor ferroelectrics that have better transforming properties and practically no hysteresis.

Ferroelectrics with partially disordered structure exhibit diffused phase transition properties. Relaxor ferroelectrics near this transition show an extraordinary softening in their dielectric and elastic properties over a wide range of temperatures. Correspondingly, dielectric permittivity ε of the relaxor shows large and broad temperature maximum where giant electrostriction is observed (because the strain x is strongly dependent on the dielectric permittivity: $x \sim \varepsilon^2$).

Relaxors are characterized by the large $\varepsilon \sim (2-6) \cdot 10^4$ and, consequently, by very big induced polarization P_i . A comparison of P_i in the relaxor ferroelectric $Pb(Mg_{1/3}Nb_{2/3})O_3$ = PMN and P_i of paraelectric material $Ba(Ti_{0.6},Sr_{0.4})O_3$ = BST (that also has rather big $\varepsilon \sim 4000$) is shown in Figure 14, *a*.

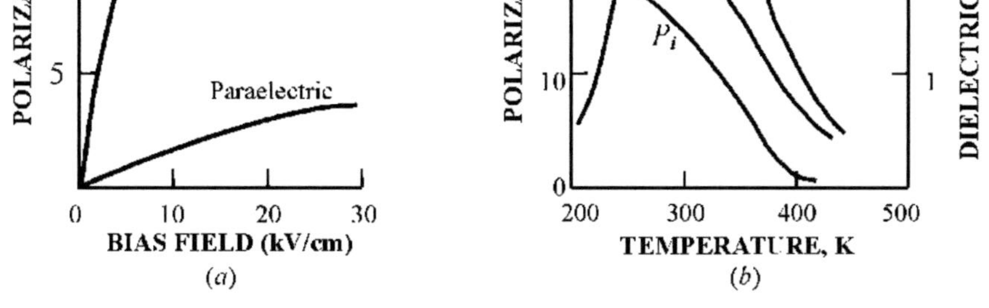

Figure 14. *a* – electrically induced polarization P_i in the relaxor of PMN and in the paraelectric BST; *b* – dielectric permittivity of PMN without (ε) and under bias field E_b=10 kV/cm (ε_b); P_i is the induced polarization in the relaxor PMN, obtained by pyroelectric measurements

Induced polarization in PMN many times exceeds one of BST. Moreover, in relaxor, the P_i depends on the temperature (like P_s of ferroelectrics), as it can be seen in Figure 14, *b*. An example is electrically induced piezoelectric effect that is explained in Figure 15.

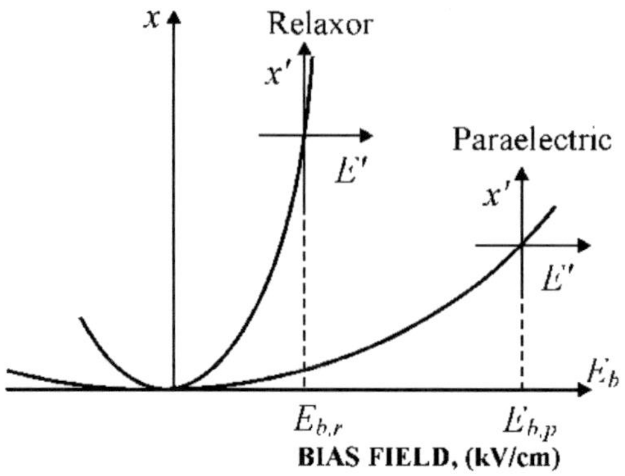

Figure 15. Electrostriction in the high-ε materials under the bias field looks like piezoelectric effect ($x' \sim E'$); $E_{b,r} < E_{b,p}$

Electric bias field E_b produces some constant internal strain x_0 at the parabolic dependency strain x on field E. Besides of steady and relatively big bias field E_b, a smaller alternating electric field E' is applied to given dielectric material. As a result, pseudo-linear "piezoelectric effect" appears that is shown in a new scale: x' - E'.

Piezoelectric effect appears instantly after the bias field is applied, and it disappears immediately after the bias field is switched off. Electrically induced piezoelectricity is large owing to giant electrostriction. Relaxor actuators can be used as precision positioner, including microwave tunable devices. Very important for device application the response time of relaxors can be estimated by the dielectric spectroscopy method.

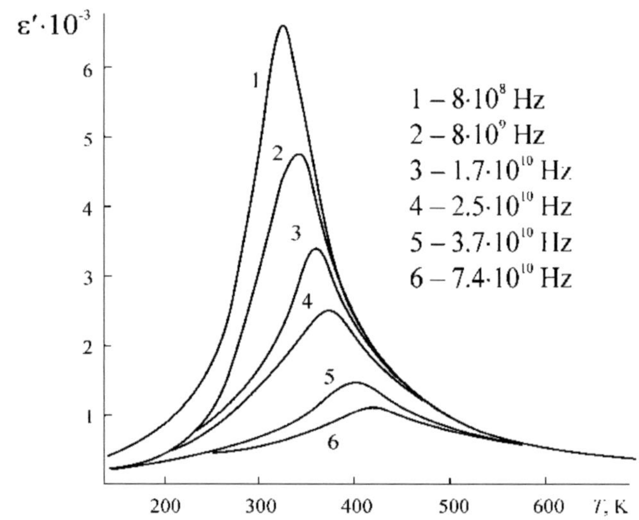

Figure 16. Dielectric spectrum of PMN at microwaves, fast dispersion of dielectric permittivity started near one gigahertz

It is obvious that response quickness is determined by the frequency dispersion of relaxor's dielectric permittivity: $\varepsilon(\nu)$. That is why dielectric dispersion in the relaxors is studied with a point of view of relaxor material applications in the fast-acting electronic devices. By the microwave dielectric spectroscopy method conducted in a broad temperature interval, the family of $\varepsilon^*(\nu,T) = \varepsilon'(\nu,T) + i\varepsilon''(\nu,T)$ curves are obtained, and one example is shown in Figure 16.

Response time of relaxor devices is determined by the mechanisms of dielectric dispersion. Electro-mechanical contribution to relaxor ε might be dominating factor so in relaxor based electronic devices the speed of response is defined by the sound speed in the relaxor, so the operating speed is dependent on the size of used relaxor element.

7. Conclusion

To achieve electromechanical control by using piezoelectric or electrostrictive actuators the dielectric- air discontinuity should create significant perturbation of the electromagnetic field. It requires a certain location of the discontinuity relatively to electromagnetic field distribution. It was demonstrated that for maximal reconfiguration of electromagnetic field by the dielectric parts displacement the border between air slot and dielectric should be perpendicular to the electric filed. In this case the displacement of dielectric parts leads to a considerable rearrangement of the electromagnetic field, and as a result to device characteristics alteration.

Effective permittivity approach not only simplifies computation but provides information about controllability of microwave structures by alteration of air slot thickness d as well. The controllability depends on frequency and dielectric thickness h. Maximal range of effective permittivity alteration increases while either frequency or thickness h reduces. At the same time, the reducing of either frequency or thickness h leads to increase of the controllability effectiveness due to decrease of required displacement of device components. Utmost controllability of effective permittivity was obtained on the assumption that either frequency or thickness of dielectric h tends to zero. Calculated dependences reflect asymptotic control over effective permittivity by alteration of air slot thickness d. Analysis of the dependences shows that the effective permittivity may be controlled in the range from permittivity of dielectric to one. Such high controllability cannot be achieved by other methods including ferroelectric permittivity control by electrical bias.

For given working frequency effectiveness of controllability increases if thickness of dielectric layer is decreased. Criterion for maximal thickness of dielectric was estimated. It is necessary to note that decrease in dielectric thickness reduces characteristic impedance of structure. That is why adding of matching sections should be considered in actual device design.

Presented method of control not only preserves high quality factor of microwave devices in the case of application low loss dielectrics but demonstrates reducing of dielectric loss during the control as well.

Effective permittivity approach significantly simplifies simulation of microwave devices. However, this approach has limitations related with high order modes excitation. That is why this technique should be carefully verified by the rigorous solution, boundary element method for instance.

Author details

Yuriy Prokopenko, Yuriy Poplavko, Victor Kazmirenko and Irina Golubeva
National Technical University of Ukraine "Kiev Polytechnic Institute", Kiev, Ukraine

8. References

[1] Yu. Poplavko, Yu. Prokopenko and V. Molchanov. Tunable Dielectric Microwave Devices with Electromechanical Control, Passive Microwave Components and Antennas, Vitaliy Zhurbenko (Ed.), In-Tech, 2010, p.367-382, ISBN: 978-953-307-083-4 (DOI: 10.5772/9416)

[2] T.-Y. Yun, K. Chang. A low loss time-delay phase shifter controlled by piezoelectric transducer to perturb microstrip line // IEEE Microwave Guided Wave Letters.– Mar. 2000.– Vol. 10, P. 96–98 (DOI: 10.1109/75.845709).

[3] H. A. Whiller. Transmission Line Properties of Parallel Wide Strip by Conformal Mapping Approximation // IEEE Transaction on Microwave Theory and Techniques.– 1964.– Vol. 12, P. 280–289 (DOI: 10.1109/TMTT.1964.1125810).

[4] E. Hammerstad, O. Jensen. Accurate Models for Microstrip Computer-Aided Design // IEEE MTT-S International Microwave Symposium Digest.– 1980, P. 407–409 (DOI: 10.1109/MWSYM.1980.1124303).

[5] C. Veyres, V. F. Hanna. Extension of the Application of Conformal Mapping Techniques to Coplanar Line with Finite Dimensions // International Journal of Electronics.– 1980.– Vol. 48, P. 47–56 (DOI: 10.1080/00207218008901066).